Algae Refinery

Algae Refinery: Up- and Downstream Processes offers complete coverage of algae refinery, including up- and downstream processes while proposing an integrated algal refinery for the advancement of existing technologies and summarizing the strategies and future perspectives of algal refinery. It provides a concise introduction to the algal science, biology, technology, and application of algae. It explains downstream and upstream steps of algal refinery for the production of algal biomass, with several social benefits.

Features:

1. Provides various aspects of algal bioprocess including upstream and downstream processes
2. Explains the major research streams of algae structures and their pathways
3. Covers algal-based CO_2 capture technology
4. Explores the potential applications of algae for socioeconomical benefits
5. Deliberates algal bioremediation approach for clean and sustainable development

Greener Technologies for Sustainable Industry and Environment
Series Editors
Pradeep Verma and Maulin P Shah

Greener technologies, such as microbial-based approaches and sustainable technologies with low net carbon output for energy generation, chemical production, bioremediation, agriculture, and so forth, are preferable to less green alternatives. This series attempts to provide space for scientists, chemical engineers, chemists, academicians, industrialists, and environmentalists to bring out the best of the literature in their area of expertise for developing future sustainable industry and environment.

Algae Refinery
Up- and Downstream Processes
Edited by Sanjeet Mehariya and Pradeep Verma

For more information about this series, please visit: www.routledge.com/Greener-Technologies-For-Sustainable-Industry-And-Environment/book-series/GTSIE

Algae Refinery

Up- and Downstream Processes

Edited by
Sanjeet Mehariya and Pradeep Verma

CRC Press
Taylor & Francis Group
Boca Raton London New York

CRC Press is an imprint of the
Taylor & Francis Group, an **informa** business

Designed cover image: © Shutterstock

First edition published 2024
by CRC Press
2385 NW Executive Center Drive, Suite 320, Boca Raton FL 33431

and by CRC Press
4 Park Square, Milton Park, Abingdon, Oxon, OX14 4RN

CRC Press is an imprint of Taylor & Francis Group, LLC

ISBN: 978-1-032-52788-8 (hbk)
ISBN: 978-1-032-59718-8 (pbk)
ISBN: 978-1-003-45595-0 (ebk)

DOI: 10.1201/9781003455950

Typeset in Times
by Apex CoVantage, LLC

Dedicated

To all contributors and future algal biotechnologists for their contributions to sustainable development of algal biorefinery.

Contents

Figures

Tables

Preface

Algae are a large and diverse group of single-celled/multi-celled autotrophic organisms existing in a variety of environments. In recent years, due to increasing industrialization and urbanization, rapid climate change is occurring. The uneven rains, seismic movements, cyclones, etc., are impacting agriculture production, thus impacting food and feed production. Also, water pollution by rapid growth has become a critical problem along with a lack of resources to meet global energy and chemical needs. The algal-based biorefinery thus can act as a potential solution to this existing problem of lack of natural resources and pollution and climate change. Thus, intensive and integrated research on algal biorefinery is highly required in the upstream and downstream processes.

With contributions from world experts, this book focuses on algae refinery from the upstream and downstream processes. In addition, it proposes an integrated algal refinery for the advancement of existing technologies. Also, it summarizes the strategies and future perspectives of the microalgal refinery to integrate with circular bioeconomy concepts.

Chapter 1 discusses the introduction to photosynthetic algae, classification, and general overview of microalgae. Moreover, a brief overview of upstream and downstream processes such as cultivation, harvesting, drying, and extraction has been deliberated. Moreover, the applications of microalgae and its biomass in bioenergy, pharmaceuticals, and food production as well as environmental benefits have been discussed. Furthermore, this book chapter focuses on the utilization of microalgae as an effective and economically feasible feedstock for biorefineries.

Chapter 2 discusses the issue related to surface water pollution associated with emerging contaminants and their algal-based removal. The concentration of these contaminants in different water bodies ranges from micrograms to milligrams per liter of wastewater. These pollutants can be either organic or inorganic including pharmaceutical compounds, industrial chemicals, endocrine disruptors, and heavy metals. Removal of these contaminants is one of the biggest challenges of the era. Many traditional and modern techniques are employed to eliminate these contaminants from wastewater, but each have their limitations. The higher-energy input requirements of traditional techniques including ion exchange, electrochemical treatment, osmosis, evaporation, and precipitation make them more expensive for application on a large scale. However, modern techniques based on biological processes (ex. Bioremediation) are bringing new hopes of a sustainable, simple, and green way of removing these pollutants from wastewater. Referring to several studies, one of the most effective strategies to eliminate emerging pollutants from wastewater is to employ microalgae for bioremediation. Microalgae are eukaryotic, photosynthetic, ubiquitous microscopic organisms that can thrive in almost all kinds of water bodies. They are reported to be bioremediate heavy metals and emerging contaminants by the mechanism of bio-assimilation and biosorption. Algae-based treatments have become an efficient, economical, and low-energy input approach compared to chemical treatments. The application of microalgae for the bioremediation of emerging organic and inorganic contaminants from wastewater will be discussed in this chapter. Further, the impact of these emerging pollutants on the environment and the phytoremediation capabilities of algae will also be discussed.

Chapter 3 describes the advances in cultivation and emerging application of microalgae *Chlorella vulgaris*. *Chlorella vulgaris* is a green microalga of great interest for the components found in its biomass; some of these components are proteins, carbohydrates, and lipids; also, the pigments it contains are considered high-value-added products. In recent years, the study and application of this microalga have increased for the energy, pharmaceutical, cosmetic, and food industry and applications. Therefore, this chapter aims to emphasize the emerging technologies developed under the biorefinery concept using various alternatives of culture media, like synthetic medium supplemented with agro-industrial residues, whose organic matter is an essential source of nutrition for biomass growth and is also necessary for the accumulation of the high-added-value components. Different strategies for recovering microalgal biomass are discussed such as centrifugation, flotation, flocculation, filtering, and sedimentation. The methods of extracting high-value-added compounds, such as the subjection of biomass to freezing cycles, are also mentioned. It also reviews the systems used for cultivation, such as open ponds, raceways, and photobioreactors, for the development of third-generation biorefineries with an impact on the circular bioeconomy and process sustainability.

Chapter 4 explores how algae are important for nutrient recovery from different waste streams. A national trend to lower the permissible nutrient levels in wastewater treatment plant (WWTP) outflow is provoking facility owners to think about new treatment approaches like algae. The assessed sensitivity of nearby water bodies presently determines the amount of fertilizer loading permitted downstream. Algae have the potential to play a significant part in helping facilities manage their side-stream treatment to meet strict downstream requirements, reduce their total effluent nutrients, and minimize overall plant operating costs. Algae, often thought to be an annoyance for WWTPs, may now be essential to resource recovery and nutrient management. Algae may be used to remediate wastewater and have several benefits. Algae are easily accessible, highly flexible, and able to significantly absorb nutrients. The technique is particularly suitable for areas with high levels of sunlight and high temperatures because photosynthesis is essential for the development of microalgae and requires minimal upkeep. Algae offer a different route, and when used, it produces high-quality treated water. The requirement for big footprints and the high cost/energy associated with algae harvesting are major obstacles to the widespread use of algae in wastewater treatment. University–industry research collaboration is coming to notice, though, for its pioneering method of cultivating and collecting the algae in a wastewater application with a less environmental impact.

Chapter 5 highlights the potential of the microalgae cell factories for developing various valuable products which find wide application in different fields. This chapter explains the potential of algae biomass application as a biorenewable feedstock to produce biofuels and their subsequent utilization as a potential source of pharmaceuticals and high-value products; it could be an option for a sustainable solution for making a better global environment and economy.

Chapter 6 discusses how algae are a group of microorganisms that are able to be used for metal remediation from contaminated sources. The fast-growing industrialization and anthropogenic activities, including fossil fuel burning, unmanaged use of agrochemicals, and the release of sewage sludge, trigger soils and water bodies to be harshly contaminated with persistent and nonbiodegradable heavy metals that threaten all life forms. These reactive metals accumulate in the food web, instigating severe health concerns. Diverse modern techniques such as ion exchange, chemical extraction, and electrolytic technologies have been adapted for the remediation of heavy metals, but these techniques are neither economical nor sustainable; instead, they need continuous monitoring and stringent control. Therefore, this chapter summarizes the information on algal strains used for eliminating metal contamination and the mode of action they followed for heavy metals remediation. Algal strains, namely *Anabaena* spp., *Chlorella* spp., *Cladophora* spp., *Spirulina* spp., *Scenedesmus* spp., *Oscillatoria* spp., *Phaeodactylum tricornutum*, etc., have exhibited the capability to eradicate heavy metals, acting as hyper-adsorbents and hyper-accumulators with high selectivity for various metals. Phycoremediation alone is not economical; therefore, integrated remediation systems must be developed, making this approach more sustainable and economical. Furthermore, this review explores current progressions in the phycoremediation of heavy metals. Also discussed is the genetic engineering approach applied to create transgenic species leading to the over-expression of metallothioneins and phytochelatins that form complexes with heavy metals and are stored within the vacuoles utilizing bioaccumulation to remove hazardous metals. Following the discussion, this chapter concluded that phycoremediation is a more bioeconomic, sustainable, and clean technology.

Chapter 7 describes how algal–bacterial interactions in the environment have various applications. In the natural environment, microalgae are not alone, as most of the time they are found in consortia with various microorganisms such as bacteria and fungi. Therefore, their interaction with other organisms such as bacteria plays a very important role in various processes such as biogeochemical cycles, nutrient recycling, wastewater treatment, and sustainable agriculture. Understanding such interaction helps us to develop processes for various biotechnological applications. In this book chapter, we will discuss our understanding of microalgae–bacteria interaction in the natural environment and its probable applications such as the production of sustainable biomass and wastewater remediation.

Chapter 8 explains the application of diatoms, which are the living factories of biogenic porous silica. The frustules have an inimitable architecture of intricate nanopatterns that are biocompatible, thermally stable, and corrosion- and pathogen resistant. Diatom silica microparticles have inspired multiple nanofabrication techniques, which are sustainable, economical, environmentally benign, and efficient in various nano-designed biosystems. Diatom-based nanoarchitectonics is a novel nano-scale approach in material science, which works under complex physiological conditions. This review focuses on such potential bioapplications and their tools to prepare hierarchical silica nanostructures to tailor multifunctional properties for improved microfabrication and functional design. A brief account of scientometric analysis on diatom silica as advanced nanoarchitectonics material to create molecular machines and micro devices is described. Moreover, the challenges faced in diatom-based nanostructure technologies are also discussed.

Chapter 9 explores the biofuel production by algal biomass; the production of algal biofuels is one such renewable alternative to overcome this global crisis. Algae being flourishing organisms are proficient of growth and expansion under variable conditions like controlled environment in closed photobioreactors as well as in open pond ecosystems. Also, alteration in their optimal growth conditions and genetic alterations result in varied secretion and aggregation of biofuels. Bio-transforming the algal biomass in distinct forms of biofuels can be a key step to solving the problem of growing energy demand. Although algal biofuel serves as a substitute to conventional fossil fuel, the production technology needs to conquer commercial barricades and various challenges that obstruct the production of such green energy solutions. A superior understanding of basic procedures involved in biofuel production is needed. In this chapter, a thorough discussion is made on different biofuels produced by algae with major challenges that obstruct productivity. In addition, it also enlightens the advancements made to meliorate the productivity of biofuels.

Chapter 10 describes a market survey and trend to increase the global demand and consumption of algal biomass for various applications. Therefore, this chapter emphasizes the developing fields of algal science with a specific focus on algal diversity, digestibility, bioavailability, active bioactive components, and applications that are required for a better assessment of the health benefits of these algae or algal by-products. Furthermore, vast opportunities exist for phycologists, other researchers, and people interested in developing new products and their industrial applications in this emerging field.

Overall, this book covers various aspects of algal biorefinery, including the upstream and downstream processes.

Sanjeet Mehariya, PhD
Pradeep Verma, PhD

Acknowledgments

First and foremost, the editors would like to express their sincere gratitude to all the distinguished authors for their thoughtful contribution to making this project successful. We appreciate their patience and diligence in revising the first draft of the chapters after assimilating the suggestions and comments. We would like to thank and acknowledge the solicitous contributions of all the reviewers who spent their valuable time in providing constructive comments to improve the quality of the chapters. Finally, we would like to acknowledge the support of our mentors, family members, friends, and colleagues for their love and encouragement; this work is dedicated to their smiles.

About the Editors

Sanjeet Mehariya (PhD) is Researcher at the Algal Technology Program, Center for Sustainable Development, College of Arts and Sciences, Qatar University, Doha, Qatar. Also, Dr Mehariya worked as Postdoctoral Researcher at the Department of Chemistry, Umeå University, Umeå, Sweden. Dr. Mehariya earned a PhD in Engineering at the University of Campania "Luigi Vanvitelli", Italy. He has collaborated with national and international stakeholders, including policymakers, industries, and prominent RTD institutes in the field of biobased economy. He has worked at CSIR-Institute of Genomics and Integrative Biology, Delhi, India; Konkuk University, Seoul, South Korea; ENEA-Italian National Agency for New Technologies, Energy and Sustainable Economic Development, Rome, Italy; Hong Kong Baptist University, Hong Kong; University of Campania "Luigi Vanvitelli", Italy; and Sapienza—University of Rome, Italy. He has published more than 80 research and review articles and has edited four books. He is a life member of various societies such as Scandinavian Plant Physiology Society (SPPS), Sweden; Biotech Research Society (BRSI), India; the Association of microbiologist of India (AMI); Life Indian Science Association (ISA); and International Forum for BOTANISTS Under the auspices of ISLS, India. Dr. Mehariya serves as a peer review member assessor for the Horizon H2020-Marie Sklodowska-Curie Individual Fellowships-Europe.

Pradeep Verma (PhD) is currently working at the Department of Microbiology, Central University of Rajasthan (CURAJ), Ajmer, India. Prof. Verma completed his PhD from Sardar Patel University, Gujarat, India, in 2002. He was selected as UNESCO fellow (2002) and joined Czech Academy of Sciences, Prague, Czech Republic. He later moved to Charles University, Prague, to work as Post-Doctoral Fellow. He joined as Visiting Scientist (2004) at UFZ Centre for Environmental Research, Halle, Germany. He was awarded a DFG fellowship (2004) to work as Post-Doctoral Fellow at Gottingen University, Germany. He moved to India in 2007 where he joined Reliance Life Sciences, Mumbai, and worked extensively on biobutanol production which attributed a few patents to his name. Later, he was awarded with JSPS Post-Doctoral Fellowship Programme and joined Laboratory of Biomass Conversion, Research Institute of Sustainable Humanosphere Kyoto University, Japan. He is also a recipient of various prestigious awards such as Ron-Cockcroft award by Swedish society and UNESCO Fellow ASCR Prague. Prof. Verma is Fellow of many prestigious societies in India, for example, BRSI, MSI, and AMI. Prof. Verma is Group Leader of Bioprocess and Bioenergy Laboratory, Department of Microbiology, Central University of Rajasthan, Bandarsindri, Kishangarh, Ajmer, Rajasthan, India, and his area of expertise involves Bioprocess Development and Lignocellulosic and Algal Biomass-Based Biorefinery.

Contributors

Getachew Tafere Abrha
Omics of Algae Group, International Centre for Genetic Engineering and Biotechnology, Aruna Asaf Ali Marg
New Delhi, India

Prachi Agrawal
Department of Biotechnology, Guru Ghasidas Central University, Bilaspur
Chhattisgarh, India

Trisha Bagchi
Department of Botany, West Bengal State University, Berunanpukuria, Malikapur, Barasat, 24 Parganas (North)
West Bengal, India

Amit K. Bajhaiya
Algal Biotechnology Lab, Department of Microbiology, School of Life Sciences, Central University of Tamil Nadu, Thiruvarur
Tamil Nadu, India

Kiran Bala
Department of Biosciences and Biomedical Engineering, Indian Institute of Technology Indore, Simrol, Indore
Madhya Pradesh, India

Regina Barboza-Rodríguez
Biorefinery Group, Food Research Department, School of Chemistry, Autonomous University of Coahuila, Saltillo
Coahuila, Mexico

Alejandra Cabello-Galindo
Biorefinery Group, Food Research Department, School of Chemistry, Autonomous University of Coahuila, Saltillo
Coahuila, Mexico

Vikas Chandra
Department of Biotechnology, Guru Ghasidas Central University, Bilaspur
Chhattisgarh, India

K. Chandrasekhar
Department of Biotechnology, Vignan's Foundation for Science, Technology and Research, Vadlamudi, Guntur
Andhra Pradesh, India

Murthy Chavali
MIT World Peace University, Kothrud, Pune
Maharashtra, India

Meenakshi Fartyal
Department of Biotechnology, Kanoria PG Mahila Mahavidyalaya, University of Rajasthan, Jaipur
Rajasthan, India

Tonmoy Ghosh
Institute for Water and Wastewater Technology, Durban University of Technology
Durban, South Africa

Rahul Kumar Goswami
Bioprocess and Bioenergy Laboratory (BPBEL), Department of Microbiology, Central University of Rajasthan, Bandarsindri, Kishangarh, Ajmer
Rajasthan, India

Neeru Gupta
Omics of Algae Group, International Centre for Genetic Engineering and Biotechnology, Aruna Asaf Ali Marg
New Delhi, India

Chitra Jain
Biomitra Life Sciences Private Limited, Jaipur
Rajasthan, India

Pannaga Pavan Jutur
Omics of Algae Group, International Centre for Genetic Engineering and Biotechnology, Aruna Asaf Ali Marg
New Delhi, India

Sahil Kapoor
Department of Botany, Goswami Ganesh Dutta S.D. College
Chandigarh, India

Surojit Kar
Department of Botany, West Bengal State University, Berunanpukuria, Malikapur, Barasat, 24 Parganas (North)
West Bengal, India

Bikash Kumar
Department of Biosciences and Biomedical Engineering, Indian Institute of Technology Indore, Simrol, Indore
Madhya Pradesh, India

Gourav Kumar
Omics of Algae Group, International Centre for Genetic Engineering and Biotechnology, Aruna Asaf Ali Marg
New Delhi, India

Samanta Machado-Cepeda
Biorefinery Group, Food Research Department, School of Chemistry, Autonomous University of Coahuila, Saltillo
Coahuila, Mexico

Abdalah Makaranga
Omics of Algae Group, International Centre for Genetic Engineering and Biotechnology, Aruna Asaf Ali Marg
New Delhi, India

Sanjeet Mehariya
Algal Technology Program, Center for Sustainable Development, College of Arts and Sciences, Qatar University
Doha, Qatar

Bijaya Nag
Omics of Algae Group, International Centre for Genetic Engineering and Biotechnology, Aruna Asaf Ali Marg
New Delhi, India

Asha Arumugam Nesamma
Omics of Algae Group, International Centre for Genetic Engineering and Biotechnology, Aruna Asaf Ali Marg
New Delhi, India

Imran Pancha
Department of Industrial Biotechnology, Gujarat Biotechnology University, Gandhinagar
Gujarat, India

Kushal Kant Pant
Department of Biotechnology, Guru Ghasidas Central University, Bilaspur
Chhattisgarh, India

Sanchita Paul
Department of Botany, West Bengal State University, Berunanpukuria, Malikapur, Barasat, 24 Parganas (North)
West Bengal, India

Sukhvinder Singh Purewal
University Centre for Research & Development (UCRD), Chandigarh University, Mohali
Punjab, India

Subhisha Raj
Algal Biotechnology Lab, Department of Microbiology, School of Life Sciences, Central University of Tamil Nadu, Thiruvarur
Tamil Nadu, India

Alka Rani
Department of Botany, School of Basic and Applied Sciences, Central University of Punjab, Bathinda
Punjab, India

Rosa M. Rodríguez-Jasso
Biorefinery Group, Food Research Department, School of Chemistry, Autonomous University of Coahuila, Saltillo
Coahuila, Mexico

Miriam L. Rosales-Aguado
Biorefinery Group, Food Research Department, School of Chemistry, Autonomous University of Coahuila, Saltillo
Coahuila, Mexico

Gilver Rosero-Chasoy
Biorefinery Group, Food Research Department, School of Chemistry, Autonomous University of Coahuila, Saltillo
Coahuila, Mexico

Héctor A. Ruiz
Biorefinery Group, Food Research Department, School of Chemistry, Autonomous University of Coahuila, Saltillo
Coahuila, Mexico

Khem Chand Saini
School of Basic & Applied Sciences, Nirwan University, Jaipur
Rajasthan, India

Preethi Selvaraj
Algal Biotechnology Lab, Department of Microbiology, School of Life Sciences, Central University of Tamil Nadu, Thiruvarur
Tamil Nadu, India

Meenakshi Singh
Department of Ecology & Biodiversity, Sustaina Greens LLP, Vadodara
Gujarat, India

Madan Sonkar
Department of Biotechnology, Guru Ghasidas Central University, Bilaspur
Chhattisgarh, India

Arathi Sreenikethanam
Algal Biotechnology Lab, Department of Microbiology, School of Life Sciences, Central University of Tamil Nadu, Thiruvarur
Tamil Nadu, India

Udaypal
Bioprocess and Bioenergy Laboratory (BPBEL), Department of Microbiology, Central University of Rajasthan, Bandarsindri, Kishangarh, Ajmer
Rajasthan, India

Pradeep Verma
Bioprocess and Bioenergy Laboratory (BPBEL), Department of Microbiology, Central University of Rajasthan, Bandarsindri, Kishangarh, Ajmer
Rajasthan, India

1 Introduction to Microalgae and Its Refinery

Udaypal, Rahul Kumar Goswami, Sanjeet Mehariya and Pradeep Verma

1.1 INTRODUCTION

Algae are a wide group of microorganisms that vary in size from microscopic blue–green algae to massive, complex seaweeds that may be a few meters in length (Singh and Saxena 2015). Compared to terrestrial plants, algae possess a wide variety of photosynthetic processes, particularly in the form of pigments and structures that capture light and fix carbon dioxide (CO_2) (Larkum et al. 2003). They are photoautotrophic, having a simple thallus and no tissue differentiation. They can be found in a variety of aquatic environments, including harsh environments and fresh as well as saline water. There is a lot of variation in their reproduction and perennation methods due to their diversity in structure and habitat (Sahoo and Baweja 2015). The term "*algae*" includes both macroalgae (large size) and the extremely broad class of microorganisms known as microalgae (microscopic size). According to estimates, there are millions of different kinds of algae, the majority of which are microalgae (Evangelista et al. 2008).

Microalgae are unicellular photosynthetic organisms that often are found in freshwater and marine aquatic habitats. The size range of microalgae is 0.2 to 2 μm (picoplankton) up to filamentous forms that are 100 μm or larger (Ghosh et al. 2021). They can be cultivated in both open and closed systems in diverse cultivation modes, that is, photoautotrophic, heterotrophic, and mixotrophic. Depending on the species, microalgae can produce a variety of co-products (Gara and Stark 2012). Due to the co-production of additional molecules like pigments, proteins, polyunsaturated fatty acids (PUFAs), and antioxidants, microalgae are gaining significant attention as potential sources to produce therapeutic agents, food, cosmetics, energy, and pharmaceuticals (Bule et al. 2018). The global market for nutraceuticals and food supplements made from microalgae is well-established, increasing by five-fold since the beginning of the century and has significant growth potential (Enzing et al. 2014). An estimated US$6.5 billion is the market value of microalgae worldwide, of which US$2.5 billion comes from the health food industry, US$1.5 billion from docosahexaenoic acid (DHA) manufacturing, and US$700 million from aquaculture.

DOI: 10.1201/9781003455950-1

About 7.5 million tonnes of microalgae biomass are produced annually throughout the globe (Mobin and Alam 2017).

For the manufacturing of human nutrition and food supplements, thousands of production systems are under operation (Benemann 2009). The high-value-added compounds (HVACs) of microalgae may provide several health advantages, including anti-inflammatory, antioxidant, antiaging, antibacterial, anti-obesity, and anticancer properties, according to published data. They can be used to manufacture medications, cosmeceuticals, and next-generation nutritional supplements (Goswami et al. 2022a). Nowadays, microalgae are being utilized to produce: (i) high-value substances such as PUFAs, pigments, and phycobiliproteins (Mehariya et al. 2021a); (ii) entire biomass used in foods, feeds, and nutraceuticals (Mehariya et al. 2021b); (iii) production of biofuel and biofertilizers from processed biomass; or (iv) bioremediation by utilizing living microalgae (Cardenas et al. 2018).

In the current situation, only a certain microalgal component is used to produce biofuels such as lipids and carbohydrates (Mehariya et al. 2022). A biorefinery strategy is needed to utilize all the fractions of biomass to get through these limitations and optimization of economics (Zhu et al. 2015). By combining bioprocessing and chemical methods, bio-refineries produce biofuels as well as valuable products from microalgal biomasses which have a lesser impact on the environment (Rizwan et al. 2018). To advance microalgae on the market, a biorefinery strategy is essential. The biorefinery strategy has drawn the attention of researchers who are keen to make use of the valuable components synthesized by microalgae (Zhuang et al. 2020).

In addition to all the beneficial compounds, they also assist in lowering atmospheric CO_2 and bioremediation. Over the past 40 years, numerous studies have been conducted on microalgae-based bioremediation to remove contaminants from wastewater (Gill et al. 2013). Similarly, it has been discovered that utilizing microalgae biomass as biosorbents to remove heavy metals is a cost-efficient, environmentally friendly, and efficient approach to do it (Leong and Chang 2020). By utilizing wastewater and CO_2 sequestration, microalgae cultivation connects two major areas of concern: global warming and water pollution control/water recycling (Patel et al. 2017). These advantages make microalgae a potential candidate for the biorefinery approach (Koyande et al. 2019).

The chapter briefly includes the classification and overview of reported microalgae. Moreover, it elaborates on the microalgal upstream and downstream biorefinery systems such as cultivation and processing, Moreover, this chapter also elaborates on main commercial products obtained from microalgal biomass and their sustainable applications.

1.2 CLASSIFICATION AND OVERVIEW OF THE MICROALGAE

Microalgae represents a wide group of organisms; they are found in a variety of sizes, shapes, and habitats. Microalgae are mainly classified on the basis of their structural diversity, sizes, pigments, habitats, and nature of stored biomolecules (Table 1.1) (Correa et al. 2017).

TABLE 1.1
Brief Classifications and Overview of Commercially Important Microalgae and Their Reported Products.

Sr. No.	Division	Structural features	Habitat	Major commercial products	Microalgal species	References
1	Chlorophyta	Cup-shaped chloroplast, contain chlorophyll a and b in a single chloroplast.	Inhabit in freshwater, marine, or perhaps in terrestrial environments.	Astaxanthin, β-carotene, protein, fatty acids	*Dunaliella salina, Haematococcus pluvialis, Chlorella vulgaris, Chlamydomonas* sp.	Heimann and Huerlimann (2015); Goswami et al. (2022a), Shah et al. (2016)
2	Rhodophyta	Cells are spherical, contain a large single chloroplast with a single central pyrenoid, pigments found in Rhodophyta are chlorophyll a, d, and phycoerythrin.	Mostly marine, freshwater species are rare.	Phycobilin, pigments, cosmetics, antioxidants	*Porphyridium cruentum, Rhodella reticulata*	Román et al. (2002); Chen et al. (2010)
3	Haptophyta	One or two pyrenoid-containing chloroplasts, antapical nucleus, possess a haptonema.	Mostly marine	As a feed in aquaculture, fatty acids	*Pavlova salina, Isochrysis galbana*	Jordan (2012)
4	Stramenopiles	Typically have two flagella, two rows of tripartite hairs found on flagella, plastid contains four outer membranes.	Mostly in Marine water and some species are terrestrial.	Aquaculture feed, protein, laminarin	*Nannochloropsis oculata, Chaetoceros muelleri, Skeletonema costatum*	Heimann and Huerlimann (2015); Blanfuné et al. (2016); Chen et al. (2021)
5	Dinophyta	Dinokont flagellation, cells might be spherical or oval, motile or not.	Mainly marine, with a few freshwater species, photosynthetic, 50% are heterotrophs.	DHA, pigments, antioxidants	*Crypthecodinium cohnii, Gymnodinium* sp.	D'Alessandro et al. (2016); Schagerl et al. (2003)

1.3 UPSTREAM AND DOWNSTREAM PROCESSING OF MICROALGAE

1.3.1 Microalgae Cultivation

Microalgae cultivation for the production of value-added products has received a lot of attention in the past few decades. To effectively improve microalgae productivity optimization of growth conditions and selection of appropriate cultivation method are crucial. Various types of cultivation systems and modes used to cultivate microalgae are mentioned in the subsequent subsections.

1.3.1.1 Types of Cultivations Systems

There are two main cultivation systems such as open systems (circular pond and raceway pond) and closed systems (photobioreactors), which are used to cultivate the microalgae both in the laboratory or commercial scale (Figure 1.1a).

(a) *Open systems*: In open systems, raceway ponds, open ponds, circular ponds, and unstirred ponds are generally used for large-scale cultivation (Mehariya et al. 2021c). Microalgae production in open ponds is perhaps the oldest and most straightforward scientific approach for cultivating algae (Jerney and Spilling 2020). Compared to closed approaches or photobioreactors, open cultivation methods have much lower production costs. They are comparatively easy to operate and scale up (Kusmayadi et al. 2020). However, there

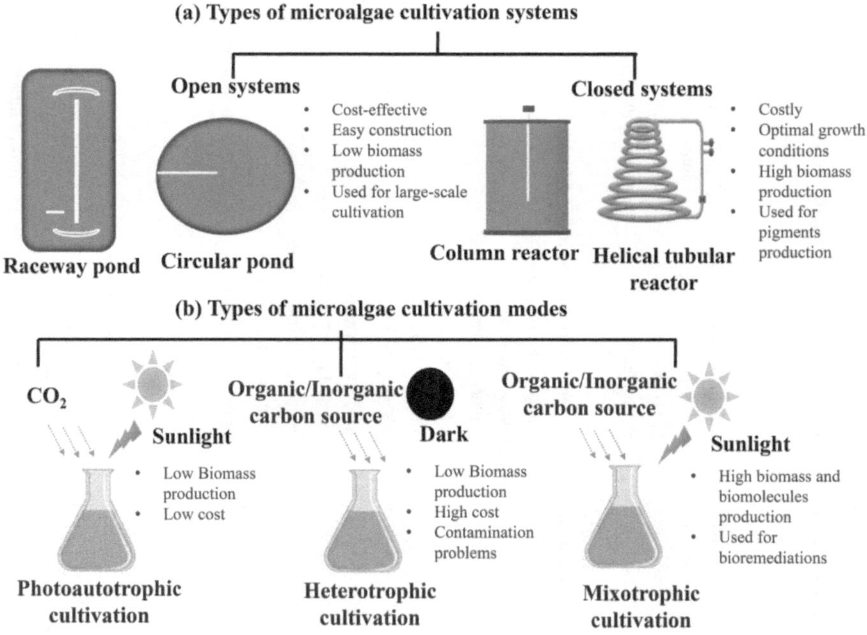

FIGURE 1.1 General representation of types of (a) microalgae cultivation systems and (b) types of microalgae cultivation modes.

are certain drawbacks in addition to these benefits, as open systems need a lot of space to scale up and are vulnerable to contamination and bad weather (Narala et al. 2016)

(b) *Closed systems:* The term "closed systems" refers to photobioreactors, which do not directly exchange gases or other contaminants with the outside environment of the cultivation systems (Yen et al. 2019). Closed photobioreactors provide benefits including improved cultivation parameter control and a lower chance of contamination (Apel and Weuster-Botz 2015). Photobioreactors (PBRs) are far more capable of supporting much higher photosynthesis effectiveness, biomass productivity, and biomass concentration than open systems (Wang et al. 2012). However, higher operational and capital expenses are still big challenges in close-system cultivation (Sirohi et al. 2022).

1.3.1.2 Modes of Cultivations

There are three major types of cultivation modes, that is, (a) photoautotrophic, (b) heterotrophic, and (c) mixotrophic (Figure 1.1b).

(a) *Photoautotrophic cultivation*: It is the most common and natural mode of cultivation where energy requirement is fulfilled by sunlight/LED and atmospheric CO_2 utilized as a carbon source (Zhan, et al. 2017).

(b) *Heterotrophic* cultivation: In this mode, organic compounds are supplied in a bioreactor that fulfills both energy and carbon requirement in a dark condition (Wang et al. 2014).

(c) *Mixotrophic* cultivation: Inorganic carbon (CO_2) and organic compounds are both utilized as carbon sources, and sunlight is utilized as an energy source (Abreu et al. 2012). This is considered as the best method for the treatment of wastewater (Cicco et al. 2021b).

1.3.1.3 Growth Parameters' Optimizations

Optimization of growth parameters such as light, temperature, carbon source, and pH of the growth medium can maximize the production of algal biomass.

(a) *Light*: It is required as an energy source for the process of photosynthesis. The light to the algae culture is provided through sunlight or artificially using light emitting diode (Wang et al. 2014). Optimum light intensity varies from species to species. However, 3,000–5,000 lux intensity is considered optimum for most of the microalgae (Goswami et al. 2022b). Day–night cycle is also an optimal for the growth of microalgae because continued illumination can lead to the photoinhibitions. It is suggested that continuous light promotes the growth of microalgae (Cicco et al. 2021a). While Goswami et al. (2023) reported that 12:12 hour day–night cycles promote the growth of *Tetraselmis indica*.

(b) Temperature: It shows a positive correlation with microalgal growth, but high temperature leads to oxidative stress (Esther et al. 2017). A temperature range of 20–35°C is considered optimum for most microalgae (Brindhadevi et al. 2021). However, some strains also survived into the low temperatures (15°C) (Goswami et al. 2022b, 2022c).

(c) pH is a key factor that limits the activity of enzymes that are involved in proteins and pigment synthesis pathways (Qiu et al. 2017). Slightly alkaline (7–9) pH is favorable for most of the microalgae (Hossain and Mahlia 2019). Moreover, it also facilitates the availability and uptake of nutrients to microalgae and promotes their growth (Goswami et al. 2023).

(d) Carbon source is a major component of algal biomass and its coproducts. Therefore, the optimization of carbon sources significantly influences algal biomass production. Both CO_2 and organic compounds (i.e., glucose, glycerol, sodium acetate, and citric acids) can be utilized as carbon sources (Zhan, et al. 2017; Goswami et al. 2022b, 2022c).

1.3.2 Downstream Processing

The downstream processing of microalgae consists of three steps: harvesting, drying, and extraction of biomolecules and conversion into different HVAC products (Figure 1.2) (Yew et al. 2019; Gu et al. 2022). Microalgae biomass was utilized for the generations of pigments and carotenoids. While lipids and fatty acid are converted into biofuels products, but production of biofuel from microalgal biomass is not commercialized yet (Goswami et al. 2022d).

1.3.2.1 Harvesting

The recovery of the cultivated microalgal biomass from their growth medium is referred to as harvesting. The ideal method for the recovery of high biomass at a low cost requires a small amount of energy, maintenance, and operational costs (Khoo et al. 2020). Centrifugation, flocculation, filtration, flotation, magnetic separation, electrolysis, ultrasound, and immobilization are common methods used for algal biomass harvesting (Kim et al. 2013). However, yet, no efficient harvesting methods are discovered. Many scientific investigations are carried out to find low-cost harvesting techniques. Among them bioflocculation-based cultivations are cheap, reliable, and energy-extensive process compared to others (Goswami et al. 2022e).

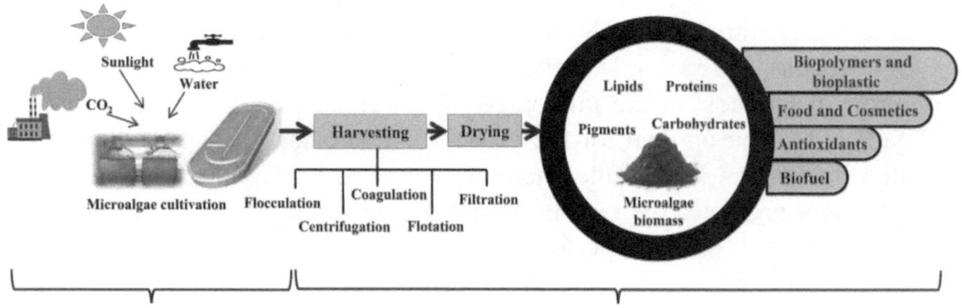

FIGURE 1.2 Upstream cultivation and downstream processing of microalgae biomass.

1.3.2.2 Drying

Dewatering/drying is a key process in downstream processing in which water is removed from the harvested cell culture to obtain biomass paste (Guldhe et al. 2014). Solar drying, convective drying, spray drying, and freeze drying are common methods of biomass drying. However, the selection of a suitable drying method is extremely dependent on the properties of the microalgal biomass (e.g., cell size, surface charge, salt concentration, pH) (Chen et al. 2015).

1.3.2.3 Extraction

The extraction process is crucial for recovering useful compounds from microalgae. Most of the by-products are encapsulated in microalgae cells (Salinas-Salazar et al. 2019). Therefore, the cell wall must be disrupted for the release of the desired molecules (Piasecka et al. 2014). Ultrasound extraction, enzymatic lysis, and chemical and mechanical techniques are some common examples of extraction procedures. Mechanical methods are utilized for cell rupture in large-scale applications because they provide full cell destruction with a high output yield (Onumaegbu et al. 2018). However, mechanical methods are more energy-intensive processes and increase the cost of extractions. Enzyme-assisted extraction through mono-based enzymes, such as cellulase, lipases, mannase, or cocktail enzymes, can lead to the cheapest methods of biomolecules' extraction from biomass. However, it is a time-consuming process as compared to other methods. While green solvent-based extraction or supercritical CO_2-based extraction was far better and cheapest, greener methods for the extractions of biomolecules (Goswami et al. 2021b).

1.4 ALGAE BIOREFINERY AND APPLICATIONS OF ITS PRODUCTS

1.4.1 Bioenergy Products

Fuels produced through algal biomass are a sustainable, possibly environmentally friendly, and affordable alternative source. Therefore, microalgae became a promising candidate to produce bioenergy.

1.4.1.1 Biodiesel

Biodiesel is a mono-alkyl ester that may be utilized in normal diesel engines with little-to-no modification. It is generated by transesterifying triglycerides or natural fatty acids using short-chain alcohols. Experimental evidence suggests that biodiesel is less harmful to the environment than petro-diesel, as it produces less gaseous pollutants overall and no net CO_2 or sulfur emissions (Rawat et al. 2013). Microalgae biomass has great potential to be utilized as feedstock for biodiesel compared to other vegetables oil because microalgae can multiply very fast and contain a significant amount of oil in their biomass. Usually, microalgae double their biomass within 24 hours, and oil content can reach up to 80% of their dry biomass weight. These features make microalgae a potential candidate for biodiesel production (Chisti 2007). The diatom *Phaeodactylum tricornutum*, several

Chlorella species, many strains and species of the eustigmatofite, *Nannochloropsis*, *Chlamydomonas* biomass are the most promising sources for biodiesel production (Gouveia et al. 2017). Four main processes required for biodiesel production from microalgae are (i) cultivation, (ii) harvesting, (iii) oil extraction, and (iv) transesterification (Wahlen et al. 2013).

However, there are still many difficulties associated with producing biodiesel from algal biomass. According to the latest data, producing biodiesel from microalgal biomass is more expensive than fossil fuels (Chowdhury and Loganathan 2019). Currently, various efforts are underway to find the most productive algae species and the generation of an in-situ transesterification directly from the biomass to reduce the time and cost of biodiesel production (Nwokoagbara et al. 2015).

1.4.1.2 Bioethanol

At present, ethanol is considered the most popular biofuel which can be generated from microalgal biomass. Utilizing microalgal biomass to produce ethanol offers benefits because microalgal starch and cellulose are similar to higher plant starch, and compared to agricultural crops, microalgal biomass has a low lignin content, which makes the pretreatment process quite easy (Ho et al. 2013). Therefore, it is recognized as an excellent raw material for the production of bioethanol (Maia et al. 2020). By employing chemical or enzymatic methods, these polysaccharides can be hydrolyzed to produce monosaccharides, and they can be fermented to get bioethanol. With their high carbohydrate content, species from the genera *Chlorella, Chlamydomonas, Scenedesmus, Chlorococcum*, and *Tetraselmis* represent ideal candidates for the production of bioethanol (Simas-Rodrigues et al. 2015). The process required for bioethanol production primarily depends on the nature of algal biomass. The process primarily consists of upstream and downstream activities that include cultivation, harvesting, pretreatment, saccharification, fermentation, and product recovery (Harun and Danquah 2011). The release and availability of fermentable sugars for the fermentation process are dependent on biomass pretreatment, which is a critical step (Ho et al. 2013).

However, like any other industrial process, the economics of algal biomass-based bioethanol production must be considered before it can be commercialized. The process of producing algal-based bioethanol faces several challenges such as the cost of pretreatment and the degradation of sugars during hydrolysis, which must be solved before it can be produced successfully on a large scale and for commercial use (Li et al. 2014). There are currently several attempts in the process to boost bioethanol production yields; including raising biomass growth rates, altering culture conditions to produce more carbohydrates, or enhancing the efficiency of converting carbohydrates to ethanol (Lakatos et al. 2019) and leftover biomass could be used for biopolymer production (Mehariya et al. 2023).

1.4.1.3 Biomethane

Currently, the primary methods for producing biofuel from microalgae biomass include extraction, hydrothermal liquefaction, pyrolysis, and anaerobic digestion. Notably, the dewatering of the microalgae slurry before the extraction, hydrothermal liquefaction, and pyrolysis procedures makes these conversion methods which are energy-expensive processes. Additionally, only lipids can be extracted during the extraction process, which leads to the wasting of proteins and carbohydrates (Xiao et al. 2019). Typically, extremely high temperatures or pressures are utilized in the hydrothermal liquefaction and pyrolysis processes, which require complicated systems and expensive operating costs (Wang et al. 2017). Anaerobic digestion can utilize wet biomass directly to generate methane, saving the energy needed for dewatering and lowering the overall utilization of energy in biogas generation; in addition, it is a carbon-neutral approach (Damtie et al. 2021).

The following four steps are involved in the production of biomethane via anaerobic digestion: (i) hydrolysis of biomass, (ii) acidogenesis, (iii) acetogenesis, and (iv) methanogenesis. The hydrolysis process involves the biodegradation of the polymeric substances of the microalgal biomass, such as proteins, lipids, and carbohydrates, by the specific microorganisms in the digester, enabling the production of simple molecules like fatty acids, glucose, and amino acids. In the acidogenic phase, simple molecules are transformed into small fatty acids and molecules into gaseous form such as CO_2, bio-H_2, and CH_3COOH. By acetogenesis, acetogenic bacteria transform fatty acids into acetic acid and H_2, which ultimately metabolize into methane by the action of methanogenic bacteria in the methanogenesis step (Zabed et al. 2019; Prajapati and Malik 2015). *Chlorella, Tetraselmis, Spirulina*, and *Scenedesmus*, are ideal microalgal genera to produce biomethane (González-Fernández et al. 2012). However, some limitations of biomethane generation put a constraint on it from being feasible on a large scale. The major challenges are that anaerobic digesters have a slow biodegradation rate and low C/N proportions in a digester due to a large concentration of proteins (Lu et al. 2019; Passos et al. 2018).

1.4.1.4 Biohydrogen

Currently, the majority of industrial H_2 production technologies are based on thermochemical processes, which typically require significant fossil fuel usage. Reports suggest that producing biohydrogen from microalgae is an extremely attractive approach to achieving zero carbon emissions and bioenergy sustainability (Li et al. 2022). With water as a by-product, hydrogen demonstrates the significant benefit of CO_2-free combustion (Limongi et al. 2021).

Microalgae produce biohydrogen via two majors procedures: (i) light-dependent pathway and (ii) light-independent pathway. The light-dependent pathway is known as biophotolysis in which the action of light energy causes the split of the H_2O into H_2 and O_2. The light-independent pathway includes dark fermentation of microalgal biomass, which is carried out by anaerobic bacteria (Singh et al. 2022; Goswami et al. 2021a). *Chlorella, Chlamydomonas Scenedesmus*, and *Tetraspora* are considered ideal microalgal genera for dark fermentation (Khetkorn et al. 2017).

However, some drawbacks limit the commercial scaling of biohydrogen production from microalgae such as the O_2-sensitivity of enzymes, lower yield, lower rate of biomass conversion, and the lack of understanding of the involved enzymes and biochemical reactions (Goswami et al. 2021b).

1.4.2 Pharmaceuticals

The health industry has traditionally used secondary metabolites derived from plants. Currently, researchers have switched to microalgae as a production platform, because of the lower yields and seasonal changes of plant sources. Microalgae cells can produce novel chemical compounds that are believed to be challenging to obtain by chemical synthesis. Microalgae are used for manufacturing many kinds of pharmaceutical products which are mentioned in the subsequent subsections (Jha et al. 2017).

1.4.2.1 Single-Cell Protein

Microalgae can be grown in controlled environments with little water consumption, and they express high levels of protein. Numerous species of microalgae have amino acid profiles that are similar to those of well-balanced proteins as described by WHO/FAO (Janssen et al. 2022). Comprehensive research and nutritional studies have shown that these algal proteins are superior to traditional vegetable proteins (Montenegro-Herrera et al. 2022). *Spirulina*, *Chlorella*, *Dunaliella*, and *Scenedesmus* genus are the most common microalgae used in the human diet and protein production at the commercial scale (Barka and Blecker 2016).

However, in addition to its beneficial characteristics, microalgal single-cell protein has some drawbacks such as high nucleic acid amount in biomass, increasing uric acid in serum that results in kidney stones formation, the cell wall of microalgae being nondigestible by humans, endotoxins, and high risk of contamination. Selection of appropriate microalgae, suitable substrate, and optimization of conditions may help to overcome these limitations (Sharif et al. 2021; Becker 2007).

1.4.2.2 Antiviral Agents

There are currently no effective vaccinations available against many common viral diseases, and the manufacturing of vaccines against other viruses, such as HIV and HCV, has shown to be ineffective (Mahendran et al. 2021). Some antiviral substances can be found naturally, and microalgae are one of the most promising sources for their synthesis. One of the earliest research studies on the antiviral action of microalgae was carried out on *Chlorella pyrenoidosa* and was found to have an inhibitory impact on mice against the vesicular stomatitis virus (VSV) (Carbone et al. 2021). Santoyo et al. (2010) studied the antiviral effect of ethanol extract of *Chlorella vulgaris* against herpes simplex virus type 1 and found significant inhibition of infection. Huleihel et al. (2001) studied the antiviral activity of red algae *Porphyridium* sp. against Herpes simplex viruses and Varicella zoster virus and reported 50% inhibition of infection.

Compounds produced by microalgae such as phycobiliprotein, fucoidan, xylomannan, and polyphenol directly inactivates viruses by binding them irreversibly, stopping them from interacting with their target cells or destroying the virion's structural integrity (Reynolds et al. 2021). Genus *Chlorella*, *Dunaliella*, *Chlamydomonas*, and *Scenedesmus* are the most popular microalgae that contain antiviral substances and show a significant inhibition of viral infections (Khavari et al. 2021).

1.4.2.3 Antibacterial Agents

An important issue that has raised the demand for new antibacterial chemicals is the growth of bacteria that are more resistant to antibiotics. Metabolites synthesized by microalgae such as phenolics, terpenes, glycolipids, fatty acids, glycolipids, and alkaloids have antimicrobial properties. Therefore, several investigations are focused on the antibacterial potential of green microalgae (Little et al. 2021). Bhuyar et al. (2020) investigated the antibacterial activity of *Oscillatoria* extract against Gram –ve bacteria *Streptomyces aureus, Bacillus subtilis,* and Gram +ve bacteria *Escherichia coli,* and the results show significant inhibition of bacterial growth. Hidhayati et al. (2022) investigated the antibacterial activity of ethyl acetate extract of *Spirulina platensis* against *Propionibacterium*, and *Enterobavter aerogenes* result shows that the extract has strong antibacterial properties against bacteria. Ethanolic extract of *Chlorella vulgaris* contains antibacterial compounds: 38.8% linoelaidic acid and 30.0% phytol that show strong antibacterial activity against *Escherichia coli* (Kim et al. 2017). Ethanol extract of *Chlorella* sp. and *Scenedesmus* sp., methanol extract of *Chlorococcum and Tetraselmis*, and ethyl acetate extract of *Nannochloropsis* show effectiveness against bacterial growth (Ognistaia et al. 2022).

1.4.2.4 Antioxidants

In past decades, scientists have been searching for a natural replacement for synthetic antioxidants; microalgae are considered a potential source of antioxidants because they stimulate antioxidant defense against oxidative stress and prevent cells from reactive oxygen species (Assunção et al. 2017). Carotenoids, pigments, and fatty acids produced by microalgae contain antioxidant properties (Gauthier et al. 2020). Microalgae *Chlorella, Tetraselmis, Phaeodactylum*, and Bot *ryococcus* contain large amounts of antioxidants. Thus, they are considered a potential source of antioxidants (Goiris et al. 2012). Ascorbic acid, carotenoids, glutathione, tocopherols, and phenolic compounds are secondary metabolites of microalgae, and they can neutralize reactive oxygen species (Coulombier et al. 2021). Guedes et al. (2013) studied that *Scenedesmus obliquus* contains a significant amount of antioxidants such as zeaxanthin, violaxanthin, neoxanthin, and luteoxanthin that protect DNA from oxidative damage. Widowati et al. (2017) evaluated the antioxidant activity of *Dunaliella salina, Isochrysis galbana*, and *Tetraselmis chuii* and found that *Isochrysis galbana* shows excellent antioxidant potential followed by *Dunaliella salina*. Bellahcen et al. (2020) carried out a comparison study of ethanolic, lipid, and aqueous extracts of *Spirulina platensis* for their antioxidant activity, and result shows that ethanolic extract has the highest antioxidant activity. Nasirian et al. (2018) examined the effect of the antioxidant activity of *Spirulina platensis* on diabetes; result shows that the activity of antioxidant enzymes was significantly increased and lowered the level of diabetes. Antioxidants produced by microalgae show anticancer, anti-inflammatory, antidiabetic, neuroprotective, and cardiovascular protective effects (Yuan et al. 2011).

1.4.3 Cosmetics

Cosmetics are products that improve the appearance of the skin (Mourelle et al. 2017). Synthetic cosmetics can have harmful effects on the skin. Due to the excellent antioxidant property of microalgae by-products, they can be utilized to produce cosmetics (Zhuang et al. 2022). Microalgal metabolites such as β-carotene, folic acid, pantothenic acid, and vitamin B_{12} have antimicrobial, UV protective, and anti-aging properties and can also repair and heal skin; due to these properties, microalgae draw the significant attention of cosmeceutical industries (Yarkent et al. 2020). *Dunaliella*, *Chlorella*, *Arthrospira*, and *Haematococcus* are considered potential microalgae to produce cosmetics at a commercial scale (Morocho et al. 2020).

1.4.4 Chemicals

Microalgae can produce a wide variety of chemicals and metabolites (Figure 1.3), and due to their photoautotrophic nature, they are considered sustainable and low-cost alternatives to synthetic chemicals.

1.4.4.1 Pigments

Due to the harmful effects of synthetic colorants, natural pigments produced by microalgae become attractive alternatives. Microalgae synthesize a wide variety

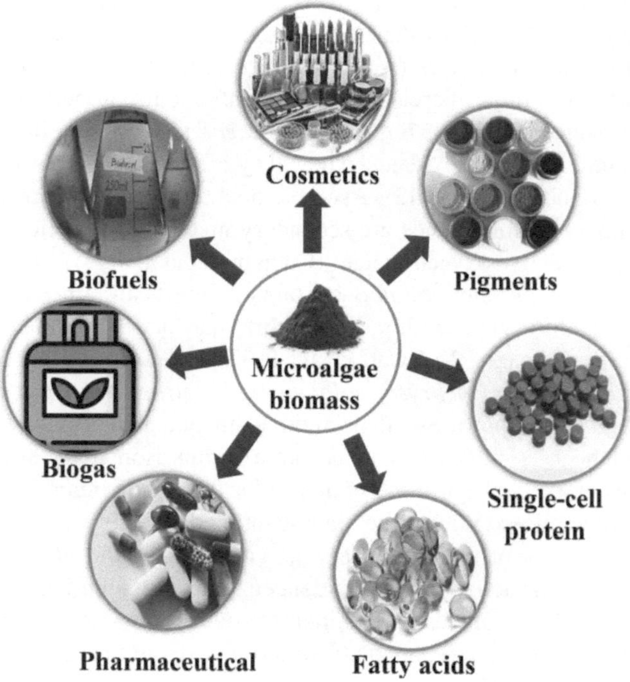

FIGURE 1.3 Microalgae biomass-based biorefineries and different high value-added products.

of pigments such as chlorophyll a, b, and c; xanthophylls; phycobiliproteins; astaxanthin; and β-carotene (Begum et al. 2016). Due to their antioxidant nature, these pigments receive significant value in the world food market (Sun et al. 2023). Natural pigments are also utilized in the production of cosmetics due to their anti-aging property (Morocho et al. 2020). *Chlorella*, *Haematococcus*, *Spirulina*, and *Dunaliella* are the most studied microalgae to produce pigments (Silva et al. 2020). However, the usage of these pigments still faces numerous obstacles such as the expensive production costs and instability of isolated pigments (Pagels et al. 2020).

1.4.4.2 Biopolymers and Bioplastics

Due to its non-degradability and generation of waste in huge amounts, synthetic plastic places harmful effects on the environment (Mastropetros et al. 2022). Biopolymers are considered sustainable alternatives to fossil-based synthetic polymers due to their biodegradability and renewability (Madadi et al. 2021). Microalgae synthesize poly-hydroxy-alkanoate (PHA) and starch used for biopolymer synthesis, and characteristics of these polymers are similar to those of petrochemical polymers, which safe for the environment (Rajpoot et al. 2022). Biopolymers are produced by the fermentation of microalgal biomass or by combining biomass by mixing some additives (Onen et al. 2020). *Spirulina, Chlorella, Neochloris,* and *Phaeodactylum* are the potential microalgae for the synthesis of biopolymers due to their starch or polyhydroxy acetone (PHA)-rich nature (Onen et al. 2020).

1.4.4.3 Nanoparticles

In recent decades, the usage of nanoparticles has significantly increased due to their wide application in the medical and pharma sectors. Due to harmful effects on the environment, physical and chemical methods are not considered sustainable for the biofabrication of nanoparticles (Jena et al. 2013). Microalgae take in metals ions during detoxification so they are considered a novel alternative to physical and chemical methods and also considered potential candidates for wastewater treatment (Mohseniazar et al. 2011; Agarwal et al. 2019). Zayadi et al. (2020) synthesized gold nanoparticles using *Chlorella* and *Spirulina* which resulted in higher stability and less reactivity of gold nanoparticles. Ebrahiminezhad et al. (2016) synthesized silver nanoparticles by utilizing *Chlorella vulgaris* that shows anticancer and antioxidant properties. Muthusamy et al. (2017) synthesized silver nanoparticles by utilizing *Spirulina platensis* that shows antibacterial activity against *Staphylococcus* and *Klebsiella*.

1.4.5 Food products

Protein, carbohydrates, lipids, and other physiologically active substances can be synthesized by microalgae in large amounts (Table 1.2). In this scenario, microalgae are becoming more and more popular in food applications (Rahman 2020).

TABLE 1.2
Protein, Lipid, and Carbohydrate Composition of Commercially Important Microalgae.

Microalgae	Protein (%)	Lipid (%)	Carbohydrate (%)	References
Chlorella vulgaris	47.82 ± 0.05	13.32 ± 0.07	8.08 ± 0.09	Tokuşoglu and Üunal (2003)
Isochrysis galbana	37	24	11	Reiriz et al. (1989)
Tetraselmis gracilis	33.02 ± 0.43	7.95 ± 0.42	29.96 ± 1.99	Gorgonio et al. (2013)
Porphyridium cruentum	34.1± 4.4	6.53 ± 0.46	32.1 ± 5.6	Fuentes et al. (2000)
Dunaliella salina	25.67	18.02	40.21	Muhaemin and Kaswadji (2010)
Spirulina platensis	63	7.53	15.35	Tokuşoglu and Üunal (2003)
Nannochloropsis sp.	28.8	18.4	35.9	Rebolloso et al. (2001a)
Phaeodactylum tricornutum	36.4	18	26.1	Rebolloso et al. (2001b)
Galdieria sulphuraria	32.5	1.77	62.9	Graziani et al. (2013)
Tetraselmis indica	n.d.	44 ± 0.47	n.d.	Goswami et al. (2023)

1.4.5.1 Fatty Acid

Polyunsaturated fatty acids (PUFAs) are not synthesized by the human body so they are known as essential fatty acids. Nowadays, PUFAs are obtained from fish oil, but their application in food additives is limited due to their bad smell, taste, and poor oxidative stability (Sayeda et al. 2015). In contrast, PUFAs obtained from microalgae have good taste and show medicinal benefits such as anticancer, anti-obesity, antidiabetic properties; reduction in inflammation; and prevention from cardiovascular diseases, and so they are considered the potential alternative to fish and vegetable oil (Kumar et al. 2019). Microalgae comprise significant amounts of PUFAs such as eicosapentaenoic acid (EPA), linolenic acid (ALA), and DHA (Santos-Sánchez et al. 2016; Mehariya et al. 2021b).

Currently, microalgae species from the genera *Chlorella*, *Isochrysis*, *Spirulina*, *Haematococcus*, *Dunaliella*, *Nostoc*, *Nannocloropsis*, *Porphyridium*, *Arthrospira*, *Schizochytrium*, and *Thalassioi* are used for the production of fatty acids at commercial scale (Maltsev and Maltseva 2021).

1.4.5.2 Protein and Amino Acids

Protein is an essential component of the human diet because it provides the majority of the required nitrogen and essential amino acids to the human body (Torres-Tiji et al. 2020). Due to its higher nutritional property, the presence of essential amino acids, and being low allergic as compared to milk and soy protein, microalgal protein is considered a novel protein supplement (Soto-Sierra et al. 2018). Due to higher protein content (55–70%), various microalgae such as *Dualiella*, *Chlorella*, and *Spirulina* are considered potential sources for the production of protein supplements. It is common and well-established fact to consume whole microalgae biomass in food and feed (Amorim et al. 2021). Nowadays, *Dunaliella* and *Sprulina* are directly sold as protein supplements, aquafeed, or animal feed only after drying without any processing (Matos 2019).

1.4.5.3 Carbohydrates

Microalgae are photoautotrophic microorganisms that convert atmospheric CO_2 into carbohydrates, where synthesized carbohydrates can be stored inside cells or may act as a structural component (Markou et al. 2012). Structural carbohydrates of microalgae do not contain lignin, so they are considered superior animal feed and substrate for bioethanol production as compared to plant-based biomass (Debnath et al. 2021). Microalgae with higher carbohydrate content are considered a potential approach for biomethane and biohydrogen production (Liu et al. 2012).

Chlamydomonas, *Dunaliella*, *Scenedesmus*, *Nannochloropsis*, and *Tetraselmis* can produce more than 50% carbohydrate under favorable conditions, so they are considered potential microalgae for commercialization (Spolaore et al. 2006).

1.4.6 Environmental Applications

1.4.6.1 Carbon Dioxide Sequestration

The atmospheric CO_2 level is rising day by day due to anthropogenic activities or the burning of fossil fuels. CO_2 is a major greenhouse gas that contributes to more than half of the greenhouse effects (Singh and Ahluwalia 2013). Due to rapid growth, higher photosynthetic efficiency, and wide environmental adaptability, microalgae-based carbon sequestration is considered the most effective, economical, and sustainable method to overcome the greenhouse effect (Xu et al. 2019). More than 25% of annual anthropogenic CO_2 is absorbed by oceans, converted into carbonic acid, and this causes acidification of the ocean; algae can take in carbonate via the carbon concentration mechanism (CCM) and prevent the ocean from acidification (Devi et al. 2013). Simultaneously, produced algal biomass is utilized in the production of biofuel, pharmaceuticals, cosmetics, nutraceuticals, pigments, and various other valuable products (Banerjee et al. 2020; Goswami et al. 2022c).

1.4.6.2 Wastewater Treatment

Word population is rising day by day that is accelerating urbanization, and urbanization leads to wastewater generation. Consumption of wastewater cause several health-related issues such as diarrhea, vomiting, and endocrine disruption (Goswami et al. 2022d). Therefore, wastewater treatment becomes essential to recycle it and resolve the problem of water shortage (Singh et al. 2022). Among various wastewater treatment techniques, microalgae-based treatment received much attention due to its low cultivation cost, sustainable nature, and higher nutrient removal capacity (Goswami et al. 2022e, 2022f).

Microalgae could potentially be able to significantly reduce the number of nutrients in wastewater, and treated wastewater fulfills the strict guidelines for discharge and reuse of the water (Li et al. 2019; Iovinella et al. 2022). Microalgae can efficiently utilize nitrogen and phosphorus, as well as organic and inorganic carbon from wastewater, and harvested biomass can be utilized in manufacturing several HVAC products (Mohsenpour et al. 2021). Microalgae can take in heavy metals from wastewater, which can be utilized in the production of nanoparticles (Goswami et al. 2022f). Based on these advantages, microalgae-based wastewater treatment is considered a potential alternative to conventional chemical-based treatment methods (Nagarajan et al. 2019).

1.5 FUTURE PROSPECTS AND CONCLUSIONS

Microalgae-based refineries make complete use of biomass while lowering production costs and environmental pollution. Microalgae have a huge potential as raw material for biorefinery applications because they contain significant amounts of biomolecules that have various energy, health, and industrial applications. In addition, the cultivation of microalgae can remediate wastewater and reduce the CO_2 from the atmosphere. However, due to high cultivation and harvesting costs (20–30% of total production cost) and low light penetration in phototrophic and contamination in heterotrophic cultivation, the industrial production of microalgal by-products is not yet cost-effective. Moreover, biomass cultivated in polluted water is not considered safe for the production of pharmaceuticals and food products. Therefore, the commercial application of microalgae is limited only to biofuel production, so there was an immediate need to integrate it with other valuable by-product production. The implementation of the macroalgal biorefinery is crucial to support a microalgae-based economy. Research on the genetic modification of microalgae strains to face harsh environmental conditions is still needed in upcoming future.

ACKNOWLEDGMENTS

Authors did not receive any specific grant from funding agencies in the public, commercial, or not-for-profit sectors.

CONTRIBUTIONS

Udaypal has contributed toward investigation, resources, and roles/writing the original draft; **Rahul Kumar Goswami** toward review and editing; **Sanjeet Mehariya** toward review and editing; and **Pradeep Verma** has contributed to conceptualization, editing, and supervision.

REFERENCES

Abreu, Ana Paula, Bruno Fernandes, António Augusto Vicente, José Teixeira, and Giuliano Dragone. 2012. "Mixotrophic cultivation of *Chlorella vulgaris* using industrial dairy waste as organic carbon source." *Bioresource Technology* 118: 61–66. https://doi.org/10.1016/j.biortech.2012.05.055.

Agarwal, Prashant, Ritika Gupta, and Neeraj Agarwal. 2019. "Advances in synthesis and applications of microalgal nanoparticles for wastewater treatment." *Journal of Nanotechnology* 2019: 739271 3. https://doi.org/10.1155/2019/7392713.

Amorim, Matheus Lopes, Jimmy Soares, Jane Sélia dos Reis Coimbra, Mauricio de Oliveira Leite, Luiz Fernando Teixeira Albino, and Marcio Arêdes Martins. 2021. "Microalgae proteins: Production, separation, isolation, quantification, and application in food and feed." *Critical Reviews in Food Science and Nutrition* 61: 1976–2002. https://doi.org/10.1080/10408398.2020.1768046.

Apel, Andreas Christoph, and Drik Weuster-Botz. 2015. "Engineering solutions for open microalgae mass cultivation and realistic indoor simulation of outdoor environments." *Bioprocess and Biosystems Engineering* 38: 995–1008. https://doi.org/10.1007/s00449-015-1363-1.

Assunção, Mariana F. G., Raquel Amaral, Clara B. Martins, Joana D. Ferreira, Sandrine Ressurreição, Sandra Dias Santos, Jorge M. T. B. Varejão, and Lília M. A. Santos. 2017. "Screening microalgae as potential sources of antioxidants." *Journal of Applied Phycology* 29: 865–877. https://doi.org/10.1007/s10811-016-0980-7.

Banerjee, Ishita, Swapnamoy Dutta, Cheryl Bernice Pohrmen, Ravikant Verma, and Dharmatma Singh. 2020. "Microalgae-based carbon sequestration to mitigate climate change and application of nanomaterials in algal biorefinery." *Octa journal of Biosciences* 8: 129–136. doi: http://sciencebeingjournal.com/octa-journal-biosciences/microalgae-based-carbon-sequestration-mitigate-climate-change-and-applicati.

Barka, Abakoura, and Christophe Blecker. 2016. "Microalgae as a potential source of single-cell proteins. A review." *Biotechnologie, Agronomie, Société et Environnement/Biotechnology, Agronomy, Society and Environment* 20: 427–426. https://doi.org/10.25518/1780-4507.13132.

Becker, Eberhard Wolfgang. 2007. "Micro-algae as a source of protein." *Biotechnology Advances* 25: 207–210. https://doi.org/10.1016/j.biotechadv.2006.11.002.

Begum, Hasina, Fatimah M. D. Yusoff, Sanjoy Banerjee, Helena Khatoon, and Mohamed Shariff. 2016. "Availability and utilization of pigments from microalgae." *Critical Reviews in Food Science and Nutrition* 56: 2209–2222. https://doi.org/10.1080/10408398.2013.764841.

Bellahcen, Touria Ould, Abderrahmane Amiri, Ikram Touam, Fouzia Hmimid, Abdelaziz El Amrani, Abdelmjid Cherif, and Mounia Cherki. 2020. "Evaluation of Moroccan microalgae: Spirulina platensis as a potential source of natural antioxidants." *Journal of Complementary and Integrative Medicine* 17: 20190036. https://doi.org/10.1515/jcim-2019-0036.

Benemann, John R. 2009. "Microalgae biofuels: A brief introduction." *Benemann Associates and MicroBio Engineering. Walnut Creek, CA* 2009. https://advancedbiofuelsusa.info/wp-content/uploads/2009/03/microalgae-biofuels-an-introduction-july23-2009-benemann.pdf.

Bhuyar, Prakash, Mohd Hasbi Ab Rahim, Gaanty Pragas Maniam, Rameshprabu Ramaraj, and Natanamurugaraj Govindan. 2020. "Exploration of bioactive compounds and antibacterial activity of marine blue-green microalgae *(Oscillatoria sp.)* isolated from coastal region of west Malaysia." *SN Applied Science s* 2: 1–10. doi: https://doi.org/10.1007/s42452-020-03698-8.

Blanfuné, Aurelie, Charles-François Boudouresque, Marc Verlaque, and Thierry Thibaut. 2016 "The fate of Cystoseira crinita, a forest-forming Fucale (Phaeophyceae, Stramenopiles), in France (North Western Mediterranean Sea)." *Estuarine, Coastal and Shelf Science* 181: 196–208. https://doi.org/10.1016/j.ecss.2016.08.049.

Brindhadevi, Kathirvel, Thangavel Mathimani, Eldon R. Rene, Sabarathinam Shanmugam, Nguyen Thuy Lan Chi, and Arivalagan Pugazhendhi. 2021. "Impact of cultivation conditions on the biomass and lipid in microalgae with an emphasis on biodiesel." *Fuel* 284: 119058. https://doi.org/10.1016/j.fuel.2020.119058.

Bule, Mohammed Hussen, Ishtiaq Ahmed, Faheem Maqbool, Muhammad Bilal, and Hafiz MN Iqbal. 2018. "Microalgae as a source of high-value bioactive compounds." *Frontier in Bioscience* 10: 197–216. https://doi.org/10.2741/s509.

Carbone, Dora Allegra, Paola Pellone, Carmine Lubritto, and Claudia Ciniglia. 2021. "Evaluation of microalgae antiviral activity and their bioactive compounds." *Antibiotics* 10: 746. https://doi.org/10.3390/antibiotics10060746.

Cardenas, Jose Antonio Garridio, Francisco Manzano-Agugliaro, Francisco Gabriel Acien-Fernandez, and Emilio Molina-Grima. 2018. "Microalgae research worldwide." *Algal Research* 35: 50–60. https://doi.org/10.1016/j.algal.2018.08.005.

Chen, Bilian, Wenlang You, Jian Huang, Ying Yu, and Weiping Chen. 2010. "Isolation and antioxidant property of the extracellular polysaccharide from Rhodella reticulata." *World Journal of Microbiology and Biotechnology* 26: 833–840. https://doi.org/10.1007/s11274-009-0240-y.

Chen, Ching-Lung, Jo-Shu Chang, and Duu-Jong Lee. 2015. "Dewatering and drying methods for microalgae." *Drying Technology* 33: 443–454. https://doi.org/10.1080/07373937.2014.997881.

Chen, Jichen, Jianchao Yang, Hong Du, Muhammad Aslam, Wanna Wang, Weizhou Chen, Tangcheng Li, Zhengyi Liu, and Xiaojuan Liu. 2021. "Laminarin, a major polysaccharide in stramenopiles." *Marine Drugs* 19: 576. https://doi.org/10.3390/md19100576.

Chisti, Yusuf. 2007. "Biodiesel from microalgae" *Biotechnology Advances* 25: 294–306. https://doi.org/10.1016/j.biotechadv.2007.02.001.

Chowdhury, Harun, and Bavin Loganathan. 2019. "Third-generation biofuels from microalgae: A review." *Current Opinion in Green and Sustainable Chemistry* 20: 39–44. https://doi.org/10.1016/j.cogsc.2019.09.003.

Cicco, Maria Rosa di, Manuela Iovinella, Maria Palmieri, Carmine Lubritto, and Claudia Ciniglia. 2021b. "Extremophilic microalgae Galdieria gen. For urban wastewater treatment: current state, the case of "POWER" system, and future prospects." *Plants* 10: 2343. https://doi.org/10.3390/plants10112343.

Cicco, Maria Rosa di, Maria Palmieri, Simona Altieri, Claudia Ciniglia, and Carmine Lubritto. 2021a. "Cultivation of the acidophilic microalgae *galdieria phlegrea* with wastewater: Process yields." *International Journal of Environmental Research and Public Health* 18: 2291. https://doi.org/10.3390/ijerph18052291.

Correa, Iago, Paulo Drews, Silvia Botelho, Marcio Silva de Souza, and Virginia Maria Tavano. 2017. "Deep learning for microalgae classification." *2017 16th IEEE International Conference on Machine Learning and Applications (ICMLA)*: 20–25. https://doi.org/10.1109/ICMLA.2017.0–183.

Coulombier, Noémie, Thierry Jauffrais, and Nicolas Lebouvier. 2021. "Antioxidant compounds from microalgae: A review." *Marine Drugs* 19: 549. https://doi.org/10.3390/md19100549.

D'Alessandro, Emmanuel Bezzera, and Nelson Roberto, Antoniosi Filho. 2016. "Concepts and studies on lipid and pigments of microalgae: A review." *Renewable and Sustainable Energy Reviews* 58: 832–841. https://doi.org/10.1016/j.rser.2015.12.162.

Damtie, Mekdimu Mezemir, Jingyeong Shin, Hyun Min Jang, Hyun UK Cho, Jinhua Wang, and Young Mo Kim. 2021. "Effects of biological pretreatments of microalgae on hydrolysis, biomethane potential and microbial community." *Bioresource Technology* 329: 124905. https://doi.org/10.1016/j.biortech.2021.124905.

Debnath, Chandrani, Tarun Kanti Bandyopadhyay, Biswanath Bhunia, Umesh Mishra, Selvaraju Narayanasamy, and Muthusivaramapandian Muthuraj. 2021. "Microalgae: Sustainable resource of carbohydrates in third-generation biofuel production." *Renewable and Sustainable Energy Reviews* 150: 111464. https://doi.org/10.1016/j.rser.2021.111464.

Devi, M. Prathima, Yerramsetti Vekata Swamy, and Venkata Mohan Srinivasula Reddy. 2013. "Nutritional mode influences lipid accumulation in microalgae with the function of carbon sequestration and nutrient supplementation." *Bioresource Technology* 142: 278–286. https://doi.org/10.1016/j.biortech.2013.05.001.

Ebrahiminezhad, Alireza, Mahboobeh Bagheri, Seyedeh-Masoumeh Taghizadeh, Aydin Berenjian, and Younes Ghasemi. 2016. "Biomimetic synthesis of silver nanoparticles using microalgal secretory carbohydrates as a novel anticancer and antimicrobial." *Advances in Natural Sciences: Nanoscience and Nanotechnology* 7: 015018. https://doi.org/10.1088/2043-6262/7/1/015018.

Enzing, Christien, Matthias Ploeg, Maria Barbosa, and Lolke Sijtsma. 2014. "Microalgae-based products for the food and feed sector: An outlook for Europe." *JRC Scientific and Policy Reports*: 19–37. https://dx.doi.org/10.2791/3339.

Esther, Posadas, Cynthia Alcántara, P A García-Encina, Luisa Gouveia, Benoit Guieysse, Zane Norvill, Gabriel Acién et al. 2017. "Microalgae cultivation in wastewater." In *Microalgae-based Biofuels and Bioproducts*, pp. 67–91. Sawston: Woodhead Publishing. https://doi.org/10.1016/B978-0-08-101023-5.00003-0.

Evangelista, Valtere, Laura Barsanti, Anna Maria Frassanito, Vincenzo Passarelli, and Paolo Gualtieri, eds. 2008. "Algal toxins: nature, occurrence, effect and detection." *Springer Science & Business Media*: 1–15. https://doi.org/10.1007/978-1-4020-8480-5.

Fuentes, M. M. Rebolloso, Acién Gabriel Fernández, Jose Antonio Sánchez Pérez, and J. L. Guil Guerrero. 2000. "Biomass nutrient profiles of the microalga *Porphyridium cruentum*." *Food Chemistry* 70: 345–353. https://doi.org/10.1016/S0308-8146(00)00101-1.

Gara, Ian O., and Stark Melissa. 2012. "An introduction to photosynthetic microalgae." *Disruptive Science and Technology* 1: 65–67. https://doi.org/10.1089/dst.2012.0017.

Gauthier, M. R., Senhorinho Gerusa Neyla Andrade, and Scott John Ashley 2020. "Microalgae under environmental stress as a source of antioxidants." *Algal Research* 52: 102104. https://doi.org/10.1016/j.algal.2020.102104.

Ghosh, Abu, Said, Zvy Dubinsky, Vitor Verdelho, and David Iluz. 2021. "Unconventional high-value products from microalgae: A review." *Bioresource Technology* 329: 124895. https://doi.org/10.1016/j.biortech.2021.124895.

Gill, Saba Shahid, Muhammad Aamer Mehmood, Umer Rashid, Muhammad Ibrahim, Anam Saqib, and Muhammad Rizwan Tabassum. 2013. "Waste-water treatment coupled with biodiesel production using microalgae: A bio-refinery approach." *Pakistan Journal of Life and Social Sciences* 11: 179–189. doi: www.pjlss.edu.pk/pdf_files/2013_3/179-189.pdf.

Goiris, Koen, Koenraad Muylaert, Ilse Fraeye, Imogen Foubert, Jos De Brabanter, and Luc De Cooman. 2012. "Antioxidant potential of microalgae in relation to their phenolic and carotenoid content." *Journal of Applied Phycology* 24: 1477–1486. https://doi.org/10.1007/s10811-012-9804-6.

González-Fernández, Cristina, Bruno Sialve, Nicolas Bernet, and Jean-Philippe Steyer. 2012. "Impact of microalgae characteristics on their conversion to biofuel. Part II: Focus on biomethane production." *Biofuels, Bioproducts and Biorefining* 6: 205–218. https://doi.org/10.1002/bbb.337.

Gorgonio, Cristiane Mesquita da Silva, Donato Alexandre Gomes Aranda, and Sonia Couri. 2013. "Morphological and chemical aspects of *Chlorella pyrenoidosa*, *Dunaliella tertiolecta*, *Isochrysis galbana* and Tetraselmis gracilis microalgae." *Natural Science* 5: 33918. https://doi.org/10.4236/ns.2013.57094.

Goswami, Rahul Kumar, Komal Agrawal, Maulin Pramod Shah, and Pradeep Verma. 2022e. "Bioremediation of heavy metals from wastewater: A current perspective on microalgae-based future." *Letters in Applied Microbiology* 75: 701–717. https://doi.org/10.1111/lam.13564.

Goswami, Rahul Kumar, Komal Agrawal, and Pradeep Verma. 2022a. "Microalgae biomass biorefinery: A sustainable renewable energy feedstock of the future." In *Micro-algae: Next-generation Feedstock for Biorefineries: Cultivation and Refining Processes*, pp. 1–29. Singapore: Springer Nature. https://doi.org/10.1007/978-981-19-0793-7_1.

Goswami, Rahul Kumar, Komal Agrawal, and Pradeep Verma. 2022c. "Microalgae *Dunaliella* as biofuel feedstock and β-carotene production: An influential step towards environmental sustainability." *Energy Conversion and Management: X* 13: 100154. https://doi.org/10.1016/j.ecmx.2021.100154.

Goswami, Rahul Kumar, Komal Agrawal, and Pradeep Verma. 2022d. "Microalgal-based remediation of wastewater: A step towards environment protection and management." *Environmental Quality Management* 32: 105–123. https://doi.org/10.1002/tqem.21850.

Goswami, Rahul Kumar, Komal Agrawal, Sanjeet Mehariya, Rajinikanth Rajagopal, Obulisamy Parthiba Karthikeyan, and Pradeep Verma. 2023. "Development of economical and sustainable cultivation system for biomass production and simultaneous treatment of municipal wastewater using Tetraselmis indica BDUG001." *Environmental Technology* 1–45. https://doi.org/10.1080/09593330.2023.2166429.

Goswami, Rahul Kumar, Sanjeet Mehariya, Obulisamy Parthiba Karthikeyan, and Pradeep Verma. 2022b. "Influence of carbon sources on biomass and biomolecule accumulation in Picochlorum sp. cultured under the mixotrophic condition." *International Journal of Environmental Research and Public Health* 19: 3674. https://doi.org/10.3390/ijerph19063674.

Goswami, Rahul Kumar, Sanjeet Mehariya, Obulisamy Parthiba Karthikeyan, Vijai Kumar Gupta, and Pradeep Verma. 2022f. "Multifaceted application of microalgal biomass integrated with carbon dioxide reduction and wastewater remediation: A flexible concept for sustainable environment." *Journal of Cleaner Production* 339: 130654. https://doi.org/10.1016/j.jclepro.2022.130654.

Goswami, Rahul Kumar, Sanjeet Mehariya, Parthiba Karthikeyan Obulisamy, and Pradeep Verma. 2021a. "Advanced microalgae-based renewable biohydrogen production systems: A review." *Bioresource Technology* 320: 124301. https://doi.org/10.1016/j.biortech.2020.124301.

Goswami, Rahul Kumar, Sanjeet Mehariya, Pradeep Verma, Roberto Lavecchia, and Antonio Zuorro. 2021b. "Microalgae-based biorefineries for sustainable resource recovery from wastewater." *Journal of Water Process Engineering* 40: 101747. https://doi.org/10.1016/j.jwpe.2020.101747.

Gouveia, Luisa, Oliveira Ana Cristina, Congestri Roberta, Bruno Laura, Soares Aline Terra, Menezes Rafael Silva, and Tzovenis Ioannis. 2017. "Biodiesel from microalgae." In *Microalgae-based Biofuels and Bioproducts*, 235–258. Sawston: Woodhead Publishing. https://doi.org/10.1016/B978-0-08-101023-5.00010-8.

Graziani, Giulia, Simona Schiavo, Maria Adalgisa Nicolai, Silvia Buono, Vincenzo Fogliano, Gabriele Pinto, and Antonino Pollio. 2013. "Microalgae as human food: chemical and nutritional characteristics of the thermo-acidophilic microalga *Galdieria sulphuraria*." *Food & Function* 4: 144–152. https://doi.org/10.1039/C2FO30198A.

Gu, Siwen, Jiaan Wang, and Yu Zhuang. 2022. "A two-tier superstructure model for optimization of microalgae-based biorefinery." *Energies* 15: 9166. https://doi.org/10.3390/en15239166.

Guedes, A. Catarina, Maria S. Gião, Rui Seabra, AC Silva Ferreira, Paula Tamagnini, Pedro Moradas-Ferreira, and Francisco Xavier Malcata. 2013. "Evaluation of the antioxidant activity of cell extracts from microalgae." *Marine Drugs* 11: 1256–1270. https://doi.org/10.3390/md11041256.

Guldhe, Abhishek, Bhaskar Singh, Ismail Rawat, Krishan Ramluckan, and Faizal Bux. 2014. "Efficacy of drying and cell disruption techniques on lipid recovery from microalgae for biodiesel production." *Fuel* 128: 46–52. https://doi.org/10.1016/j.fuel.2014.02.059.

Harun, Razif, and Michael Kobina Danquah. 2011. "Influence of acid pre-treatment on microalgal biomass for bioethanol production." *Process Biochemistry* 46: 304–309. https://doi.org/10.1016/j.procbio.2010.08.027.

Heimann, Kirsten, and Roger Huerlimann. 2015. "Microalgal classification: major classes and genera of commercial microalgal species." In *Handbook of Marine Microalgae*, pp. 25–41. Cambridge: Academic Press. https://doi.org/10.1016/B978-0-12-800776-1.00003-0.

Hidhayati, Noor, Ni Wayan Sri Agustini, Marsiti Apriastini, and Dhea Peby Ananda Diaudin. 2022. "Bioactive compounds from microalgae *Spirulina platensis* as antibacterial candidates against pathogen bacteria." *Jurnal Kimia Sains dan Aplikasi* 25: 41–48. https://doi.org/10.14710/jksa.25.2.41-48.

Ho, Shih-Hsin, Shu-Wen Huang, Chun-Yen Chen, Tomohisa Hasunuma, Akihiko Kondo, and Jo-Shu Chang. 2013. "Bioethanol production using carbohydrate-rich microalgae biomass as feedstock." *Bioresource Technology* 135: 191–198. https://doi.org/10.1016/j.biortech.2012.10.015.

Hossain, Nazia, and Teuku Meurah Indra Mahlia. 2019. "Progress in physicochemical parameters of microalgae cultivation for biofuel production." *Critical Reviews in Biotechnology* 3: 835–859. https://doi.org/10.1080/07388551.2019.1624945.

Huleihel, Mahmoud, Vladimir Ishanu, Jacov Tal, and Shoshana Arad. 2001. "Antiviral effect of red microalgal polysaccharides on Herpes simplex and Varicella zoster viruses." *Journal of Applied Phycology* 13: 127–134. https://doi.org/10.1023/A:1011178225912.

Iovinella, Manuela, Francesco Lombardo, Claudia Ciniglia, Maria Palmieri, Maria Rosa Di Cicco, Marco Trifuoggi, Marco Race, Carla Manfredi, Carmine Lubritto, Massimiliano Fabbricino, Mario De Stefano, and Seth Jon Davis. 2022. "Bioremoval of yttrium (III), cerium (III), europium (III), and terbium (III) from single and quaternary aqueous solutions using the extremophile Galdieria sulphuraria (Galdieriaceae, Rhodophyta)." *Plants* 11: 1376. https://doi.org/10.3390/plants11101376.

Janssen, Marcel, Rene H. Wijffels, and Maria J. Barbosa. 2022. "Microalgae based production of single-cell protein." *Current Opinion in Biotechnology* 75: 102705. https://doi.org/10.1016/j.copbio.2022.102705.

Jena, Jayashree, Nilotpala Pradhan, Bisnu Prasad Dash, Lala Behari Sukla, and P. Kumar Panda. 2013. "Biosynthesis and characterization of silver nanoparticles using microalga *Chlorococcum humicola* and its antibacterial activity." *International Journal of Nanomaterial and Biostructure* 3: 1–8. doi: www.urpjournals.com.

Jerney, Jacqueline, and Kristian Spilling. 2020. "Large scale cultivation of microalgae: open and closed systems." In *Biofuels from Algae: Methods and Protocols Humana*, pp. 1–8. New York: Springer. https://doi.org/10.1007/7651_2018_130.

Jha, Durga, Vishakha Jain, Brinda Sharma, Anil Kant, and Vijay Kumar Garlapati. 2017. "Microalgae-based pharmaceuticals and nutraceuticals: an emerging field with immense market potential." *ChemBioEng Reviews* 4: 257–272. https://doi.org/10.1002/cben.201600023.

Jordan, Richard W. 2012. "Haptophyta." *Wiley Online Library*. https://doi.org/10.1002/9780470015902.a0001981.pub2.

Khavari, Fatemeh, Massoud Saidijam, Mohammad Taheri, and Fatemeh Nouri. 2021. "Microalgae: Therapeutic potentials and applications." *Molecular Biology Reports* 48: 4757–4765. https://doi.org/10.1007/s11033-021-06422-w.

Khetkorn, Wanthanee, Rajesh Prasad Rastogi, Aran Incharoensakdi, Peter Lindblad, Datta Madamwar, Ashok Pandey, and Christian Larroche. 2017. "Microalgal hydrogen production–A review." *Bioresource Technology* 243: 1194–1206. https://doi.org/10.1016/j.biortech.2017.07.085.

Khoo, Kuan Shiong, Kit Wayne Chew, Guo Yong Yew, Wai Hong Leong, Yee Ho Chai, Pau Loke Show, and Wei-Hsin Chen. 2020. "Recent advances in downstream processing of microalgae lipid recovery for biofuel production." *Bioresource Technology* 304: 122996. https://doi.org/10.1016/j.biortech.2020.122996.

Kim, Jungmin, Gursong Yoo, Hansol Lee, Juntaek Lim, Kyochan Kim, Chul Woong Kim, Min S. Park, and Ji-Won Yang. 2013. "Methods of downstream processing for the production of biodiesel from microalgae." *Biotechnology Advances* 31: 862–876. https://doi.org/10.1016/j.biotechadv.2013.04.006.

Kim, Yun-Jung, Sang-Chul Ha, Dae Uk Kim, and Il-Shik Shin. 2017. "Antibacterial activity of ethanol extracts from marine micro-algae." *Korean Journal of Food Science and Technology* 49: 390–395.

Koyande, Apurav Krishna, Pau-Loke Show, Ruixin Guo, Bencan Tang, Chiaki Ogino, and Jo-Shu Chang. 2019. "Bio-processing of algal bio-refinery: A review on current advances and future perspectives." *Bioengineered* 10: 574–592. https://doi.org/10.1080/21655979.2019.1679697.

Kumar, Banothu. Ramesh, Garlapati Deviram, Thangavel Mathimani, Pham Anh Duc, and Arivalagan Pugazhendhi. 2019. "Microalgae as rich source of polyunsaturated fatty acids." *Biocatalysis and Agricultural Biotechnology* 17: 583–588. https://doi.org/10.1016/j.bcab.2019.01.017.

Kusmayadi, Adi, Eko Agus Suyono, Dillirani Nagarajan, Jo-Shu Chang, and Hong-Wei Yen. 2020. "Application of computational fluid dynamics (CFD) on the raceway design for the cultivation of microalgae: a review." *Journal of Industrial Microbiology and Biotechnology* 47: 373–382. https://doi.org/10.1007/s10295-020-02273-9.

Lakatos, Gergely Ernő, Karolína Ranglová, João Câmara Manoel, Tomáš Grivalský, Jiří Kopecký, and Jiří Masojídek. 2019. "Bioethanol production from microalgae polysaccharides." *Folia Microbiologica* 64: 627–644. https://doi.org/10.1007/s12223-019-00732-0.

Larkum, Anthony W. D., Johan A. Raven, and Susan E. Douglas. 2003. "The algae and their general characteristics." *Photosynthesis in Algae* 14: 1–10. https://doi.org/10.1007/978-94-007-1038-2_1.

Leong, Yoong Kit, and Jo-Shu Chang. 2020. "Bioremediation of heavy metals using microalgae: Recent advances and mechanisms." *Bioresource Technology* 303: 122886. https://doi.org/10.1016/j.biortech.2020.122886.

Li, Kexun, Shun Liu, and Xianhua Liu. 2014. "An overview of algae bioethanol production." *International Journal of Energy Research* 38: 965–977. https://doi.org/10.1002/er.3164.

Li, Shengnan, Fanghua Li, Xun Zhu, Qiang Liao, Jo-Shu Chang, and Shih-Hsin Ho. 2022. "Biohydrogen production from microalgae for environmental sustainability." *Chemosphere* 291: 132717. https://doi.org/10.1016/j.chemosphere.2021.132717.

Li, Tian-Tian, Ai-Jun Tong, Yuan-Yuan Liu, Zi-Rui Huang, Xu-Zhi Wan, Yu-Yang Pan, Rui-Bo Jia, Bin Liu, Xin-Hua Chen, and Chao Zhao. 2019. "Polyunsaturated fatty acids from microalgae Spirulina platensis modulates lipid metabolism disorders and gut microbiota in high-fat diet rats." *Food and Chemical Toxicology* 131: 110558. https://doi.org/10.1016/j.fct.2019.06.005.

Limongi, Antonina Rita, Emanuele Viviano, Maria De Luca, Rosa Paola Radice, Giuliana Bianco, and Giuseppe Martelli. 2021. "Biohydrogen from microalgae: Production and applications." *Applied Sciences* 11: 1616. https://doi.org/10.3390/app11041616.

Little, Shannon M., Gerusa N. A. Senhorinho, Mazen Saleh, Nathan Basiliko, John A. Scott, Shannon M. Little, Gerusa N. A. Senhorinho, Mazen Saleh, Nathan Basiliko, and John A. Scott. 2021. "Antibacterial compounds in green microalgae from extreme environments: A review." *Algae* 36: 61–72. https://doi.org/10.4490/algae.2021.36.3.6.

Liu, Chien-Hung, Chin-Yen Chang, Chieh-Lun Cheng, Duu-Jong Lee, and Jo-Shu Chang. 2012. "Fermentative hydrogen production by *Clostridium butyricum* CGS5 using carbohydrate-rich microalgal biomass as feedstock." *International Journal of Hydrogen Energy* 37: 15458–15464. https://doi.org/10.1016/j.ijhydene.2012.04.076.

Lu, Dingnan, Xiao Liu, Onur G. Apul, Lin Zhang, David K. Ryan, and Xiaoqi Zhang. 2019. "Optimization of biomethane production from anaerobic co-digestion of microalgae and septic tank sludge." *Biomass and Bioenergy* 127: 105266. https://doi.org/10.1016/j.biombioe.2019.105266.

Madadi, Rozita, Hamid Maljaee, Luísa S. Serafim, and Sónia P. M. Ventura. 2021. "Microalgae as contributors to produce biopolymers." *Marine Drugs* 19: 466. https://doi.org/10.3390/md19080466.

Mahendran, Manishaa Sri, Sinouvassane Djearamane, Ling Shing Wong, Govindaraju Kasivelu, Anto Cordelia Tanislaus Antony Dhanapal, and A. Dhanapal. 2021. "Antiviral properties of microalgae and cyanobacteria." *Journal of Experimental Biology and Agricultural Sciences* 9: S43–S48. doi: http://dx.doi.org/10.18006/2021.9.

Maia, Jorge Lucas Da, Jéssica Soares Cardoso, Duna Joanol da Silveira Mastrantonio, Caroline Krause Bierhals, Juliana Botelho Moreira, Jorge Alberto Vieira Costa, and Michele Greque de Morais. 2020. "Microalgae starch: A promising raw material for the bioethanol production." *International Journal of Biological Macromolecules* 165: 2739–2749. https://doi.org/10.1016/j.ijbiomac.2020.10.159.

Maltsev, Yevhen, and Kateryna Maltseva. 2021. "Fatty acids of microalgae: Diversity and applications." *Reviews in Environmental Science and Bio/Technology* 20: 515–547. https://doi.org/10.1007/s11157-021-09571-3(0123456789.

Markou, Giorgos, Irini Angelidaki, and Dimitris Georgakakis. 2012. "Microalgal carbohydrates: an overview of the factors influencing carbohydrates production, and of main bioconversion technologies for production of biofuels." *Applied Microbiology and Biotechnology* 96: 631–645. https://doi.org/10.1007/s00253-012-4398-0.

Mastropetros, Savvas Giannis, Konstantinos Pispas, Dimitris Zagklis, Sameh S. Ali, and Michael Kornaros. 2022. "Biopolymers production from microalgae and cyanobacteria cultivated in wastewater: Recent advances." *Biotechnology Advances* 107999. https://doi.org/10.1016/j.biotechadv.2022.107999.

Matos, Ângelo Paggi. 2019. "Microalgae as a potential source of proteins." In *Proteins: Sustainable Source, Processing and Applications*, pp. 63–96. Cambridge: Academic Press. https://doi.org/10.1016/B978-0-12-816695-6.00003-9.

Mehariya, Sanjeet, Francesca Fratini, Roberto Lavecchia, and Antonio Zuorro. 2021a. "Green extraction of value-added compounds form microalgae: A short review on natural deep eutectic solvents (NaDES) and related pre-treatments." *Journal of Environmental Chemical Engineering* 9: 105989. https://doi.org/10.1016/j.jece.2021.105989.

Mehariya, Sanjeet, Martin Plöhn, Antonio Leon-Vaz, Alok Patel, and Christiane Funk. 2022. "Improving the content of high value compounds in Nordic *Desmodesmus* microalgal strains." *Bioresource Technology* 359: 127445. https://doi.org/10.1016/j.biortech.2022.127445

Mehariya, Sanjeet, Martin Plöhn, Piotr Jablonski, Stefan Stagge, Leif J. Jönsson, and Christiane Funk. 2023. "Biopolymer production from biomass produced by Nordic microalgae grown in wastewater." *Bioresource Technology* 376: 128901. https://doi.org/10.1016/j.biortech.2023.128901

Mehariya, Sanjeet, Rahul Kumar Goswami, Obulisamy Parthiba Karthikeysan, and Pradeep Verma. 2021b. "Microalgae for high-value products: A way towards green nutraceutical and pharmaceutical compounds." *Chemosphere* 280: 130553. https://doi.org/10.1016/j.chemosphere.2021.130553.

Mehariya, Sanjeet, Rahul Kumar Goswami, Pradeep Verma, Roberto Lavecchia, and Antonio Zuorro. 2021c. "Integrated approach for wastewater treatment and biofuel production in microalgae biorefineries." *Energies* 14: 2282. https://doi.org/10.3390/en14082282.

Mobin, Saleh, and Firoz Alam. 2017. "Some promising microalgal species for commercial applications: A review." *Energy Procedia* 110: 510–517. https://doi.org/10.1016/j.egypro.2017.03.177.

Mohseniazar, Mahdi, Mohsen Barin, Habib Zarredar, Siamak Alizadeh, and Dariush Shanehbandi. 2011. "Potential of microalgae and lactobacilli in biosynthesis of silver nanoparticles." *BioImpacts: BI* 1: 149. doi:10.5681/bi.2011.020.

Mohsenpour, Seyedeh Fatemeh, Sebastian Hennige, Nicholas Willoughby, Adebayo Adeloye, and Tony Gutierrez. 2021. "Integrating micro-algae into wastewater treatment: A review." *Science of the Total Environment* 752: 142168. https://doi.org/10.1016/j.scitotenv.2020.142168.

Montenegro-Herrera, Carlos A., Francisco Vera-López Portillo, Georgina T. Hernández-Chávez, and Alfredo Martinez. 2022. "Single-cell protein production potential with the extremophilic red microalgae *Galdieria sulphuraria*: Growth and biochemical characterization." *Journal of Applied Phycology* 34: 1341–1352. https://doi.org/10.1007/s10811-022-02733-y.

Morocho-Jácome, Ana Lucía, Nadia Ruscinc, Renata Miliani Martinez, João Carlos Monteiro de Carvalho, Tânia Santos de Almeida, Catarina Rosado, João Guilherme Costa, Maria Valéria Robles Velasco, and André Rolim Baby. 2020. "(Bio) Technological aspects of microalgae pigments for cosmetics." *Applied Microbiology and Biotechnology* 104: 9513–9522. https://doi.org/10.1007/s00253-020-10936-x.

Mourelle, Mourelle Lourdes, Carmen Paula Gómez, and José L. Legido. 2017. "The potential use of marine microalgae and cyanobacteria in cosmetics and thalassotherapy." *Cosmetics* 4: 46. https://doi.org/10.3390/cosmetics4040046.

Muhaemin, Moh Muhaemin Moh, and Richardus F. Kaswadji. 2010. "Biomass nutrient profiles of marine microalgae *Dunaliella salina*." *Jurnal Penelitian Sains* 13: 13313. https://doi.org/10.56064/jps.v13i3.142.

Muthusamy, Govarthanan, Selvankumar Thangasamy, Mythili Raja, Sudhakar Chinnappan, and Selvam Kandasamy. 2017. "Biosynthesis of silver nanoparticles from Spirulina microalgae and its antibacterial activity." *Environmental Science and Pollution Research* 24: 19459–19464. doi: 10.1007/s11356-017-9772-0.

Nagarajan, Dillirani, Adi Kusmayadi, Hong-Wei Yen, Cheng-Di Dong, Duu-Jong Lee, and Jo-Shu Chang. 2019. "Current advances in biological swine wastewater treatment using microalgae-based processes." *Bioresource Technology* 289: 121718. https://doi.org/10.1016/j.biortech.2019.121718.

Narala, Rakesh Reddy, Sourabh Garg, Kalpesh K. Sharma, Skye R. Thomas-Hall, Miklos Deme, Yan Li, and Peer M. Schenk. 2016. "Comparison of microalgae cultivation in photobioreactor, open raceway pond, and a two-stage hybrid system." *Frontiers in Energy Research* 4: 29. https://doi.org/10.3389/fenrg.2016.00029.

Nasirian, Fariba, Masoumeh Dadkhah, Nasrollah Moradi-Kor, and Zia Obeidavi. 2018. "Effects of Spirulina platensis microalgae on antioxidant and anti-inflammatory factors in diabetic rats." *Diabetes, Metabolic Syndrome and Obesity: Targets and Therapy* 11: 375–380. https://doi.org/10.2147/DMSO.S172104.

Nwokoagbara, Ezinne, Akeem K Olaleye, and Meihong Wang. 2015. "Biodiesel from microalgae: The use of multi-criteria decision analysis for strain selection." *Fuel* 159: 241–249. https://doi.org/10.1016/j.fuel.2015.06.074.

Ognistaia, Albina, Zhanna Markina, and Tatiana Orlova. 2022. "Antimicrobial activity of marine microalgae." *Russian Journal of Marine Biology* 48: 217–230. https://doi.org/10.1134/S1063074022040071.

Onen Cinar, Senem, Zhi Kai Chong, Mehmet Ali Kucuker, Nils Wieczorek, Ugur Cengiz, and Kerstin Kuchta. 2020. "Bioplastic production from microalgae: A review." *International Journal of Environmental Research and Public Health* 17: 3842. https://doi.org/10.3390/ijerph17113842.

Onumaegbu, Chaukwuma, James Mooney, Abed Alaswad, and Abdul Ghani Olabi. 2018. "Pre-treatment methods for production of biofuel from microalgae biomass." *Renewable and Sustainable Energy Reviews* 93: 16–26. https://doi.org/10.1016/j.rser.2018.04.015.

Pagels, Fernando, Daniel Salvaterra, Helena Melo Amaro, and Ana Catarina Guedes. 2020. "Pigments from microalgae." In *Handbook of Microalgae-based Processes and Products* 465–492. Cambridge: Academic Press. https://doi.org/10.1016/B978-0-12-818536-0.00018-X.

Passos, Fabiana, Cesar Mota, Andrés Donoso-Bravo, Sergi Astals, David Jeison, and Raúl Muñoz. 2018. "Biofuels from microalgae: Biomethane." *Energy from Microalgae* 247–270. Cham: Springer. https://doi.org/10.1007/978-3-319-69093-3_12.

Patel, Akash, Bharat Gami, Pankaj Patel, and Beena Patel. 2017. "Microalgae: Antiquity to era of integrated technology." *Renewable and Sustainable Energy Reviews* 71: 535–547. https://doi.org/10.1016/j.rser.2016.12.081.

Piasecka, Agata, Izabela Krzemiñska, and Jerzy Tys. 2014. "Physical methods of microalgal biomass pretreatment." *International Agrophysics* 28: 341–348. https://doi.org/10.2478/intag-2014-0024.

Prajapati, Sanjeev Kumar, and Anushree Malik. 2015. "Algal biomass as feedstock for biomethane production: An introduction." *Journal of Environmental and Social Sciences* 2: 103. do i: www.opensciencepublications.com/fulltextarticles/ESS-2454-5953-2-103.html.

Qiu, Renhe, Song Gao, Paola A. Lopez, and Kimberly L. Ogden. 2017. "Effects OF pH on cell growth, lipid production and CO_2 addition of microalgae *Chlorella sorokiniana*." *Algal Research* 28: 192–199. https://doi.org/10.1016/j.algal.2017.11.004.

Rahman, Khondokar Mizanur 2020. "Food and high value products from microalgae: Market opportunities and challenges. "*Microalgae Biotechnology for Food, Health and High Value Products*, pp. 3–27. Singapore: Springer. https://doi.org/10.1007/978-981-15-0169-2_1.

Rajpoot, Aman Singh, Tushar Choudhary, H. Chelladurai, Tikendra Nath Verma, and Vikas Shende. 2022. "A comprehensive review on bioplastic production from microalgae." *Materials Today: Proceedings* 56: 171–178. https://doi.org/10.1016/j.matpr.2022.01.060.

Rawat, Ismail, R. Ranjith Kumar, Taurai Mutanda, and Faizal Bux. 2013. "Biodiesel from microalgae: A critical evaluation from laboratory to large scale production." *Applied Energy* 103: 444–467. https://doi.org/10.1016/j.apenergy.2012.10.004.

Rebolloso-Fuentes, M. M., Juan Antonio Navarro-Pérez, Felipe García-Camacho, Jose Joaquin Ramos-Miras, and Jose Luis Guil-Guerrero. 2001a. "Biomass nutrient profiles of the microalga *Nannochloropsis."Journal of Agricultural and Food Chemistry* 49: 2966–2972. https://doi.org/10.1021/jf0010376.

Rebolloso-Fuentes, M. M., Juan Antonio Navarro-Pérez, Jose Joaquin Ramos-Miras, and Jose Luis Guil-Guerrero. 2001b. "Biomass nutrient profiles of the microalga *Phaeodactylum tricornutum*." *Journal of Food Biochemistry* 25: 57–76. https://doi.org/10.1111/j.1745-4514.2001.tb00724.x.

Reiriz, Fernández María José, Perez-Camacho Alejandro, M. J. Ferreiro, Juan Blanco, Miguel Planas, Campos Maria Jesus, and Uxío Labarta. 1989. "Biomass production and variation in the biochemical profile (total protein, carbohydrates, RNA, lipids and fatty acids) of seven species of marine microalgae." *Aquaculture* 83: 17–37. https://doi.org/10.1016/0044-8486(89)90057-4.

Reynolds, Daman, Michael Huesemann, Scott Edmundson, Amy Sims, Brett Hurst, Sherry Cady, Nathan Beirne, Jacob Freeman, Adam Berger, and Song Gao. 2021. "Viral inhibitors derived from macroalgae, microalgae, and cyanobacteria: A review of antiviral potential throughout pathogenesis." *Algal Research* 57: 102331. https://doi.org/10.1016/j.algal.2021.102331.

Rizwan, Muhammad, Ghulam Mujtaba, Sheraz Ahmed Memon, Kisay Lee, and Naim Rashid. 2018. "Exploring the potential of microalgae for new biotechnology applications and beyond: A review." *Renewable and Sustainable Energy Reviews* 92: 394–404. https://doi.org/10.1016/j.rser.2018.04.034.

Román, R. Bermejo, Jose Maria Alvarez-Pez, Francisco Gabriel Acién Fernández, and Emilio Molina Grima. 2002. "Recovery of pure B-phycoerythrin from the microalga *Porphyridium cruentum*." *Journal of Biotechnology* 93: 73–85. https://doi.org/10.1016/S0168-1656(01)00385-6.

Sahoo, Dinabandhu, and Pooja Baweja. 2015. "General characteristics of algae." In *The Algae World*, pp. 3–29. Dordrecht: Springer. https://doi.org/10.1007/978-94-017-7321-8_1.

Salinas-Salazar, Carmen, J. Saul Garcia-Perez, Rashmi Chandra, Carlos Castillo-Zacarias, Hafiz M. N. Iqbal, and Roberto Parra-Saldívar. 2019. "Methods for extraction of valuable products from microalgae biomass." In *Microalgae Biotechnology for Development of Biofuel and Wastewater Treatment*, pp. 245–263. Singap ore: Springer. doi:10.1007/978-981-13-22648_11.

Santos-Sánchez, Norma Francenia, R. Valadez-Blanco, B. Hernández-Carlos, Alejandra Torres-Ariño, P. C. Guadarrama-Mendoza, and Raul Salas-Coronado. 2016. "Lipids rich in ω-3 polyunsaturated fatty acids from microalgae." *Applied Microbiology and Biotechnology* 100: 8667–8684. https://doi.org/10.1007/s00253-016-7818-8.

Santoyo, Susana, Merichel Plaza, Laura Jaime, Elena Ibanez, Guillermo Reglero, and Francisco J. Senorans. 2010. "Pressurized liquid extraction as an alternative process to obtain antiviral agents from the edible microalga *Chlorella vulgaris*." *Journal of Agricultural and Food Chemistry* 58: 8522–8527. https://doi.org/10.1021/jf100369h.

Sayeda, M. Abdo, Gamila H. Ali, and Farouk K. El-Baz. 2015. "Potential production of omega fatty acids from microalgae." *International Journal of Pharmaceutical Sciences Review and Research* 34: 210–215. do i: www.globalresearchonline.net.

Schagerl, Michael, Christian. Pichler, and Karl Donabaum. 2003. "Patterns of major photosynthetic pigments in freshwater algae. 2. *Dinophyta, Euglenophyta, Chlorophyceae* and *Charales*." *Annales de Limnologie-International Journal of Limnology* 9: 49–62. https://doi.org/10.1051/limn/2003005.

Shah, Md Mahfuzur Rahman, Yuanmei Liang, Jay Jiayang Cheng, and Maurycy Daroch. 2016. "Astaxanthin-producing green microalga *Haematococcus pluvialis*: from single cell to high value commercial products." *Frontiers in Plant Science* 7: 531. https://doi.org/10.3389/fpls.2016.00531.

Sharif, Muhammad, Muhammad Hammad Zafar, Amjad Islam Aqib, Muhammad Saeed, Mayada R. Farag, and Mahmoud Alagawany. 2021. "Single cell protein: Sources, mechanism of production, nutritional value and its uses in aquaculture nutrition." *Aquaculture* 53: 735885. https://doi.org/10.1016/j.aquaculture.2020.735885.

Silva, Samara C., Isabel C. F. R. Ferreira, Madalena M. Dias, and Maria Filomena Barreiro. 2020. "Microalgae-derived pigments: A 10-year bibliometric review and industry and market trend analysis." *Molecules* 25: 3406. https://doi.org/10.3390/molecules25153406.

Simas-Rodrigues, Cíntia, Helena Villela, Aline P. Martins, Luiza G. Marques, Pio Colepicolo, and Angela P. Tonon. 2015. "Microalgae for economic applications: Advantages and perspectives for bioethanol." *Journal of Experimental Botany* 66: 4097–4108. https://doi.org/10.1093/jxb/erv130.

Singh, Divya, Rahul Kumar Goswami, Komal Agrawal, Venkatesh Chaturvedi, and Pradeep Verma. 2022. "Bio-inspired remediation of wastewater: A contemporary approach for environmental clean-up." *Current Research in Green and Sustainable Chemistry* 5: 100261. https://doi.org/10.1016/j.crgsc.2022.100261.

Singh, Jasvinder, and Rakesh Chandra Saxena. 2015. "An introduction to microalgae: Diversity and significance." In *Handbook of Marine Microalgae*, pp. 11–24. Cambridge: Academic Press. https://doi.org/10.1016/B978-0-12-800776-1.00002-9.

Singh, Uday Bhan, and Amrik Singh Ahluwalia. 2013. "Microalgae: A promising tool for carbon sequestration." *Mitigation and Adaptation Strategies for Global Change* 18: 73–95. https://doi.org/10.1007/s11027-012-9393-3.

Sirohi, Ranjna, Ashutosh Kumar Pandey, Panneerselvam Ranganathan, Shikhangi Singh, Aswathy Udayan, Mukesh Kumar Awasthi, Anh Tuan Hoang, Chaitanya Reddy Chilakamarry, Sang Hyoun Kim, and Sang Jun Sim. 2022. "Design and applications of photobioreactors-A review." *Bioresource Technology* 349: 126858. https://doi.org/10.1016/j.biortech.2022.126858.

Soto-Sierra, Laura, Petya Stoykova, and Zivko L. Nikolov. 2018. "Extraction and fractionation of microalgae-based protein products." *Algal Research* 36: 175–192. https://doi.org/10.1016/j.algal.2018.10.023.

Spolaore, Pauline, Claire Joannis-Cassan, Elie Duran, and Arsène Isambert. 2006. "Commercial applications of microalgae." *Journal of Bioscience and Bioengineering* 101: 87–96. https://doi.org/10.1263/jbb.101.87.

Sun, Han, Yuxin Wang, Yongjin He, Bin Liu, Haijin Mou, Feng Chen, and Shufang Yang. 2023. "Microalgae-derived pigments for the food industry." *Marine Drugs* 21: 82. https://doi.org/10.3390/md21020082.

Tokuşoglu, Özlenm, and M. K. Üunal. 2003. "Biomass nutrient profiles of three microalgae: *Spirulina platensis, Chlorella vulgaris*, and *Isochrisis galbana*." *Journal of Food Science* 68: 1144–1148. https://doi.org/10.1111/j.1365-2621.2003.tb09615.x.

Torres-Tiji, Yasin, Francis J. Fields, and Stephen Patrick Mayfield. 2020. "Microalgae as a future food source." *Biotechnology Advances* 41: 107536. https://doi.org/10.1016/j.biotechadv.2020.107536.

Wahlen, Bradley D., Michael R. Morgan, Alex T. McCurdy, Robert M. Willis, Michael D. Morgan, Daniel J. Dye, Bruce Bugbee, Byard D. Wood, and Lance C. Seefeldt. 2013. "Biodiesel from microalgae, yeast, and bacteria: Engine performance and exhaust emissions." *Energy & Fuels* 27: 220–228. https://doi.org/10.1021/ef3012382.

Wang, Bei, Christopher Q. Lan, and Mark Horsman. 2012. "Closed photobioreactors for production of microalgal biomasses." *Biotechnology Advances* 30: 904–912. https://doi.org/10.1016/j.biotechadv.2012.01.019.

Wang, Jinghan, Haizhen Yang, and Feng Wang. 2014. "Mixotrophic cultivation of microalgae for biodiesel production: status and prospects." *Applied Biochemistry and Biotechnology* 172: 3307–3329. https://doi.org/10.1007/s12010-014-0729-1.

Wang, Meng, Eunyoung Lee, Merrill P. Dilbeck, Matthew Liebelt, Qiong Zhang, and Sarina J. Ergas. 2017. "Thermal pretreatment of microalgae for biomethane production: Experimental studies, kinetics and energy analysis." *Journal of Chemical Technology & Biotechnology* 92: 399–407. https://doi.org/10.1002/jctb.5018.

Wang, Shi-Kai, Amanda R. Stiles, Chen Guo, and Chun-Zhao Liu. 2014. "Microalgae cultivation in photobioreactors: An overview of light characteristics." *Engineering in Life Sciences* 14: 550–559. https://doi.org/10.1002/elsc.201300170.

Widowati, Ita, Muhammad Zainuri, Hermien Pancasakti Kusumaningrum, Ragil Susilowati, Yann Hardivillier, Vincent Leignel, Nathalie Bourgougnon, and Jean-Luc Mouget. 2017. "Antioxidant activity of three microalgae *Dunaliella salina, Tetraselmis chuii* and *Isochrysis galbana* clone Tahiti." *IOP Conference Series: Earth and Environmental Science* 55: 012067. https://doi.org/10.1088/1755-1315/55/1/012067.

Xiao, Chao, Qiang Liao, Qian Fu, Yun Huang, Hao Chen, Hong Zhang, Ao Xia, Xun Zhu, Alissara Reungsang, and Zhidan Liu. 2019. "A solar-driven continuous hydrothermal pretreatment system for biomethane production from microalgae biomass." *Applied Energy* 236: 1011–1018. https://doi.org/10.1016/j.apenergy.2018.12.014.

Xu, Xianzhen, Xiaoguang Gu, Zhongyang Wang, William Shatner, and Zhenjun Wang. 2019. "Progress, challenges and solutions of research on photosynthetic carbon sequestration efficiency of microalgae." *Renewable and Sustainable Energy Reviews* 110: 65–82. https://doi.org/10.1016/j.rser.2019.04.050.

Yarkent, Çağla, Ceren Gürlek, and Suphi S. Oncel. 2020. "Potential of microalgal compounds in trending natural cosmetics: A review." *Sustainable Chemistry and Pharmacy* 17: 100304. https://doi.org/10.1016/j.scp.2020.100304.

Yen, Hong-Wei, I-Chen Hu, Chun-Yen Chen, Dillirani Nagarajan, and Jo-Shu Chang. 2019. "Design of photobioreactors for algal cultivation." In *Biofuels from Algae*, pp. 225–256. Amsterdam: Elsevier. https://doi.org/10.1016/B978-0-444-64192-2.00010-X.

Yew, Guo Yong, Sze Ying Lee, Pau Loke Show, Yang Tao, Chung Lim Law, Thi Trung Chinh Nguyen, and Jo-Shu Chang. 2019. "Recent advances in algae biodiesel production: from upstream cultivation to downstream processing." *Bioresource Technology Reports* 7: 100227. https://doi.org/10.1016/j.biteb.2019.100227.

Yuan, Jian-Ping, Juan Peng, Kai Yin, and Jiang-Hai Wang. 2011. "Potential health-promoting effects of astaxanthin: a high-value carotenoid mostly from microalgae." *Molecular Nutrition & Food Research* 55: 150–165. https://doi.org/10.1002/mnfr.201000414.

Zabed, Hossain M., Xianghui Qi, Junhua Yun, and Huanhuan Zhang. 2019. "Anaerobic digestion of microalgae biomass for methane production." *Microalgae Biotechnology for Development of Biofuel and Wastewater Treatment*: 397–421. https://doi.org/10.1007/978-981-13-2264-8_16.

Zayadi, Rabiatul Adawiyah, and Faridah Abu Bakar. 2020. "Comparative study on stability, antioxidant and catalytic activities of bio-stabilized colloidal gold nanoparticles using microalgae and cyanobacteria." *Journal of Environmental Chemical Engineering* 8: 103843. https://doi.org/10.1016/j.jece.2020.103843.

Zhan, Jiao, Junfeng Rong, and Qiang Wang. 2017. "Mixotrophic cultivation, a preferable microalgae cultivation mode for biomass/bioenergy production, and bioremediation, advances and prospect." *International Journal of Hydrogen Energy* 42: 8505–8517. https://doi.org/10.1016/j.ijhydene.2016.12.021.

Zhu, Liandong, Shuhao Huo, and Lei Qin. 2015. "A microalgae-based biodiesel refinery: sustainability concerns and challenges." *International Journal of Green Energy* 12: 595–602. https://doi.org/10.1080/15435075.2013.867406.

Zhuang, Dingling, Ning He, Kuan Shiong Khoo, Eng-Poh Ng, Kit Wayne Chew, and Tau Chuan Ling. 2022. "Application progress of bioactive compounds in microalgae on pharmaceutical and cosmetics." *Chemosphere* 291: 132932. https://doi.org/10.1016/j.chemosphere.2021.132932.

Zhuang, Lin-Lan, Mengting Li, and Huu Hao Ngo. 2020. "Non-suspended microalgae cultivation for wastewater refinery and biomass production." *Bioresource Technology* 308: 123320. https://doi.org/10.1016/j.biortech.2020.123320.

2 Phycoremediation
A Sustainable Alternative for Removing Emerging Contaminants from Wastewater

Preethi Selvaraj, Arathi Sreenikethanam, Subhisha Raj and Amit K. Bajhaiya

2.1 INTRODUCTION

Surface water pollution is one of the most serious environmental issues today, mainly due to the direct discharge of contaminated water from various industries and other sources. It leads to major changes in physical, chemical, and biological properties of water. This water pollution/contamination is caused by the introduction of different kinds of pollutants into the water bodies. Emerging contaminants (ECs) are one such pollutants which have received recent attention for their adverse effects on human health and ecosystems (Gogoi et al. 2018). ECs are manmade or naturally occurring substances or microbes that are not regularly observed in the environment but have the ability to infiltrate and cause negative ecological and health consequences in humans. As a result, the impact of ECs in wastewater has increased and caused various adverse effects on environment such as the emergence of drug-resistant bacteria, marine life explosions, neurotoxic effects, endocrine disruption, and tumours (Ojha and Tiwary 2021). Different types of ECs exist in water which include organic contaminants like pharmaceutical compounds, industrial chemicals, endocrine disruptors and inorganic contaminants like heavy metals (Kumar et al. 2020). Both the organic and inorganic contaminants have harmful effects on environment. Therefore, it is necessary to eliminate these pollutants from wastewater. In order to remove these contaminants from wastewater, various traditional and modern methods are used. However, traditional treatment methods are not efficient in complete degradation of ECs. In these methods, the water is treated to remove contaminants based on their physical, chemical, and biological properties. Examples of traditional methods are ion-exchange, electrochemical treatment, osmosis, evaporation, and precipitation. But these techniques are not cost-effective and require higher energy inputs as they involve multiple processes for carbon, nitrogen, and phosphorus removal (Rodriguez-Narvaez et al. 2017). On the other hand, modern

DOI: 10.1201/9781003455950-2

biological methods are quite efficient, practical, and cost-effective in removing ECs from wastewater. Biological treatment technologies are mainly used to remove ECs primarily through biodegradation pathways. Large molecular weight ECs undergo a process called biodegradation where microorganisms break them down into smaller molecules and even biomineralize them into organic compounds like water and carbon dioxide (Rosenfeld and Feng 2011). During the classical biodegradation process, organic complexes serve as the first substrate for the cell development and induction of microorganisms (Rajasulochana and Preethy 2016). Further, in biological treatment methods, algal bioremediation has been recognized in recent years as the best technique for remediating ECs in wastewater. The microalgae, which are being used in algal bioremediation methods, are ubiquitous microscopic organisms and are main primary producers in our ecosystem which can thrive in all kinds of environments. They also have very high photosynthetic efficiency. In addition to their ecological importance, algae have several applications as they are utilized in biofuel, cosmeceuticals, nutraceutical, and bio-pharmaceuticals industries. In recent years, algae have generated a lot of interest in using them as feedstock for biofuel production. Along with that, microalgae-based method is one of the most promising techniques for the wastewater treatment and nutrient recovery from various sources including industrial, municipal, and agricultural effluents (Maryjoseph and Ketheesan 2020). Microalgae are widely utilized to treat effluents because they can adapt to a varied range of environments and also produce potentially valuable biomass coupled with tertiary bio-treatment. Many studies report that microalgae have the ability to use organic and inorganic substances as nutrients for cell growth (Sousa et al. 2022). This capability of uptaking nutrients allows them to grow in tolerance to varied environmental conditions. Additionally, in the biological processes, co-culturing methods also play a vital role in water treatment. These co-culture techniques such as the algal–bacterial symbiotic association serve as a sustainable and economic approach which is characterized by the symbiotic relationship of microalgae with other microorganisms. Co-culturing methods also include algae-activated sludge and algae co-cultured with nanoparticles. Phycoremeditation and associated co-cultivation techniques are therefore particularly efficient at removing pollutants from wastewater. This book chapter offers a thorough analysis of the occurrence of EC, their ecotoxicological impacts, and algal-based bioremediation of emerging pollutants from waste water against this context.

2.2 EMERGING CONTAMINANTS (ECs)

ECs can be microorganisms or any synthetic or naturally occurring substances that are not often found in the environment but have the ability to infiltrate and cause harmful impacts on human health (Rosenfeld and Feng 2011). ECs are not always entirely novel substances. They might be compounds that have been around in the environment for a while but whose existence and significance were not recognized until they cause any serious effect on the environment. ECs threaten the aquatic environment due to their persistence and ability to harm organisms on prolonged exposure. The increased detection of different types of ECs in different water bodies has facilitated the study of their toxic effects and disposal management (Ojha and Tiwary 2021). But developing appropriate regulations based on permitted limits for EC in water and comprehending how different treatment technologies might lower their concentrations to non-toxic levels is a field that needs a lot of focus. The effectiveness of various treatment technologies was assessed on the basis of their capacity to maintain EC concentrations below predetermined regulatory thresholds. Unlike most conventional treatments, very advanced combination treatments like modern methods can reduce EC concentrations below recommended values. Technologies with removal rates above 90% include membrane bioreactors coupled with reverse osmosis, membrane distillation, and ozonation (Sengupta et al. 2021). However, data on ECs is often sparse, and field testing may not exist or is in the early stages of development. Thus, it is important to research more on the adverse effects of ECs on the environment and aquatic ecosystem. Generally, these ECs are divided into two categories: organic contaminants and inorganic contaminants.

2.2.1 Organic Contaminants

Pesticides, medicines, surfactants, personal care items, food additives, and other organic chemicals are utilized on a regular basis in a wide range of global sectors (Table 2.1). These substances are categorized as emerging organic contaminants

TABLE 2.1
Types of Emerging Contaminants and the Phycoremediation Efficiency.

Emerging contaminants	Algal species involved in ECs' removal	Contaminants (removal)	Percentage of removal	Reference
Personal healthcare products	*Chlamydomonas* sp. Tai-03	Sulfadiazine	54.53%	(Pan et al. 2018; Xie et al., 2020)
		Ciprofoxacin	100%	
	Scenedesmus obliquus	Climbazole	>88%	
Industrial chemicals	*Rhodomonas* sp. JZB-2	Para-xylene	100%	(Li et al., 2020; Papazi et al., 2019)
	Scenedesmus obliquus	Various chlorinated phenolic compounds	90%	
Pesticides	*Scenedesmus obliquus*	2,6 dichlorophenol	50%	(Lindner and Pleissner 2019)
	Chlorella pyrenoidosa	α-naphthol	100%	
		β-naphthol		
Heavy metals	*Chlorella sorokiniana*	Ammonium, nitrates, phosphates, etc.	80%	(Guldhe et al. 2017)
Plasticizers	*Cylindrotheca closterium*	Diethyl phthalate	81.2%	(Gao and Chi 2015)
Hormones	*Chlorella pyrenoidosa*	Progesterone	95%	(Peng et al. 2014)
Surfactants	*Chlorella vulgaris*	Nonylphenol	>80%	(Gao et al. 2011)

(EOCs) when they are found in the environment in harmful concentrations. These organic contaminants resist environmental degradation (Sengupta et al. 2021). Therefore, organic compounds strongly pollute various types of water bodies, which in turn can lead to bioaccumulation. Even community drinking water supplies can become contaminated by these EOCs from a variety of sources, such as through runoff from farms, artificial recharge, or discharge from wastewater treatment plants. Several EOCs have recently been reported in community drinking water sources around the world (Xie et al. 2020). EOCs such as carbamazepine, atrazine, and caffeine are commonly detected contaminants in drinking water sources, which pose serious human health risks (García et al. 2020). In drinking water treatment facilities, these EOCs and other organic elements lead to the generation of carcinogenic disinfection by-products (DBPs). But the removal of these EOCs from water cannot be effectively accomplished by conventional methods, whereas modern methods like biological processes are efficient in removing those contaminants (Maryjoseph and Ketheesan 2020). Some of the most important organic contaminants are pharmaceuticals, endocrine-disrupting compounds, pesticides, and personal care products.

2.2.1.1 Pharmaceuticals

Pharmaceutical contaminants (PCs) are pharmaceutically active components which are one among the most important emerging contaminant groups, causing adverse effects on both living organisms and the environment even at very low levels. These kinds of contaminants remain in the environment as it takes a long time for these substances to break down (Samal et al. 2022). But with the growing demand for pharmaceuticals, the presence of these contaminants in water bodies has increased dramatically. This is mostly because traditional water treatment plants are ineffective at processing and getting rid of these pharmaceutical compounds. Hence, it is crucial to identify the appropriate drug categories and devise methods toeradicate them (Egorova et al. 2017). When PCs are released into the water bodies, it can have genotoxic, mutagenic, and ecotoxicological effects on all the living beings. The continuous human exposure to these PCs can lead to changes in genetic traits and biological behaviours. For example oestrogen is one of the most found pharmaceutical waste, and its presence in the drinking water can reduce male fertility. Further, it can also raise the prevalence of testicular and breast cancers. In addition, newborns, the elderly, and those with renal or liver disorders are harmed by such contaminants in drinking water (Samal et al. 2022). Therefore, it is of great importance to develop an effective treatment method to eliminate PCs from wastewater (Ebele et al. 2017). So, various mechanisms and technologies are being investigated for pharmaceutical contaminants' removal and pharmaceutical wastewater treatment. But the most effective of them all are biological methods (Arya et al. 2016). Some common biological treatment options for the removal of PCs include waste stabilization ponds, membrane bioreactors, and algae photobioreactors (Table 2.1) (Abdallat et al. 2022).

2.2.1.2 Endocrine Disrupting Compounds (EDCs)

Endocrine system is made up of several ductless glands that secrete hormones into the bloodstream to control different bodily processes. These hormonal molecules known as xenohormones influence our physiological processes and behaviours by flowing through the blood to far-off cells and tissues (Kasonga et al. 2021). But the presence of some chemicals called the endocrine disruptors can disrupt the body's hormonal balance. This group of chemicals includes exogenous and endogenous substances that interfere with the natural hormonal function of the body. Considering the complexity and significance of hormones in the functions of living organisms, it should be emphasized that endocrine-disrupting chemicals have a diverse effect on the environment and aquatic ecosystem (Petrovic et al. 2004). These EDCs have been found to contaminate the surface and groundwater, marine environmental effluents, and water reservoirs of rivers and lakes. In recent years, endocrine disruptors have entered the aquatic environment through various routes such as through the disposal of medical pharmaceutical waste, dissolution of chemicals used in sewage treatment plants, industrial and domestic products, and residues from pesticide and agricultural activities (Kasonga et al. 2021; Pironti et al. 2021). Ozonation, UV irradiation, photolysis, reverse osmosis, peroxidation, and ultrasonic procedures are some of the classic ways used to remove EDCs from wastewater. But compared to these methods, some studies show that phycoremediation gave efficient results in EDCs treatment from wastewater. Microalgal-based treatment was able to remove EDCs with more than 90% effectiveness in 12 hours, which validates the capacity of quick initial absorption, bioaccumulation, and biodegradation properties of microalgae (Gao et al. 2011).

2.2.1.3 Pesticides

Chemical substances employed to eradicate pests, such as insects, rodents, fungi and weeds, are referred as pesticides. They are mainly classified into fungicides, herbicides, bactericides, and insecticides according to their use. But they are the main sources of water contamination through surface runoff from agricultural areas (Smital et al. 2004). And they can cause acute and chronic health impacts like neonatal methemoglobinemia (caused by nitrate in groundwater), acute and chronic neurotoxicity (caused by insecticides, fungicides, and fumigants), lung damage (caused by paraquat), and chemical burns (caused by anhydrous ammonia). A number of cancers, particularly haematological tumours, have also been connected to different pesticides' exposure. Pesticides have also been linked to immunologic abnormalities and detrimental effects on growth, development, and reproduction. Thus, it is important to remove these contaminants from water bodies effectively (Sankhla 2018) to control the severe effects on environment and human population.

2.2.1.4 Personal Care Products

Personal care products (PCPs) are a wide range of typical household items used for hygiene and cleaning purposes. These include antiseptics, disinfectants, insect repellents, perfumes, and UV filters. Some of them are classified as emerging chemicals due to their detrimental impacts on aquatic ecosystems, and some are found to be linked to a higher risk of ailments, endocrine and reproductive disorders, as well as cancers. For instance, common hair care products may contain formaldehyde, which can cause irritation, skin rashes, and lung disorders on their exposure. Mostly the entry of these chemicals into water bodies happens through effluent discharges from sewage treatment plants. As the removal of these chemicals is incomplete or ineffective (Montes-Grajales et al. 2017) by conventional traditional methods, modern treatment methods using sustainable biological organisms are required to bioremediate the PCPs.

2.2.2 Inorganic Contaminants

Heavy metals, which are hazardous even at their low concentrations, make up the majority of inorganic contaminants. They can create health problems by entering our body through food, water, or air. They can also enter the environment via various uncontrolled human activities (Kurwadkar 2019). Nowadays, the waste and wastewater generated by industries and construction companies contain inorganic acids (sulphuric acid, hydrochloric acid, nitric acid), heavy metals, and trace elements (Cd, Hg, Pb, As, Cr, etc.), which reduce the quality of common water sources (Nidheesh et al. 2022). These contaminants can negatively affect the aquatic flora and fauna, disrupting the ecosystems and environment. They can also cause various human health problems. Traditional methods like chemical treatments, activated sludge processes, membrane filtration advanced oxidation processes, and electro Fenton procedures are utilized to remove these inorganic pollutants from waste water (Krishnan et al. 2021). However, it has been shown that the treatment of effluents with complex compositions is ineffective when utilizing these physicochemical methods solely. Additionally, these methods have a number of limitations, such as high energy costs, aeration-related expenses, and sludge control. Traditional treatment also uses a lot of resources and produces a lot of harmful wastes. Thus, as an environmentally benign method, phycoremediation is gaining prominent interest to reduce the production of these environmental pollutants (Dayana Priyadharshini et al. 2021). *Spirulina*, *Chlorella*, *Chlamydomona*s, *Oscillatoria*, *Nostoc*, and *Scenedesmus* are some of the non-pathogenic algal species used for wastewater treatment. These microalgae showed efficient results in eradicating contaminants from wastewater (Emparan et al. 2019).

2.2.2.1 Heavy Metals

Heavy metals are a frequent contaminant that enters municipal wastewater through industrial effluent discharges into the sewage lines or through household discharge of chemicals. Lead, copper, nickel, cadmium, zinc, mercury, arsenic, and chromium are the heavy metals which are most frequently identified in waste water (Francisco et al. 2019). These dangerous heavy metals can seriously harm the lively ecosystem (Du et al. 2022). Heavy metals have been shown to have an impact on a number of cellular organelles and enzymes functioning in metabolism, detoxification, and DNA repair. Apoptosis, carcinogenesis, and cell cycle regulation can be triggered by these metal ions and their interactions with nuclear proteins and DNA in the cells. So, removal of these heavy metals is important, and it can be done via phycoremediation. This is because microalgal cells have the ability to attach with the heavy metals through their cell membrane as they have carboxyl (COO^-) and hydroxyl (OH) functional groups. Hence, microalgae can take in the heavy metals as nutrients for their cell growth.

2.3 METHODS FOR REMOVING EMERGING CONTAMINANTS FROM WASTEWATER

The need for clean water is growing as a result of atmospheric adaption, industrialization, population growth, and environmental devastation. But the presence of organic and inorganic contaminants in wastewater presents a significant obstacle to water regeneration (Grassi et al. 2012; Sengupta et al. 2021). In order to identify and remove these contaminants, only a few treatment methods are available today. For this reason, the development of qualified and efficient treatment technologies is a major area in environmental studies these days. Broadly speaking, wastewater treatment methods are divided into three types; biological, physical, and chemical methods (Rodriguez-Narvaez et al. 2017). Biological treatments are easy and inexpensive but are ineffective for synthetic dyes because they resist aerobic biodegradation (Norfazilah Wan Ismail and Umairah Mokhtar 2021). Chemical treatments generate toxic by-products and are also not effective, while physical treatments are often effective. However, these traditional methods are less efficient, non-economical, and do not remove pollutants from contaminated water. However, a combination of modern and biological techniques showed efficient results in removing contaminants from water (Mohsenpour et al. 2021). In this, microorganisms are co-cultured with other microorganisms for the effective removal of contaminants. Also, they were found to be less expensive and more sensitive. Some of the commonly used methods for wastewater remediation/treatment are explained later in the chapter (Figure 2.1).

2.3.1 Traditional Methods

Traditional methods are common and widely used treatment methods used to remove contaminants from wastewater. But some of these techniques were found to

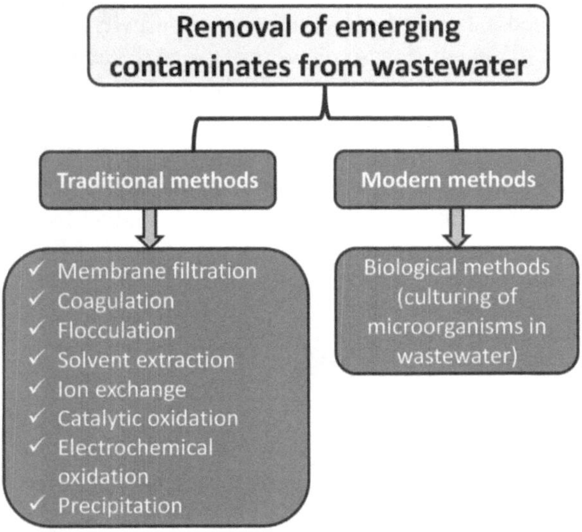

FIGURE 2.1 Methods for removing emerging contaminants from wastewater.

be inefficient in completely removing the contaminants from wastewater. However, these methods are still used in primary and secondary treatments. Some of the important traditional wastewater treatment techniques/processes are coagulation, ion exchange, electrochemical methods, osmosis, precipitation, and evaporation.

2.3.1.1 Coagulation

Coagulation is a crucial step in the treatment of wastewater. It is a method used to treat industrial wastewater in order to reduce suspended particles and colloidal turbidity (the et al. 2016). In this instance the coagulant is used to stabilize the colloidal particles, which are repulsing from each other due to their negative charge. This coagulant will bind to the colloidal particles and will cut down their repulsive force. Hence, the colloidal solids will settle down as agglomerates. These agglomerates can be separated at the end (Nidheesh et al. 2022). Generally, these coagulants are classified into chemical and natural coagulants. Chemical coagulants are hydrolyzing metallic salts such as ferric sulphate, ferric chloride, poly ferric chloride, magnesium chloride, and polyalkylene. Natural coagulants include microorganisms such as bacteria, microalgae, and fungi. Advantages of the coagulation method is that it will effectively remove metal contaminants and increase water clarity to reduce turbidity. However, this method is not very efficient in removing certain other emerging contaminants. Besides, it requires more energy (electricity) and money to operate.

2.3.1.2 Ion Exchange

Ion exchange method (Figure 2.2) employs synthetic and industrially produced ion exchange resins to remove the contaminants from wastewater. It is made up of small, micro porous beads that won't dissolve in water or other organic solvents. Polystyrene and polyacrylate are often used as substrates which act as backbone in ion exchange. A microporous exchange resin that has been supersaturated with a free solution is the principal element of the ion exchange apparatus and for water softening, and a sodium supersaturated sulfonated polystyrene bed is usually used to cover the surface of the bed. The diameter of these resins is between 0.3 and 1.3 millimeters. The material's gel-structured compartments include beads that are scattered throughout and contain around 50% water (Weidlich et al. 2001). The water is evenly distributed in the beads, and the water-soluble substances can flow in and out freely. Ions bind to these resin beads when water flows through them and releases a free solution into water (Chen et al. 2016). Water-soluble molecules, especially ions, can interact with the functional groups linked to the polymer backbone. These ions have either positive

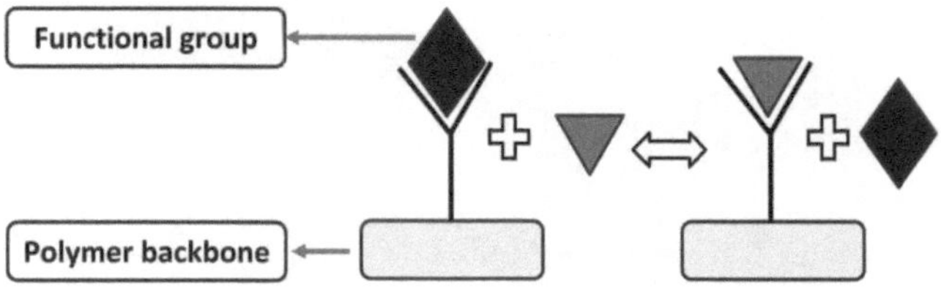

FIGURE 2.2 Ion exchange method to remove emerging contaminants from wastewater.

(cations) or negative (anions) charges. The interaction among ions and functional groups is mediated by electrostatic forces since functional groups are also charged. Functional groups with positive charges bind and interact with anions, while negative charges interact with cations (Figure 2.2). The ions that are connected to the functional groups form a weak bond with them. By passing a different ion across the functional group, the exchange can be reversed. An exchange reaction next to another can be used to repeat this process again (Entezari and Tahmasbi 2009). It is one of the most suitable technologies for the effective removal of dissolved inorganic ions. This technique is frequently used to soften water, remove or recover metals in the chemical sector, and prepare highly pure water for industrial uses. However, this technique is not that effective in killing bacteria and it has very high long-term operating costs.

2.3.1.3 Electrochemical Treatments

Electrochemical methods are commonly employed to extract domestic and industrial contaminants from wastewater. These are often used to eliminate contaminants from pretreated water. Compared to all conventional methods, these method give better results in pretreatment of wastewater (Chen et al. 2016). There are several electrochemical techniques for removing contaminants from wastewater (Figure 2.3). They are separation method, conventional method, sono-electro catalysis method, and the combination methods (Radjenovic and Sedlak 2015).

2.3.1.3.1 Separation Methods

Electrochemical treatment can be used as a separation method to separate different components in a mixture. The separation process involves applying an electrical current to the mixture, which triggers electrochemical reactions that cause the components to move towards different electrodes, based on their chemical properties. There are different electrochemical separation methods that can be used, depending on the

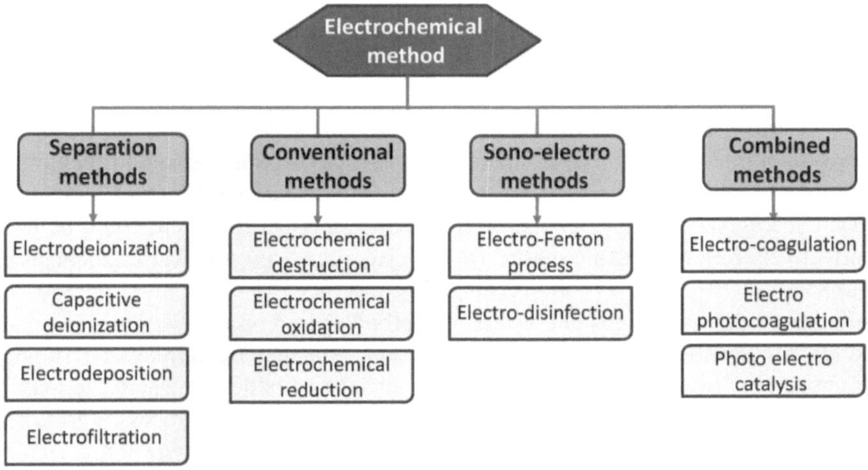

FIGURE 2.3 Electrochemical methods to remove emerging contaminants from wastewater.

type of mixture and components being separated. Some common methods include: electrodeionization (EDI), which is a water purification technology that combines ion exchange and electrodialysis processes to remove dissolved ions from water. It is often used in industrial and pharmaceutical applications where high-purity water is required. The EDI process involves passing a stream of water between two electrodes, one of which is positively charged (the anode) and the other negatively charged (the cathode). Between the electrodes, there will be ion exchange membranes that allow only certain ions to pass through them. When an electric current is applied to the electrodes, the ions in the water migrate towards the electrodes of opposite charge and are removed from the water. A recent study reported that EDI was effective in removing contaminants such as silica, sulfate, and chloride, and it can operate at a high recovery rate with minimal scaling and fouling (Thangamani et al. 2021). The next method is capacitive deionization, which is one of the latest methods employed in salt water purification. The principle of absorption and de-absorption is employed in this technique (Y. Chen et al. 2016). Ions are absorbed in the cathode and anode as current flows through the salt water, and these ions are then released from the electrodes. Another separation technique is electrodeposition in which an electric current is applied to the contaminated water, causing the heavy metals to get attracted to the oppositely charged electrode. This causes the metals to accumulate on the electrode surface, forming a solid layer that can be easily removed from the water. This process can be optimized by controlling various operational parameters such as the applied voltage, electrode material, and flow rate (Urbina-Suarez et al. 2021). The final technique is the electro-filtration technique, which is a water treatment process that combines electrochemical and filtration technologies to remove suspended particles and organic compounds from water. This process involves applying an electric field to the water, which causes the particles and contaminants to move towards the oppositely charged electrodes and to agglomerate and precipitate. The water then passes through a filter to remove the agglomerates and other remaining particles.

2.3.1.3.2 Conventional Method

The conventional method of electrochemical treatment is a process that converts organic compounds in the wastewater into harmless or less harmful by-products through electrochemical reactions. When wastewater is exposed to an electric current, highly reactive species like hydroxyl radicals and ozone are formed. These reactive species then interact with the organic components in the wastewater to break down them into simple compounds. A study published in the *Journal of Environmental Chemical Engineering* in 2021 found that electrochemical treatment could be used to effectively remove organic matter from wastewater, with removal efficiencies of up to 85%.

There are different types of conversion methods based on the contaminants. The first technique is the electrochemical destruction. This method employs electrolytes to remove ions from wastewater. The anode and the cathode are where the ions in the water are captured, which make water devoid of ions. This process is followed by passing the water with high force through the membrane where ion exchange happens. The next technique is electrochemical oxidation in which chemical reactions occur where an atom or a molecule loses electrons. This occurs because the conversion process uses an anode made of a catalytic substance (De Battisti and Martínez-Huitle 2018). In this approach, oxidation might happen either by anodic or indirect oxidation. The electrochemical reduction is another type of conversion in which an atom or molecule's electron or electrons deposit on the surface of the cathode as a result of an electric current flowing through the electrochemical system. This method is widely used to remove heavy metals and organic contaminants. Although it is a cost-effective technique, the components of the waste water have a significant impact on how well it works. For the most part, this technique works well to remove hazardous metals.

2.3.1.3.3 Sono-electro Catalysis

Sono-electro catalysis is a process that combines ultrasound (sonication), electrochemistry, and catalysis to promote chemical reactions. This technique can be used in wastewater treatments. It is divided into two types: electro-Fenton process and electro-disinfection. The advanced oxidation process known as the electro-Fenton process, which uses radical reactions to treat organic contaminants like pharmaceuticals, pesticides, dyes, phenols, and phenolic compounds, is named after Hendry John Horstman Fenton who discovered the transfer of electrons in specific metals. This method involves an oxidation method to remove contaminants that have low degradable ability in wastewater. In comparison to the traditional chemical disinfection procedure, the second technique called the electro-disinfection technique is more effective. It operates on the theory that during water electrolysis, strong oxidizing agents like O_2, O_3, and hypochlorite are produced via anodic generation. It uses oxygen gas to remove contaminants. In some methods, it uses chlorine gas and hypochlorite ions as well (Z. Chen et al. 2022).

2.3.1.3.4 Combined Methods

Electro-coagulation is a combined technique by which a variety of contaminants are eliminated from wastewater. This method is widely used to remove the contaminants from industrial wastewater before releasing them into any water bodies (Nidheesh et al. 2022). This method is an economical and efficient technique for decontamination. It requires low energy input and operating cost which make it more economical. Another combination technique is electro flotocoagulation, which involves the removal of contaminants and fine particles. This method deals with the particle size and uses light energy to drive chemical reactions at the surface of an electrode. In this process, a semiconductor material is used as an electrode that absorbs light to produce electrons and holes. These electrons and holes can then react with chemicals at the surface of the electrode, promoting or catalysing a chemical reaction. It exhibits great efficiency in removing emerging contaminants like harmful ions and heavy metals.

2.3.1.4 Osmosis

Osmosis is the diffusion of a solvent through a deferentially permeable membrane. Water is typically the solvent in biological systems. Osmosis will take place when there is differing water concentrations on opposite sides of a differentially permeable membrane (Lee et al. 2010). In simple words, osmosis is a process by which water molecules pass through a semipermeable membrane from a region with higher water concentration to an area with lower water concentration. However, even though osmosis can remove many harmful contaminants from the water, a large amount of water wastage happens during this process (Yang et al. 2019).

2.3.1.5 Reverse Osmosis

Reverse osmosis (RO) is a method of water purification that removes impurities like chlorine, salt, and dirt from drinking water using a system consisting of semi-permeable membrane (Venkatesan and Wankat 2011). In this process, instead of infiltrating naturally, water is allowed to pass through the membrane with the assistance of applied pressure (Lee et al. 2010). This technique removes suspended and dissolved solids, including bacteria, from water (Sherwood et al. 1967). Additionally, it can purify drinking water of ions, mineral chemicals, and other contaminants (Lee et al. 2010; Sherwood et al. 1967). Reverse osmosis also removes around 99% of dissolved salt particles, colloids, and other pathogenic bacteria and pyrogens from the feed water. Contaminants are eliminated by RO membranes based not only on the size but also the charge they carry. Pollutants are more likely to flow through the reverse osmosis membrane if they have a lower electric charge. For example sodium (monovalent) and calcium (divalent) can cross the membrane easily because of their lower charge. Similarly, since gases like carbon dioxide are weakly ionized, they cannot be eliminated via reverse osmosis (Wenten and Khoiruddin 2016). Even though RO can remove all kinds of pollutants such as pesticides, pharmaceuticals, heavy metals, and odour from the wastewater, it also removes essential minerals from the drinking water. Moreover, RO requires a costly installation process and expensive maintenance to work properly.

2.3.1.6 Evaporation

Evaporation is a separation process that makes use of the physical transformation of water from liquid to vapour phase. In contrast to previous separation techniques, this method separates the water from the contaminants instead of filtering the adulterants from the water. Prior to the latest advances in technology, evaporation system's capital and energy requirements had been too high for wastewater treatment and recycling processes (Kumar et al. 2022). The provident use of evaporation was limited to product recovery operations or when all other treatment styles had failed. It is based on the concept that there exists a finite "vapor pressure" above any material. Even though this method results in high-purity final product, poor step coverage and lower throughput due to low vacuum are its drawbacks.

2.3.1.7 Precipitation

Chemical precipitation is a frequently used technique in the industry and is also regarded as one of the most potent and established techniques for treating wastewaters. To assist in their settling, it transforms the dissolved metal ions and pollutants into solid particles (Kazadi Mbamba et al. 2015). Coagulant precipitates the metal ions and other contaminants by altering pH, electro oxidation potential, or co-precipitation. In this method, the upper layer contains the purified water, and the lower part contains the contaminants. So, the purified water can be removed by separating the upper portion. The usage of excessive chemicals is a significant disadvantage associated with this method.

2.3.2 Modern Methods

Besides conventional methods, there are certain other wastewater treatment methods called modern methods, and they widely utilize biological processes to remove emerging contaminants from wastewater. The high maintenance and upkeep expenses of current treatment technologies make them commercially unviable. Additionally, the overuse of chemicals may contribute to environmental imbalances. So, these modern methods are engaged to treat the wastewater effectively. Several studies proved that the modern biological treatment processes can effectively remove ECs from wastewater. According to Lutzu et al., *Scenedesmus dimorphus* can remove more than 99% of the nutrients like nitrogen (N) and phosphorous (P) from the brewery effluents within a week (Lutzu et al. 2016).

2.3.2.1 Biological Processes

The microorganisms including bacteria, algae, and fungi used in the biological treatment of wastewater mimic the ability of natural ecosystems to reduce the concentration of contaminants from water in an economic and equitable manner (Morin-Crini et al. 2022). Various drugs, such as pharmaceutical beta-blockers (sotalol, propranolol, and atenolol), cancer drugs and gastroesophageal drugs (famotidine, cimetidine, citalopram, acridine, metformin), anti-inflammatories (acetaminophen, including the stimulant butalbital), and antibiotics (sulfamethazine, sulfamethazine, sulfapyridine, azithromycin, and erythromycin) can be removed efficiently by fungi generators. Usually, wastewater treatment by fungal-based biological processes utilizes aerobic and anaerobic degradation processes (Norfazilah Wan Ismail and Umairah Mokhtar 2021). And most recent research demonstrates that microalgae can also efficiently remove emerging contaminants from wastewater and their properties like biosorption, bioaccumulation, bio-uptake, and photodegradation can help them in this bioremediation process (Karadjova et al. 2008). For example *Chlorella vulgaris* displayed a 85% and 89% efficiency of total phosphorous (TP) and total nitrogen (TN) removal from waste water, respectively (Wang et al. 2017).

2.3.2.2 Phycoremediation

Microalgae are known to be effective in removing ECs such as organic and inorganic contaminants from wastewater through biological processes (phycoremediation) (Phang et al. 2015). They are highly adaptive to different kinds of environments and can grow autotrophically, mixotrophically, or heterotrophically (Sousa et al. 2022). They can even survive in very stressful or extreme environmental conditions as well. During their cell growth, they can perform some mechanisms like biodegradation, bioaccumulation, bio-uptake, biosorption, and photodegradation. These mechanisms play an important role in the uptake of contaminants into the algal cell for further metabolic reactions. As microalgae are photosynthetic organisms, they can use light energy to take in the contaminants. Moreover, the cell walls of microalgae have different functional groups, and these groups can act as receptors for the ECs

(Abdelfattah et al. 2023). Therefore, the contaminants will bind to those receptors, and they can be taken inside by microalgal cells.

2.4 MECHANISM USED BY MICROALGAE FOR BIOREMEDIATION

Different mechanisms used by microalgae in wastewater bioremediation such as biosorption, bio-uptake, biodegradation, bioaccumulation, photodegradation, and volatilization are explained later in the chapter.

2.4.1 Microalgal Biosorption of ECs

Microalgae can remove organic contaminants like pesticides from wastewater by a passive process known as biosorption (Varsha et al. 2022). It is a resultant of combined physical, chemical, and mechanical methods comprising different techniques like chelation, adsorbtion, electrostatic interaction, surface precipitation, and ion exchange (Hassani et al. 2022; Rathi and Kumar 2021). In the process of biosorption, pores existing on the microalgal cell surface and chemical groups present on microalgal cell walls like OH^-, SO_4^{2-}, NH_2^- and COO– act as binding sites for the organic and inorganic contaminants (Priya et al. 2022). The cell wall structure includes sulphated polysaccharides, carbohydrates, and febrile matrix which gives the cells high mechanical strength. Hence, the cell walls will be more flexible because of the amorphous fraction, and this can enhance biosorption. Biosorption is reversible and can be influenced by factors such as the pH of the water, the concentration of the pollutants, and the characteristics of the algae cells. Moreover, biosorption mainly occurs in two steps: a metabolism-independent process which involves quick and reversible binding of adsorbate to active sites on the outer layer of microalgal cells and a positive intracellular diffusion step, which is a common metabolic activity of microalgae.

2.4.2 Bio-Uptake of ECs

In the process of bio-uptake, microalgal cells can transport the contaminants into the cells by passing via the cell wall. Within the cell, these contaminants attach to the intracellular proteins and other components. This process of bio-uptake can take hours to several days. Microalgae are capable of taking in various nutrients and trace elements from their surrounding water, including nitrogen, phosphorus, iron, and other minerals. This can occur through three major pathways, which are passive diffusion, passive-facilitated diffusion, and energy-dependent/active uptake (Karadjova et al. 2008). Passive diffusion doesn't need energy for the process of uptake of emerging contaminants. ECs can diffuse into the cell membrane from the higher to lower concentration. This passive diffusion creates changes in the cell membrane permeability. Microalgae cannot mediate this mechanism, as ECs can cause depolarization or hyper-polarization in the cell wall of the microalgae. This interference with the membrane integrity leads to the increased diffusion of the emerging contaminants into the cells through the cell membrane. The second pathway called passive-facilitated diffusion is also a cell membrane diffusion process. Without the use of cellular energy, passive-facilitated processes involve the movement of molecules through a protein channel or carrier protein across a membrane. Passive facilitated processes play a role in the absorption of necessary nutrients and the outflow of waste in microalgae. For example microalgae can take up carbon dioxide (CO_2) through a passive-facilitated mechanism involving protein carbonic anhydrase. This enzyme catalyses the conversion of CO_2 to bicarbonate (HCO^{3-}) and protons (H^+), which can then be transported across the cell membrane through bicarbonate transporters. Microalgae depend on passive-facilitated mechanisms for their development and survival because they enable effective nutrient uptake and waste product excretion. The final pathway of nutrient uptake is active transport. Living cells use active transport, which requires ATP energy input, to move molecules across their membrane in contrast to a concentration gradient. Active transport is a critical process for microalgae to take in nutrients that are present in low concentrations in the surrounding water. One well-studied example of active transport in microalgae is the uptake of iron. Although iron is present in extremely low concentrations in the surrounding waters, microalgae have developed a specialized mechanism for active iron uptake. They synthesize siderophores, which are small molecules that chelate iron and increase its availability for uptake. Additionally, specialized transporters are actively used by microalgae to carry iron across their membrane. ECs can be taken in by a similar process. But this can be influenced by the physiochemical environmental parameters like temperature, pH, and metabolic state. Another challenge associated with the bio-uptake mechanism is that the accumulated ECs in the microalgal cell create ROS (reactive oxygen species) that may damage the biomolecules produced by microalgae (Matamoros et al. 2016; Monteiro et al. 2012; Sutherland and Ralph 2019).

2.4.3 Photodegradation and Volatilization

In some cases, mechanisms like biosorption, bio-uptake, or biodegradation are effortless in removing ECs. But still microalgae play an important role in their successful remediation by other two mechanisms, which are photodegradation and volatilization (Gholkar et al. 2019).

2.4.3.1 Photodegradation

Photodegradation process involves the absorption of light by the pigments in microalgae, which can then trigger a series of chemical reactions that will break down the contaminants.

Photo-oxidative degradation of the contaminants occurs through hydroxyl radicals. These hydroxyl molecules or other analogous oxidants are generated when either soluble organic molecules or nitrate reacts with light. The mechanism of photodegradation depends on several factors like physiochemical properties of the contaminants and water bodies. The intensity of the light and wavelength also influence this mechanism (Maryjoseph and Ketheesan 2020).

2.4.3.2 Volatilization

Volatilization of microalgae refers to the process of converting microalgae biomass into a gaseous state, typically by heating the biomass at high temperatures in the presence of a gas, such as air or nitrogen. This process can also be employed to recover valuable compounds from microalgae, such as lipids, proteins, and carbohydrates (Gholkar et al. 2019). Volatilization of microalgae in wastewater treatment can be used to convert organic compounds into gaseous form. This process can be used to remove nutrients, such as nitrogen and phosphorus, from wastewater and produce biogas that can be used as an energy source.

2.4.4 Biodegradation of ECs by Microalgae

Biodegradation is considered as one among the most effective methods of bioremediation of contaminants present in wastewater. Microalgae can carry out biodegradation efficiently. They can break down complex molecules into simpler molecules through their metabolism and catalytic activities (Wang et al. 2022). This ability of microalgae can be utilized to degrade the accumulated contaminants. This biodegradation process by microalgae occurs in three stages. In the first stage, cytochrome P_{450} enzymes comprising hydroxylase, carboxylase, decarboxylase, and monooxygenase enzymes are essential. In the second step, enzymes like glutathione-S-transferase (GST) stimulate the conjugation of glutathione with different compounds that have electrophilic centers in order to defend against oxidative damages in the cell. The third stage of biodegradation consists of various enzymes which bio-transform the molecules into safer or non-harmful intermediates (Pinto et al. 2002).

2.4.5 Bioaccumulation

Bioaccumulation is an active metabolic process, which uses wide range of substrates present in the cell lumen of microalgae to accumulate toxic substances in them (Rocha et al. 2018). In short, microalgal cells will take up the toxic substances and either store them or utilize them for metabolism. This process appears to be an important one in the removal of organic and inorganic contaminants from wastewater. Microalgae can accumulate different contaminants along with the nutrients and trace elements present in the environment. Bioaccumulation is a preceding step to biodegradation. It is an efficient mechanism to remove pharmaceuticals like antibiotics and drugs present in the wastewater (Rodríguez-Mozaz et al. 2015).

2.4.6 Co-culturing of Microalgae to Remove ECs

Co-culturing of microalgae with other microorganisms refers to growing them in the presence of other microbes, such as bacteria or fungi, which can break down and metabolize emerging contaminants. Microalgae facilitate with nutrients in this association, while other microorganisms help in removing the ECs from wastewater (Chu et al. 2021). There are several advantages of using this co-culturing technique for the removal of ECs. For example the use of multiple types of microorganisms enhances the overall efficiency of the wastewater treatment process and can reduce the potential of antibiotic-resistance development in bacteria. In addition, the use of microalgae can provide a sustainable source of biomass for other microorganisms, which can be used for biogas production and other applications (Ayele et al. 2021).

Different co-culturing techniques using microalgae to remove ECs from wastewater are explained later in the chapter.

2.4.6.1 Microalgae–Fungi Co-culture

Fungi–microalgae consortia have shown great potential in wastewater treatment. Besides microalgae, the application of fungi in treating wastewater has also been the subject of extensive research. The special surface characteristics and high degradative enzyme resources of fungus help in the removal of contaminants from wastewater (Ayele et al. 2021; Wang et al., 2022). Currently, monocultures of fungi or microalgae for wastewater treatment are extensively studied, while fungi-supported microalgal cultures may receive increased attention than monocultures in future (Chu et al. 2021).

2.4.6.1.1 Fungi-Assisted Microalgae Bio-Flocculation

In submerged cultures, some of the filamentous fungi have a tendency to clump together and turn into granules as shown in Figure 2.4. This process can be roughly separated into three stages; the first stage involves the introduction of fungal spores into the reactor after which they will germinate to form the first embryonic mycelium. This is known as the micromorph growth stage. In the second stage, the mycelium begins to branch and form visible granules. This is known as the macromorphological growth stage. During this stage, the hyphae of the mycelium interact with each other, and the granules continue to grow and mature. In the third stage, when the growth conditions deteriorate, the mycelium starts to undergo autolysis/self-digestion. This stage is known as granular autolysis, and it results in the release of nutrients that can be used by other microorganisms such as microalgae in the bioreactor (Chu et al. 2021).

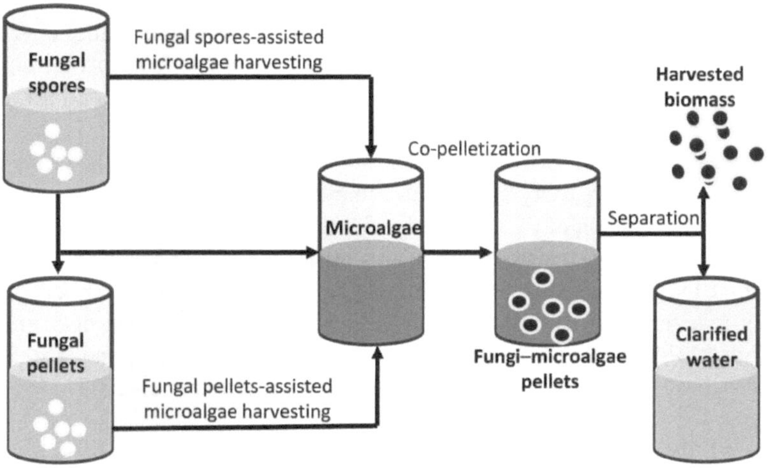

FIGURE 2.4 Fungi-assisted microalgae bio-flocculation method.

In the co-culturing method, fungal spores or whole fungal cells are cultured with microalgae, so that algal cells can be entrapped within the fungal cells, and co-pelletization can be performed. The contaminants present in the wastewater can be extracted by this microalgae-fungal pellets and they will help to clear the water (Gultom and Hu 2013).

2.4.6.2 Microalgae-Activated Sludge Co-culture Method

In this process, oxygen is supplied into a mixture of primary treated sewage or industrial waste water combined with microorganisms to reduce the organic content in the sewage (Perera et al. 2022). Here, wastewater and biomass are commonly called mixed liquor. Usually, in all activated sludge treatments, the wastewater will receive sufficient treatments like preliminary, primary, and secondary treatments before mixing. In this method, the excess mixed liquor is discharged into the settling tank, and the supernatant is treated before discharge. Usually, bacteria are used to treat wastewater, especially for municipal and industrial waste. But activated bacterial sludge has its own limitation in removing sufficient nitrogen, phosphorus, and heavy metals from the wastewater. Thus, chemical precipitation is essential for removing contaminants in this case. However, this chemical precipitation may generate certain toxic substances in water. In this scenario, the significance of microalgae-activated sludge-based wastewater treatment comes into picture. This microalgae-based treatment technique using photobioreactors shows efficient results in the removal of ECs (Hussein et al. 2020)). As microalgae are photoautotrophs, they need sufficient light energy and CO_2. Thus, microalgae-activated sludge technique can improve the efficiency of wastewater treatment while producing biomass that can be harvested for other applications. This activated sludge system can be done in two ways, which are by using either suspended wastewater system or immobilized wastewater system.

2.4.6.3 Suspended Wastewater System

Pond systems are common in wastewater treatments. They are also the most widely used type of large reactors in microalgae cultivation due to their simple structure and low capital cost. High-velocity algal ponds attempt to circumvent some of these problems by improving mixing efficiency with a pad agitator and air supply. Insufficient CO_2 supply limits algal biomass production due to unfavorable C:N:P ratios in wastewater. But specific aeration and CO_2 addition have been shown to increase biomass productivity and the elimination of undesirable components from the water. Co-cultivation of bacteria can promote heterotrophic oxidation of organic compounds in wastewater by microorganisms utilizing increased oxygen levels, as indicated by the induction of photoautotrophs. The efficiency of microalgae in removing total nitrogen and phosphorus from wastewater has been determined to range from 10% to 97% and is highly dependent on culture mode, pond size, and the type of wastewater (Delgadillo-Mirquez et al. 2016; Whitton et al. 2015).

2.4.6.4 Immobilized Wastewater Systems

Immobilized wastewater systems are a type of wastewater treatment system that uses microorganisms, such as bacteria or microalgae, which are immobilized in a support matrix such as beads or fibers. The support matrix provides a surface area for the microorganisms to grow and attach to, which helps to enhance the treatment process (Sial et al. 2021). However, it is crucial that low biomass concentrations are typical in photoautotrophic conditions, which means that large-scale cultivation of microalgae can still be challenging and costly. Additionally, the use of immobilized microorganisms may require specialized equipment and materials, which can add to the overall cost of the system.

2.4.6.5 Microalgae–Nanoparticle Co-culture Method

Qualified nanofibers are favoured over other standard supporting materials for the immobilization and encapsulation of microorganisms, including microalgae, to form bio-integrated hybrid materials with greater efficiency of pollutants' removal. Their simple and convenient nature of application and reusability and advantages of high surface-to-volume ratio, porosity, and biocompatibility make this coculture approach the most effective (Hussein et al. 2020). For example, in a study conducted in 2014, a cross-linked chitosin nanofibre mat was cocultured with *C. vulgaris*. This nanofibre mat immobilized the *C. vulgaris* and helped in the removal of NO^{3-} efficiently from wastewater (Eroglu et al. 2015).

2.5 CONCLUSION

Emerging contaminants (ECs) refer to a variety of pollutants that are not traditionally monitored in wastewater treatment, which include pharmaceuticals, personal care products, and endocrine-disrupting chemicals. These contaminants can have detrimental effects on the environment and human health, and thus their removal is a growing concern in efficient wastewater treatment (Sutherland and Ralph 2019). The removal of wastewater contaminants using algae, which is also known as phycoremediation, has shown promising results in the removal of these ECs from wastewater lately. Studies have shown that microalgae can take in these contaminants and can facilitate wastewater treatment through a variety of mechanisms, including adsorption, absorption, and biodegradation. Thus, phycoremediation has the ability to be an economical and sustainable alternative to traditional wastewater treatment methods, as it uses algae to remove contaminants from wastewater, after which these algae can be cultivated and harvested for various applications, including biofuels, fertilisers, and animal feed. So, phycoremediation has the potential to be scaled up to meet the growing demand for wastewater treatment, particularly in developing countries, where the availability of pure water is a major challenge (Abdelfattah et al. 2023).

However, there are many challenges associated with phycoremediation as well. Difficulty in harvesting needs, great demand for certain nutrients, contamination from other microorganism, immature downstream processing, complex operation, and system monitoring are some of the major challenges (Maryjoseph and Ketheesan 2020). Also, it requires a careful management of algal growth and harvesting, as there is a potential risk associated with the release of these contaminants back into the environment by microalgae if not properly managed. In conclusion, even though phycoremediation is efficient in the removal of emerging contaminants from wastewater, further research is needed to explore its full potential and to optimize its use in wastewater treatment.

REFERENCES

Abdallat, G. A., Salameh, E., Shteiwi, M., & Bardaweel, S. (2022). Pharmaceuticals as Emerging pollutants in the reclaimed wastewater used in irrigation and their effects on plants, soils, and groundwater. *Water, 14*(10), 156 0. https://doi.org/10.3390/w14101560

Abdelfattah, A., Ali, S. S., Ramadan, H., El-Aswar, E. I., Eltawab, R., Ho, S.-H., Elsamahy, T., Li, S., El-Sheekh, M. M., Schagerl, M., Kornaros, M., & Sun, J. (2023). Microalgae-based wastewater treatment: Mechanisms, challenges, recent advances, and future prospects. *Environmental Science and Ecotechnology, 13*, 100205. https://doi.org/10.1016/j.ese.2022.100205

Arya, V., Philip, L., & Murty Bhallamudi, S. (2016). Performance of suspended and attached growth bioreactors for the removal of cationic and anionic pharmaceuticals. *Chemical Engineering Journal, 284*, 1295–1307. https://doi.org/10.1016/j.cej.2015.09.070

Ayele, A., Getachew, D., Kamaraj, M., & Suresh, A. (2021). Phycoremediation of synthetic dyes: An effective and eco-friendly algal technology for the dye abatement. *Journal of Chemistry, 2021*, 1–14. https://doi.org/10.1155/2021/9923643

Chen, Y., Davis, J. R., Nguyen, C. H., Baygents, J. C., & Farrell, J. (2016). Electrochemical ion-exchange regeneration and fluidized bed crystallization for zero-liquid-discharge water softening. *Environmental Science & Technology, 50*(11), 5900–5907. https://doi.org/10.1021/acs.est.5b05606

Chen, Z., Wei, W., Liu, X., & Ni, B.-J. (2022). Emerging electrochemical techniques for identifying and removing micro/nanoplastics in urban waters. *Water Research, 221*, 118846. https://doi.org/10.1016/j.watres.2022.118846

Chu, R., Li, S., Zhu, L., Yin, Z., Hu, D., Liu, C., & Mo, F. (2021). A review on co-cultivation of microalgae with filamentous fungi: Efficient harvesting, wastewater treatment and biofuel production. *Renewable and Sustainable Energy Reviews, 139*, 11068 9. https://doi.org/10.1016/j.rser.2020.110689

Dayana Priyadharshini, S., Suresh Babu, P., Manikandan, S., Subbaiya, R., Govarthanan, M., & Karmegam, N. (2021). Phycoremediation of wastewater for pollutant removal: A green approach to environmental protection and long-term remediation. *Environmental Pollution, 290*, 117989. https://doi.org/10.1016/j.envpol.2021.117989

De Battisti, A., & Martínez-Huitle, C. A. (2018). Electrocatalysis in wastewater treatment. In *Electrochemical Water and Wastewater Treatment* (pp. 119–131). Elsevier. https://doi.org/10.1016/B978-0-12-813160-2.00005-5

Delgadillo-Mirquez, L., Lopes, F., Taidi, B., & Pareau, D. (2016). Nitrogen and phosphate removal from wastewater with a mixed microalgae and bacteria culture. *Biotechnology Reports, 11*, 18–26. https://doi.org/10.1016/j.btre.2016.04.003

Du, M., Zheng, M., Liu, A., Wang, L., Pan, X., Liu, J., & Ran, X. (2022). Effects of emerging contaminants and heavy metals on variation in bacterial communities in estuarine sediments. *Science of The Total Environment, 832*, 155118. https://doi.org/10.1016/j.scitotenv.2022.155118

Ebele, A. J., Abou-Elwafa Abdallah, M., & Harrad, S. (2017). Pharmaceuticals and personal care products (PPCPs) in the freshwater aquatic environment. *Emerging Contaminants, 3*(1), 1–16. https://doi.org/10.1016/j.emcon.2016.12.004

Egorova, K. S., Gordeev, E. G., & Ananikov, V. P. (2017). Biological activity of ionic liquids and their application in pharmaceutics and medicine. *Chemical Reviews, 117*(10), 7132–7189. https://doi.org/10.1021/acs.chemrev.6b00562

Emparan, Q., Harun, R., & Danquah, M. K. (2019). Role of phycoremediation for nutrient removal from wastewaters: A review. *Applied Ecology and Environmental Research, 17*(1), 889–915. https://doi.org/10.15666/aeer/1701_889915

Entezari, M. H., & Tahmasbi, M. (2009). Water softening by combination of ultrasound and ion exchange. *Ultrasonics Sonochemistry, 16*(3), 356–360. https://doi.org/10.1016/j.ultsonch.2008.09.008

Eroglu, E., Smith, S. M., & Raston, C. L. (2015). *Application of Various Immobilization Techniques for Algal Bioprocesses* (pp. 19–44). https://doi.org/10.1007/978-3-319-16640-7_2

Francisco, L. F. V., do Amaral Crispim, B., Spósito, J. C. V., Solórzano, J. C. J., Maran, N. H., Kummrow, F., do Nascimento, V. A., Montagner, C. C., De Oliveira, K. M. P., & Barufatti, A. (2019). Metals and emerging contaminants in groundwater and human health risk assessment. *Environmental Science and Pollution Research*, *26*(24), 24581–24594. https://doi.org/10.1007/s11356-019-05662-5

Gao, J., & Chi, J. (2015). Biodegradation of phthalate acid esters by different marine microalgal species. *Marine Pollution Bulletin*, *99*(1–2), 70–75. https://doi.org/10.1016/j.marpolbul.2015.07.061

Gao, Q. T., Wong, Y. S., & Tam, N. F. Y. (2011). Removal and biodegradation of nonylphenol by immobilized *Chlorella vulgaris*. *Bioresource Technology*, *102*(22), 10230–10238. https://doi.org/10.1016/j.biortech.2011.08.070

García, J., García-Galán, M. J., Day, J. W., Boopathy, R., White, J. R., Wallace, S., & Hunter, R. G. (2020). A review of emerging organic contaminants (EOCs), antibiotic resistant bacteria (ARB), and antibiotic resistance genes (ARGs) in the environment: Increasing removal with wetlands and reducing environmental impacts. *Bioresource Technology*, *307*, 123228. https://doi.org/10.1016/j.biortech.2020.123228

Gholkar, P., Shastri, Y., & Tanksale, A. (2019). Catalytic reactive flash volatilisation of microalgae to produce hydrogen or methane-rich syngas. *Applied Catalysis B: Environmental*, *251*, 326–334. https://doi.org/10.1016/j.apcatb.2019.03.082

Gogoi, A., Mazumder, P., Tyagi, V. K., Tushara Chaminda, G. G., An, A. K., & Kumar, M. (2018). Occurrence and fate of emerging contaminants in water environment: A review. *Groundwater for Sustainable Development*, *6*, 169–180. https://doi.org/10.1016/j.gsd.2017.12.009

Grassi, M., Kaykioglu, G., Belgiorno, V., & Lofrano, G. (2012). Removal of emerging contaminants from water and wastewater by adsorption process. In *Emerging Compounds Removal from Wastewater*. SpringerBriefs in Molecular Science. Springer. https://doi.org/10.1007/978-94-007-3916-1_2

Guldhe, A., Ansari, F. A., Singh, P., & Bux, F. (2017). Heterotrophic cultivation of microalgae using aquaculture wastewater: A biorefinery concept for biomass production and nutrient remediation. *Ecological Engineering*, *99*, 47–53. https://doi.org/10.1016/j.ecoleng.2016.11.013

Gultom, S., & Hu, B. (2013). Review of Microalgae Harvesting via Co-Pelletization with Filamentous Fungus. *Energies*, *6*(11), 5921–5939. https://doi.org/10.3390/en6115921

Hassani, A., Malhotra, M., Karim, A. V., Krishnan, S., & Nidheesh, P. V. (2022). Recent progress on ultrasound-assisted electrochemical processes: A review on mechanism, reactor strategies, and applications for wastewater treatment. *Environmental Research*, *205*, 112463. https://doi.org/10.1016/j.envres.2021.112463

Hussein, H. A., Syamsumir, D. F., Radzi, S. A. M., Siong, J. Y. F., Zin, N. A. M., & Abdullah, M. A. (2020). Phytochemical screening, metabolite profiling and enhanced antimicrobial activities of microalgal crude extracts in co-application with silver nanoparticle. *Bioresources and Bioprocessing*, *7*(1), 39. https://doi.org/10.1186/s40643-020-00322-w

Karadjova, I., Slaveykova, V., & Tsalev, D. (2008). The biouptake and toxicity of arsenic species on the green microalga Chlorella salina in seawater. *Aquatic Toxicology*, *87*(4), 264–271. https://doi.org/10.1016/j.aquatox.2008.02.006

Kasonga, T. K., Coetzee, M. A. A., Kamika, I., Ngole-Jeme, V. M., & Benteke Momba, M. N. (2021). Endocrine-disruptive chemicals as contaminants of emerging concern in wastewater and surface water: A review. *Journal of Environmental Management*, *277*, 111485. https://doi.org/10.1016/j.jenvman.2020.111485

Kazadi Mbamba, C., Batstone, D. J., Flores-Alsina, X., & Tait, S. (2015).astewaysed chemical precipitation modelling approach in wastewater treatment applied to calcite. *Water Research*, *68*, 342–353. https://doi.org/10.1016/j.watres.2014.10.011

Krishnan, R. Y., Manikandan, S., Subbaiya, R., Biruntha, M., Govarthanan, M., & Karmegam, N. (2021). Removal of emerging micropollutants originating from pharmaceuticals and personal care products (PPCPs) in water and wastewater by advanced oxidation processes: A review. *Environmental Technology & Innovation*, *23*, 101757. https://doi.org/10.1016/j.eti.2021.101757

Kumar, M., Borah, P., & Devi, P. (2020). Priority and emerging pollutants in water. In *Inorganic Pollutants in Water* (pp. 33–49). Elsevier. https://doi.org/10.1016/B978-0-12-818965-8.00003-2

Kumar, R., Qureshi, M., Vishwakarma, D. K., Al-Ansari, N., Kuriqi, A., Elbeltagi, A., & Saraswat, A. (2022). A review on emerging water contaminants and the application of sustainable removal technologies. *Case Studies in Chemical and Environmental Engineering*, *6*, 100219. https://doi.org/10.1016/j.cscee.2022.100219

Kurwadkar, S. (2019). Occurrence and distribution of organic and inorganic pollutants in groundwater. *Water Environment Research*, *91*(10), 1001–1008. https://doi.org/10.1002/wer.1166

Lee, S., Boo, C., Elimelech, M., & Hong, S. (2010). Comparison of fouling behavior in forward osmosis (FO) and reverse osmosis (RO). *Journal of Membrane Science*, *365*(1–2), 34–39. https://doi.org/10.1016/j.memsci.2010.08.036

Li, H., Li, H., Meng, F., Zhang, B., Lin, Y., Wu, J., Wang, G., & Zheng, Y. (2020). The biodegradation of Para-xylene in seawater by a newly isolated oceanic microalga Rhodomonas sp. JZB-2. *Journal of Water Process Engineering*, *36*, 101311. https://doi.org/10.1016/j.jwpe.2020.101311

Lindner, A. V., & Pleissner, D. (2019). Utilization of phenolic compounds by microalgae. *Algal Research*, *42*, 101602. https://doi.org/10.1016/j.algal.2019.101602

Lutzu, G. A., Zhang, W., & Liu, T. (2016). Feasibility of using brewery wastewater for biodiesel production and nutrient removal by *Scenedesmus dimorphus*. *Environmental Technology*, *37*(12), 1568–1581. https://doi.org/10.1080/09593330.2015.1121292

Maryjoseph, S., & Ketheesan, B. (2020). Microalgae based wastewater treatment for the removal of emerging contaminants: A review of challenges and opportunities. *Case Studies in Chemical and Environmental Engineering*, *2*, 100046. https://doi.org/10.1016/j.cscee.2020.100046

Matamoros, V., Uggetti, E., García, J., & Bayona, J. M. (2016). Assessment of the mechanisms involved in the removal of emerging contaminants by microalgae from wastewater: a laboratory scale study. *Journal of Hazardous Materials*, *301*, 197–205. https://doi.org/10.1016/j.jhazmat.2015.08.050

Mohsenpour, S. F., Hennige, S., Willoughby, N., Adeloye, A., & Gutierrez, T. (2021). Integrating micro-algae into wastewater treatment: A review. *Science of The Total Environment*, *752*, 142168. https://doi.org/10.1016/j.scitotenv.2020.142168

Monteiro, C. M., Castro, P. M. L., & Malcata, F. X. (2012). Metal uptake by microalgae: Underlying mechanisms and practical applications. *Biotechnology Progress*, *28*(2), 299–311. https://doi.org/10.1002/btpr.1504

Montes-Grajales, D., Fennix-Agudelo, M., & Miranda-Castro, W. (2017). Occurrence of personal care products as emerging chemicals of concern in water resources: A review. *Science of The Total Environment*, *595*, 601–614. https://doi.org/10.1016/j.scitotenv.2017.03.286

Morin-Crini, N., Lichtfouse, E., Fourmentin, M., Ribeiro, A. R. L., Noutsopoulos, C., Mapelli, F., Fenyvesi, É., Vieira, M. G. A., Picos-Corrales, L. A., Moreno-Piraján, J. C., Giraldo, L., Sohajda, T., Huq, M. M., Soltan, J., Torri, G., Magureanu, M., Bradu, C., & Crini, G. (2022). Removal of emerging contaminants from wastewater using advanced treatments. A review. *Environmental Chemistry Letters*, *20*(2), 1333–1375. https://doi.org/10.1007/s10311-021-01379-5

Nidheesh, P. V., Khan, F. M., Kadier, A., Akansha, J., Bote, M. E., & Mousazadeh, M. (2022). Removal of nutrients and other emerging inorganic contaminants from water and wastewater by electrocoagulation process. *Chemosphere*, *307*, 135756. https://doi.org/10.1016/j.chemosphere.2022.135756

Norfazilah Wan Ismail, W., & Umairah Mokhtar, S. (2021). Various methods for removal, treatment, and detection of emerging water contaminants. In *Emerging Contaminants*. IntechOpen. https://doi.org/10.5772/intechopen.93375

Ojha, A., & Tiwary, D. (2021). Organic pollutants in water and its health risk assessment through consumption. In *Contamination of Water* (pp. 237–250). Elsevier. https://doi.org/10.1016/B978-0-12-824058-8.00039-6

Pan, C.-G., Peng, F.-J., & Ying, G.-G. (2018). Removal, biotransformation and toxicity variations of climbazole by freshwater algae Scenedesmus obliquus. *Environmental Pollution*, *240*, 534–540. https://doi.org/10.1016/j.envpol.2018.05.020

Papazi, A., Karamanli, M., & Kotzabasis, K. (2019). Comparative biodegradation of all chlorinated phenols by the microalga Scenedesmus obliquus—The biodegradation strategy of microalgae. *Journal of Biotechnology*, *296*, 61–68. https://doi.org/10.1016/j.jbiotec.2019.03.010

Peng, F.-Q., Ying, G.-G., Yang, B., Liu, S., Lai, H.-J., Liu, Y.-S., Chen, Z.-F., & Zhou, G.-J. (2014). Biotransformation of progesterone and norgestrel by two freshwater microalgae (Scenedesmus obliquus and Chlorella pyrenoidosa): Transformation kinetics and products identification. *Chemosphere*, *95*, 581–588. https://doi.org/10.1016/j.chemosphere.2013.10.013

Perera, I. A., Abinandan, S., Panneerselvan, L., Subashchandrabose, S. R., Venkateswarlu, K., Naidu, R., & Megharaj, M. (2022). Co-culturing of microalgae and bacteria in real wastewaters alters indigenous bacterial communities enhancing effluent bioremediation. *Algal Research*, *64*, 102705. https://doi.org/10.1016/j.algal.2022.102705

Petrovic, M., Eljarrat, E., Lopez de Alda, M. J., & Barcel, D. (2004). Endocrine disrupting compounds and other emerging contaminants in the environment: A survey on new monitoring strategies and occurrence data. *Analytical and Bioanalytical Chemistry*, *378*(3), 549–562. https://doi.org/10.1007/s00216-003-2184-7

Phang, S.-M., Chu, W.-L., & Rabiei, R. (2015). *Phycoremediation* (pp. 357–389). https://doi.org/10.1007/978-94-017-7321-8_13

Pinto, G., Pollio, A., Previtera, L., & Temussi, F. (2002). Biodegradation. *Biotechnology Letters*, *24*(24), 2047–2051. https://doi.org/10.1023/A:1021367304315

Pironti, C., Ricciardi, M., Proto, A., Bianco, P. M., Montano, L., & Motta, O. (2021). Endocrine-disrupting compounds: An overview on their occurrence in the aquatic environment and human exposure. *Water*, *13*(10), 1347. https://doi.org/10.3390/w13101347

Priya, A. K., Gnanasekaran, L., Dutta, K., Rajendran, S., Balakrishnan, D., & Soto-Moscoso, M. (2022). Biosorption of heavy metals by microorganisms: Evaluation of different underlying mechanisms. *Chemosphere*, *307*, 135957. https://doi.org/10.1016/j.chemosphere.2022.135957

Radjenovic, J., & Sedlak, D. L. (2015). Challenges and opportunities for electrochemical processes as next-generation technologies for the treatment of contaminated water. *Environmental Science & Technology*, *49*(19), 11292–11302. https://doi.org/10.1021/acs.est.5b02414

Rajasulochana, P., & Preethy, V. (2016). Comparison on efficiency of various techniques in treatment of waste and sewage water – A comprehensive review. *Resource-Efficient Technologies*, *2*(4), 175–184. https://doi.org/10.1016/j.reffit.2016.09.004

Rathi, B. S., & Kumar, P. S. (2021). Application of adsorption process for effective removal of emerging contaminants from water and wastewater. *Environmental Pollution*, *280*, 116995. https://doi.org/10.1016/j.envpol.2021.116995

Rocha, A. C., Camacho, C., Eljarrat, E., Peris, A., Aminot, Y., Readman, J. W., Boti, V., Nannou, C., Marques, A., Nunes, M. L., & Almeida, C. M. (2018). Bioaccumulation of persistent and emerging pollutants in wild sea urchin Paracentrotus lividus. *Environmental Research, 161*, 354–363. https://doi.org/10.1016/j.envres.2017.11.029

Rodríguez-Mozaz, S., Huerta, B., & Barceló, D. (2015). *Bioaccumulation of Emerging Contaminants in Aquatic Biota: Patterns of Pharmaceuticals in Mediterranean River Networks* (pp. 121–141). https://doi.org/10.1007/698_2015_5015

Rodriguez-Narvaez, O. M., Peralta-Hernandez, J. M., Goonetilleke, A., & Bandala, E. R. (2017). Treatment technologies for emerging contaminants in water: A review. *Chemical Engineering Journal, 323*, 361–380. https://doi.org/10.1016/j.cej.2017.04.106

Rosenfeld, P. E., & Feng, L. G. H. (2011). Emerging contaminants. In *Risks of Hazardous Wastes* (pp. 215–222). Elsevier. https://doi.org/10.1016/B978-1-4377-7842-7.00016-7

Samal, K., Mahapatra, S., & Hibzur Ali, M. (2022). Pharmaceutical wastewater as Emerging Contaminants (EC): Treatment technologies, impact on environment and human health. *Energy Nexus, 6*, 100076. https://doi.org/10.1016/j.nexus.2022.100076

Sankhla, M. S. (2018). Water contamination through pesticide & their toxic effect on human health. *International Journal for Research in Applied Science and Engineering Technology, 6*(1), 967–970. https://doi.org/10.22214/ijraset.2018.1146

Sengupta, A., Jebur, M., Kamaz, M., & Wickramasinghe, S. R. (2021). Removal of emerging contaminants from wastewater streams using membrane bioreactors: A review. *Membranes, 12*(1), 60. https://doi.org/10.3390/membranes12010060

Sherwood, T. K., Brian, P. L. T., & Fisher, R. E. (1967). Desalination by reverse osmosis. *Industrial & Engineering Chemistry Fundamentals, 6*(1), 2–12. https://doi.org/10.1021/i160021a001

Sial, A., Zhang, B., Zhang, A., Liu, K., Imtiaz, S. A., & Yashir, N. (2021). Microalgal–bacterial synergistic interactions and their potential influence in wastewater treatment: A review. *BioEnergy Research, 14*(3), 723–738. https://doi.org/10.1007/s12155-020-10213-9

Smital, T., Luckenbach, T., Sauerborn, R., Hamdoun, A. M., Vega, R. L., & Epel, D. (2004). Emerging contaminants—Pesticides, PPCPs, microbial degradation products and natural substances as inhibitors of multixenobiotic defense in aquatic organisms. *Mutation Research/Fundamental and Molecular Mechanisms of Mutagenesis, 552*(1–2), 101–117. https://doi.org/10.1016/j.mrfmmm.2004.06.006

Sousa, H., Sousa, C. A., Simões, L. C., & Simões, M. (2022). Microalgal-based removal of contaminants of emerging concern. *Journal of Hazardous Materials, 423*, 127153. https://doi.org/10.1016/j.jhazmat.2021.127153

Sutherland, D. L., & Ralph, P. J. (2019). Microalgal bioremediation of emerging contaminants—Opportunities and challenges. *Water Research, 164*, 114921. https://doi.org/10.1016/j.watres.2019.114921

Teh, C. Y., Budiman, P. M., Shak, K. P. Y., & Wu, T. Y. (2016). Recent advancement of coagulation–flocculation and its application in wastewater treatment. *Industrial & Engineering Chemistry Research, 55*(16), 4363–4389. https://doi.org/10.1021/acs.iecr.5b04703

Thangamani, R., Vidhya, L., & Varjani, S. (2021). Electrochemical technologies for wastewater treatment and resource reclamation. In *Microbe Mediated Remediation of Environmental Contaminants* (pp. 381–389). Elsevier. https://doi.org/10.1016/B978-0-12-821199-1.00028-6

Urbina-Suarez, N. A., Machuca-Martínez, F., & Barajas-Solano, A. F. (2021). Advanced oxidation processes and biotechnological alternatives for the treatment of tannery wastewater. *Molecules, 26*(11), 3222. https://doi.org/10.3390/molecules26113222

Varsha, M., Senthil Kumar, P., & Senthil Rathi, B. (2022). A review on recent trends in the removal of emerging contaminants from aquatic environment using low-cost adsorbents. *Chemosphere, 287*, 132270. https://doi.org/10.1016/j.chemosphere.2021.132270

Venkatesan, A., & Wankat, P. C. (2011). Simulation of ion exchange water softening pretreatment for reverse osmosis desalination of brackish water. *Desalination*, *271*(1–3), 122–131. https://doi.org/10.1016/j.desal.2010.12.022

Wang, J., Tian, Q., Cui, L., Cheng, J., Zhou, H., Zhang, Y., Peng, A., & Shen, L. (2022). Synergism and mutualistic interactions between microalgae and fungi in fungi-microalgae symbiotic system. *Bioresource Technology*, *361*, 127728. https://doi.org/10.1016/j.biortech.2022.127728

Wang, L., Chen, X., Wang, H., Zhang, Y., Tang, Q., & Li, J. (2017). *Chlorella vulgaris* cultivation in sludge extracts from 2,4,6-TCP wastewater treatment for toxicity removal and utilization. *Journal of Environmental Management*, *187*, 146–153. https://doi.org/10.1016/j.jenvman.2016.11.020

Wang, Y., He, Y., Li, X., Nagarajan, D., & Chang, J.-S. (2022). Enhanced biodegradation of chlortetracycline via a microalgae-bacteria consortium. *Bioresource Technology*, *343*, 126149. https://doi.org/10.1016/j.biortech.2021.126149

Weidlich, C., Mangold, K.-M., & Jüttner, K. (2001). Conducting polymers as ion-exchangers for water purification. *Electrochimica Acta*, *47*(5), 741–745. https://doi.org/10.1016/S0013-4686(01)00754-X

Wenten, I. G., & Khoiruddin. (2016). Reverse osmosis applications: Prospect and challenges. *Desalination*, *391*, 112–125. https://doi.org/10.1016/j.desal.2015.12.011

Whitton, R., Ometto, F., Pidou, M., Jarvis, P., Villa, R., & Jefferson, B. (2015). Microalgae for municipal wastewater nutrient remediation: mechanisms, reactors and outlook for tertiary treatment. *Environmental Technology Reviews*, *4*(1), 133–148. https://doi.org/10.1080/21622515.2015.1105308

Xie, P., Chen, C., Zhang, C., Su, G., Ren, N., & Ho, S.-H. (2020). Revealing the role of adsorption in ciprofloxacin and sulfadiazine elimination routes in microalgae. *Water Research*, *172*, 115475. https://doi.org/10.1016/j.watres.2020.115475

Yang, Z., Feng, Y, Rui, X., Zhang, T., & Zhang, Z. (2019). A review on reverse osmosis and nanofiltration membranes for water purification. *Polymers*, *11*(8), 1252. https://doi.org/10.3390/polym11081252

3 Advances in Cultivation and Emerging Application of *Chlorella vulgaris*

A Sustainable Biorefinery Approach

Miriam L. Rosales-Aguado, Rosa M. Rodríguez-Jasso, Samanta Machado-Cepeda, Gilver Rosero-Chasoy, Regina Barboza-Rodríguez, Alejandra Cabello-Galindo and Héctor A. Ruiz

3.1 INTRODUCTION

It is estimated that nonrenewable resources, such as fossil fuels, crude oil, natural gas, and coal, drive the world´s primary production. Since the recent pandemic of COVID-19, all the existing problems around the world have increased in an immeasurable way. Global warming has been increasing due to the deterioration of climate change, causing the most frequent occurrence of events such as extreme climates, alteration of the productivity of agriculture caused by high temperatures, increased droughts, and reduced precipitation. Also, the Food and Agriculture Organization (FAO) forecasts that by 2030, an estimated 669 million people will be affected by hunger (Khanal et al. 2021; Thornton et al. 2018; Burki 2022).

The root of the aforementioned problems goes hand in hand with the scarcity and depletion of natural resources and overpopulation (Lara-Flores et al. 2018; Schwerhoff and Stuermer 2019). The primary goal nowadays is to reach a way of living that needs to be sustainable over the future years; be able to reduce the consumption of fresh water, fossil, and natural resources; and find other alternatives ways to meet the needs of the population and achieve the zero waste production in which a biorefinery can be defined as processing of biomass, in this case agro-industrial resources, maximizing profitability, and minimizing waste generation (Rosero-Chasoy 2022; Lara-Flores et al. 2018).

The term biorefinery translates to the integration of a resource from plant origin, in this case, raw materials and biomasses, which has been considered as a

DOI: 10.1201/9781003455950-3

future sustainable source for the replacement of fossil resources, transforming them into their essential components such as proteins, carbohydrates, and lipids and obtaining these processes' products such as biofuels and compounds of high added value like vitamins, antioxidants, enzymes, polymers, pigments, and chemicals. The process of biorefinery has been considered a sustainable future source for replacing fossil resources used for industrial purposes, alleviating the environmental pollution (Singh et al. 2019; González-Gloria et al. 2021; Cherubini 2010; Sirohi et al. 2022).

To better understand the biorefinery concept, this is classified into first, second, and third generations, depending on the type of raw material and biomass used for processing. The first generation of the biorefinery uses crops with the highest glucose and starch content, for example, cane sugar, corn starch, and vegetable oils as feedstock. The disadvantage of using this raw material is that these resources are not used to satisfy the demand of the food sector but rather to satisfy the energy demand (Jambo et al. 2016; Naik et al. 2010). The second generation of biorefinery uses as raw material agro-industrial residues that are not of interest to the food sector, such as lignocellulose biomass, that is, agricultural residues such as sawdust, corn stover, agave bagasse, sugar cane, husks-rice, wheat straw, witchgrass, sunflower stalks, among others (Jambo et al. 2016; Hassan et al. 2019; Ruiz et al. 2013).

Furthermore, the third generation of biorefinery implements the use of aquatic biomass such as macroalgae and microalgae in which microalgae are photosynthetic unicellular or filamentous microorganisms found in any habitat. They can grow in almost any environment absorbing nutrients from any water bodies like fresh, salt, and wastewater or in any culture medium chosen for their growth. They can absorb carbon dioxide from the air by photosynthesis for their biomass and compound conversion. They are well known for their fast growth rate compared to traditional agriculture or forestry (Ishika et al. 2017; Tang et al. 2020; Ali et al. 2022).

Microalgae depend on abiotic factors such as light, which could be either sunlight or artificial light; optimal temperature between 20°C and 30°C; and the nutrients in their culture medium found in different availability. Other factors that need to be considered, like the intensity and wavelength of the light, optimal pH range, and salinity, also affect biomass and pigment growth (Sharma et al. 2022; Mehariya et al. 2022; Begum et al. 2016).

On the other hand, *Chlorella vulgaris* is a microalga of great interest due to its potential use in different areas at the laboratory and industry levels. The focus is on the application and advantage of its biomass richness, such as its macronutrients and micronutrients. This chapter aims to emphasize the emerging technologies developed under the biorefinery concept using various alternatives of culture media, like synthetic medium being replaced with agro-industrial residues, whose organic matter is an essential source of nutrition for biomass growth and is also necessary for the accumulation of the high-added value components. Figure 3.1 shows how the general process of biomass growth for *C. Vulgaris* is carried on.

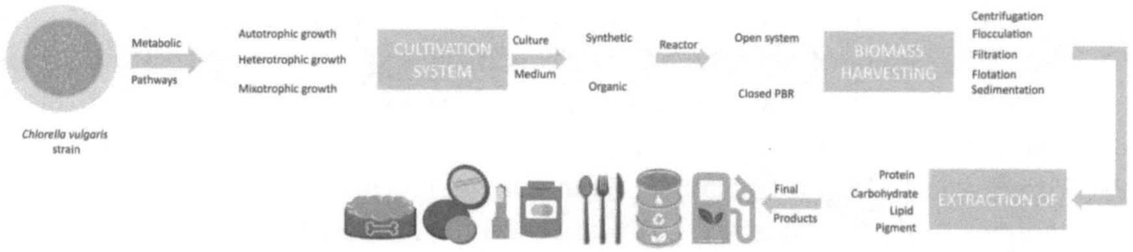

FIGURE 3.1 General diagram process of *Chlorella vulgaris* production.

3.2 *CHLORELLA VULGARIS*

Back in time, algae had been part of the food system of humanity since our ancestors. One example is the Mesoamericans and Aztecs, who consumed microalgae as a food source without even knowing the health benefits it brought to their system. They recollected this "new food" from lakes or ponds and added it to their meals. Some of the microalgae already studied have been grown to be part of food and function as high nutritional value additives for food for humans, animals, and aquaculture (Safi et al. 2014).

Microalgae have a more significant number of cellular compounds, such as pigments, antioxidants, lipids, carbohydrates, and proteins, compared to crops. Over time, they have been called a functional food, generating interest in the food sector due to the phytochemical characteristics of their biomass and the increasing demand for obtaining alternatives to food production (Wan Mahari et al. 2022; Wells et al. 2017). Figure 3.2 shows a photograph taken of Chlorella in the biorefinery laboratory using a microscope.

The primary nutrients that need to be present in a vast amount are nitrogen and phosphorus, essential for microalgae development; these sources can be either from organic or synthetic influences. A new perspective to make more sustainable cultivation for this kind of biomass is the emergence of the biorefinery approach. The agro-industrial wastes or biomasses from the agro-food industries are disposed into the environment, polluting the air, water, and soil. This process releases many nutrients such as nitrogen and phosphorus, inorganic elements considered pollutants. However, these types of nutrients serve for the growth of the algal biomass (Sharma et al. 2022; Melo et al. 2018).

Moreover, the interest in the production of high-added value compounds has been increasing from microalgal biomass. These components are pigments, such as carotenoid, phycocyanin, astaxanthin, xanthophyll, and phycobilin, which can vary depending on the type of microalgae. Either the biomass or the pigments have a significant and valuable effect on the energy, pharmaceutical, cosmetic, nutraceutical, and food industries (Begum et al. 2016; Sharma et al. 2022). Table 3.1 shows the main compounds produced and accumulated by the microalga – *Chlorella*.

One of the most exciting and remarkable microalgae is the green unicellular eukaryotic microorganism *Chlorella vulgaris*. The recorded data from the discovery of *Chlorella* date from 1890 when the Dutch researcher, Martinus Willem Beijerinck, identified it as the first microalgae with a well-looking defined nucleus (Safi et al. 2014). The principal components in the cell of *Chlorella vulgaris* have been reported

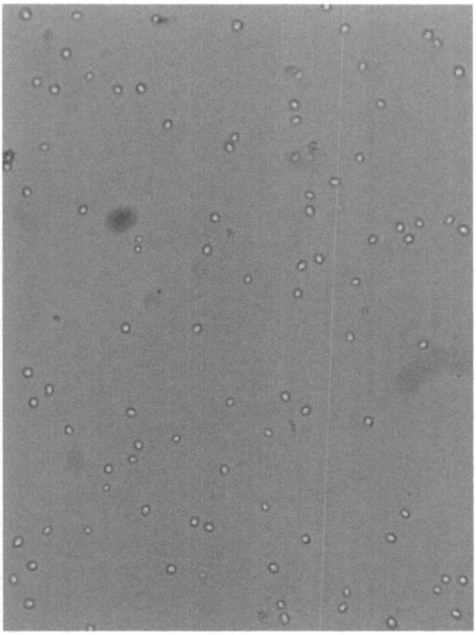

FIGURE 3.2 A typical photograph taken of *Chlorella* using a microscope.

by Safi et al. (2014). They reported the different organelles contained in a single cell of *Chlorella vulgaris*, such as the cell wall, nucleus, chlorophyll and carotenoids, chloroplast, chloroplast envelope, mitochondrion, cytoplasm, lipid droplets, Golgi body, vacuole, starch, pyrenoid and thylakoids. Also, Safi et al. (2014) reported the scientific classification as shown in Figure 3.3.

3.2.1 Growth Factors

Studies have shown that the growth rate of *Chlorella vulgaris* is more as compared to other microalgae, and it has a more extraordinary ability to adapt to different modifications to its cultivation. It can survive rough conditions such as deprivation, starvation, shock, and stress conditions, making it the ideal option for production on small and big scales (Safi et al. 2014).

On the other hand, the abiotic factors are not the only ones impacting the growth rate of microalgae. Zuccaro et al. (2020) reported that the operational parameters also need to be considered in these factors. They reported the following factors: fluid and hydrodynamic stress, mixing, culture depth, dilution rate, and harvest frequency. Countless varieties of cultivation methods are specific to the productivity and biochemical composition of microalgae (Assunção and Malcata 2020; Seo et al. 2012), considering conditions like environmental factors, metabolic pathway, and cultivation system, such as the one reported by Zuccaro et al. (2020). There are still ongoing studies and development of different growth techniques to target a specific goal in producing macronutrients and micronutrients. Figure 3.4 shows some factors that can influence the upgrowth of microalgae.

TABLE 3.1
***Chlorella* spp. within Different Approaches.**

Microalgae type	Reactor	Culture medium	Operational conditions	Produced compounds	Reference
Chlorella vulgaris	LGem Photobioreactor	M8a medium	Volume: 1,300 L pH: 6.7 Temperature: 25–30°C Light intensity: natural light and high-pressure mercury greenhouse lamps (18 hours light, 8 hours dark)	Protein, carbohydrate, and lipid quantification	(Alavijeh et al. 2020)
Chlorella vulgaris	Horizontal tubular photobioreactor	Modified Bold's basal medium	Volume: 5 L pH: 7.0 ± 0.5 Temperature: n/d Light intensity: 40 W fluorescent lamps	Protein, carbohydrate, and lipid quantification	(Carullo et al. 2018)
Chlorella vulgaris	Sequential-flow bubble column photobioreactor	Bold's basal medium	Volume: 5 L pH: n/r Temperature: 25 ± 5°C Light intensity: cool-white, fluorescent light 60–70 μmol m^{-2} s^{-1}	Biomass improvement	(Dasan et al. 2020)
Chlorella sorokiniana	Bubble column photobioreactor	Tap medium	Volume: 1 L pH: 7.0 ± 0.5 Temperature: 30 ± 2°C Light intensity: white CFL lamps 3,500, 7,000 and 10,500 lx (18 h light, 6 h dark) Air rate: 0.5 and 1 LMP	Increase biomass yield	(Nair and Chakraborty 2020)
Chlorella sorokiniana	Closed bioreactor	Tamiya liquid medium	Volume: 2 L pH: n/r Temperature: 25 ± 2°C Light intensity: 130 μmol m^{-2} s^{-1} (12 h light, 12 h dark) Air rate: 20 L h^{-1} and 1.5% CO_2-enriched sterile air	Analyze environmental conditions	(Stirk et al. 2021)
Chlorella zofingiensis	Flat Plate outdoor photobioreactor	BG-11 medium	Volume: 60 L pH: n/r Temperature: <40°C Light intensity: 200 μmol m^{-2} s^{-1}	Biomass and lipid productivity	(Feng et al. 2011)
Chlorella minutissima	Vertical tubular photobioreactor	BMM medium	Volume: 2 L pH: n/r Temperature: 30°C Light intensity: fluorescent lamps 130 μmol m^{-2} s^{-1} (12 h light, 12 h dark)	Biomass development using CO_2 and pentoses	(Freitas et al. 2017)

Chlorella protothecoides	Photobioreactor with VAP configuration	Inorganic medium	Volume: 2.6 L pH: 6.5 Temperature: 22°C Light intensity: cool light 180 $\mu Em^{-2}\ s^{-1}$ Air rate: 1 vvm	Biomass, lipid, and carotenoid content	(Santos et al. 2014)
Chlorella zofingiensis	Vertical tubular outdoor photobioreactor	BG-11 medium	Volume: 40 L pH: 7.5 ± 0.3 Temperature: 25 ± 1°C Light intensity: white light 200 ± 50 $\mu Em^{-2}\ s^{-1}$ Air rate: 0.5 vvm enriched with 6% CO_2	Biomass production	(Huo et al. 2018)
Chlorella pyrenoidosa	Open Reactor Outdoor	Modified basal heterotrophic seed medium	Volume: 50 L pH: n/r Temperature and light: depend on weather condition Air rate: 22.5 L min^{-1} enriched with 2% CO_2	Biomass concentration, lipid, and fatty acids content	(Han et al. 2013)

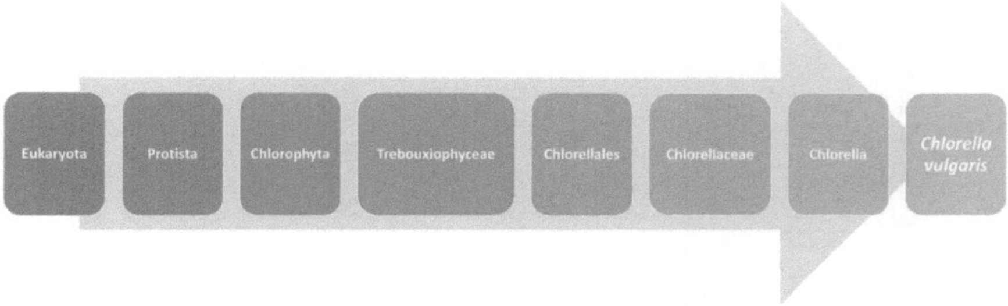

FIGURE 3.3 Scientific classification of *Chlorella vulgaris*.

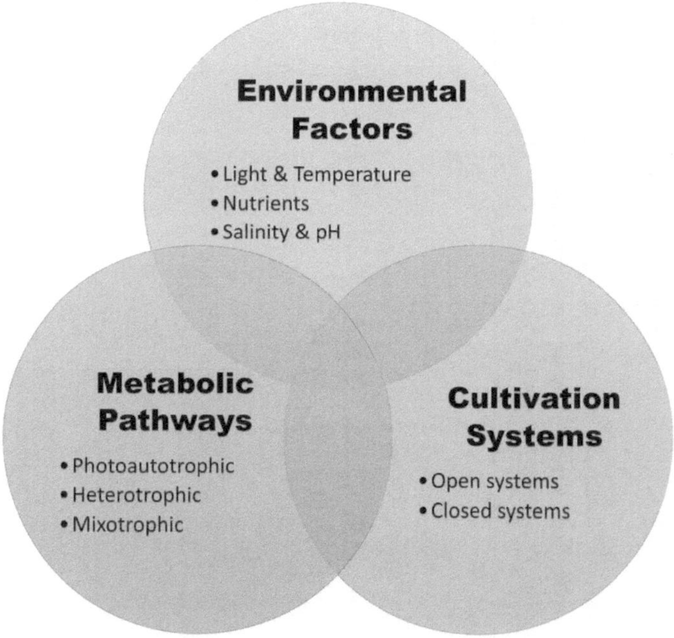

FIGURE 3.4 Principal growth factors that affect *Chlorella vulgaris*.

3.2.2 Environmental Factors

The environmental factors that affect microalgae development are interdependent, and optimizing these factors is essential for achieving high biomass productivity.

3.2.2.1 Light and Temperature

The productivity of phototrophic microalgal cultures is primarily influenced by two factors—light and temperature. Additionally, the alternating patterns of light and dark cycles also significantly impact the synthesis of organic compounds and nutrient metabolism in which light is arguably the most crucial factor predominating microalgal culture's growth and yield. However, an excess exposure to light can lead to light stress, which induces the production of carotenoids to protect the cells from photoinhibition. Overall, the significance of light in biomass production and the organic synthesis of microalgae cannot be overstated (Mehariya et al. 2022; Patel et al. 2019; Begum et al. 2016).

Each species of *Chlorella* has a unique temperature requirement. It is essential to comprehend how temperature fluctuations affect microalgae metabolism, especially in outdoor production systems that face significant variations in temperature on a daily and seasonal basis due to solar radiation. Microalgae can typically thrive within a broad temperature range of 15 to 35°C, with a response that varies significantly between species. Therefore, thoroughly understanding species-specific temperature requirements is crucial for optimizing microalgal growth in various environmental conditions (Patel et al. 2019; Serra-Maia et al. 2016).

3.2.2.2 Salinity, pH, and Other Nutrients

The role of salinity in the growth and metabolism of algal biomass must be considered. Salinity stress is considered a viable approach for enhancing biomass production, carotenoids, and lipids, with the added benefit of using seawater for large-scale microalgae cultivation. Several microalgae species have shown increased lipid content when subjected to salinity stress treatment. Although *Chlorella vulgaris* is typically a freshwater alga, some strains can grow in full-strength seawater, while others may not tolerate high salinity levels. Therefore, exploiting the potential benefits of salinity stress treatment requires a careful consideration of the specific microalgal species and their tolerance levels to salt (Patel et al. 2019; Mehariya et al. 2022; Yun et al. 2019; Luangpipat and Chisti 2017).

Maintaining the appropriate pH level is crucial for successful microalgae cultivation. THE pH affects various physiological processes of microalgae, including photosynthesis, respiration, nutrient uptake, and enzyme activity. Most microalgal species are sensitive TO pH, with only a limited number capable of thriving within the range tolerated by *Chlorella vulgaris*. Additionally, research has indicated that subjecting microalgae to high pH stress can impede the cell cycle while inducing lipid accumulation. Therefore, controlling and monitoring the pH level are essential to ensure optimal microalgal growth and productivity in culture systems (Zuccaro et al. 2020).

Although nitrogen and phosphorus are the primary nutrients necessary for optimal microalgal biomass production, other nutrients are equally essential and cannot be overstated. Magnesium, sulfur, calcium, and iron are among the micronutrients required in the cultivation medium. Research has shown that carbon storage occurs in lipids instead of starch when microalgae experience a shortage of significant nutrients such as nitrogen or phosphorus. Yun et al. (2019) reported that *Chlorella vulgaris* cultivated under nutrient-sufficient conditions typically has a low oil content, ranging from 14 to 30% of dry weight, while nutrient-deficient conditions can increase the lipid content to 70%. However, nutrient starvation reduces microalgal biomass production, particularly nitrogen or phosphate deficiency. Therefore, providing an adequate supply of all necessary nutrients is essential to ensure optimal microalgal growth (Zuccaro et al. 2020; Patel et al. 2019).

3.2.3 Metabolic Pathways

3.2.3.1 Autotrophic Growth

This type of metabolic pathway, autotrophic growth, offers a sustainable solution due to the ability of an organism to synthesize organic compounds from inorganic substances, such as carbon dioxide and water, using energy from the environment. Light can be provided to microalgae in two forms: sunlight or artificial light. Using sunlight as a light source has the added advantage of being a cost-effective option. However, artificial light may be necessary in areas where sunlight is not constant or intense enough to support microalgal growth. It is considered the primary and most common production mode for microalgae growth (Romero-Villegas et al. 2018; Pawar 2016; Zuccaro et al. 2020).

3.2.3.2 Heterotrophic Growth

This type of metabolic pathway, heterotrophic growth, offers the use of organic components mainly as carbon sources, such as the one reported by Pawar (2016): sugars (glucose, galactose, mannose, fructose, sucrose, and lactose); alcohols (methanol, ethanol, and glycerin); and short-chain fatty acid (formate, acetate, and propionate). Moreover, this pathway is carried mainly in closed photobioreactors (PBRs), and the major issues of this type of growth mode are the high cost and availability of the equipment and carbon sources (Safi et al. 2014).

3.2.3.3 Mixotrophic Growth

This type of metabolic pathway, mixotrophic growth, can be used in simultaneously organic and inorganic compounds as carbon sources (Pawar 2016). Patel et al. (2019) describes this kind of cultivation as the microalgae can simultaneously take in inorganic and organic carbon, such as CO_2 and glucose, respectively, for photosynthesis and respiration-mediated carbon capture (via ATP generation), obtaining more biomass than the other two types of metabolic pathways. This growth mode is considered a curious and fascinating type of cultivation as it provides higher biomass production and is more environmentally friendly due to the flexibility of the microalgae adaptation to organic and inorganic sources (Patel et al. 2019).

In order to make mixotrophic production economically viable, selecting an inexpensive organic carbon source is vital. One way to reduce costs and enhance the economic efficiency of the process is to combine mixotrophic culture with the recycling of industrial water waste. This approach supports the concept of the circular economy (León-Vaz et al. 2019).

3.2.4 Cultivation Systems

The equipment required for microalgae development involves photobioreactors (open and closed systems), and their operation depends on the factors that need to be monitored depending on the goals established for microalgae growth (Romero-Villegas et al. 2018).

3.2.4.1 Open Pond System

Open reactors, like circular and raceway ponds, are equipment that can be used on small and large scales and involve using open-air ponds or raceways, typically made of concrete or plastic, to grow and harvest algae; therefore, they are usually employed at the industry level since 90% of the production of microalgae needs to be at a large scale to meet the global demand, and it is considered the cheapest method for cultivation (Assunção and Malcata 2020; Bani et al. 2021).

In biorefinery terms, the open pond raceway is the best and most used due to its low price in construction; it has easy maintenance and can be an attractive option for small-scale or experimental algae cultivation (Bani et al. 2021; Pawar 2016). One of the issues of open reactors is the vulnerability to contamination due to changes in weather conditions because this type of equipment tends to be outdoors, and it is exposed to bacteria, predators, and other microorganisms polluting the original strain (Bani et al. 2021).

Compared to the closed bioreactors, the open reactors are easy to handle and are low maintenance in cost, have culture media requirement, where agro-industrial wastes can be used, and low power consumption is generated. Open pond systems offer a promising approach to the sustainable production of algae-based products, but further research is needed to optimize the technology and improve its efficiency and scalability (Romero-Villegas et al. 2018).

3.2.4.2 Closed Photobioreactors

Sirohi et al (2022) described that a photobioreactor is a closed structure that is sealed from the outside environment and has no exchange of gases or pollutants; also, it has an illuminated culture vessel. The PBR has many advantages over the open reactor systems, such as low contamination from unwanted microorganisms. They are designed to create a controlled environment for the growth of these organisms providing more control over the essential variables in microalgae cultivation, such AS pH, temperature, light, and nutrients. This kind of system allows monoculture growth and enhances the concentration of cells of the microalgae; however, this control can lead to more consistent growth rates and higher biomass yields (Sirohi et al. 2022).

The most employed PBRs at the laboratory and industrial scale are the tubular, flat-plate, and column types. They come in various shapes and sizes and can be made from various materials, including glass, plastic, and metal. This type of equipment can use different light sources such as artificial illumination, fluorescent lamps, radiation, and sunlight. The carbon dioxide produced by the organisms during photosynthesis can be circulated back into the photobioreactor, reducing the need for external sources of carbon dioxide and lowering the overall carbon footprint. There are still ongoing modifications to meet the needs and overcome the bottlenecks of the primary configurations of closed photobioreactors (Sirohi et al. 2022; Assunção and Malcata 2020).

3.3 CULTURE MEDIUM SYSTEM

Different ways of cultivating microalgae exist mainly depending on the target expected. The ability of microalgae to use or tolerate certain pollutants and organic compounds depends on the species and the specific conditions of the culture system. Moreover, *Chlorella spp.* is a robust microalga extensively employed for mitigating nutrient levels and minimizing toxicity in diverse by-product categories and wastewater streams (Lu et al. 2015; Melo et al. 2018).

3.3.1 Synthetic Mediums

Obtaining the reagents for synthetic culture media used for microalgal cultivation is often expensive, impacting primarily small-scale aquaculture farmers. This limitation is challenging for developing cities when the microalgae biomass growth is necessary for commercial products since the primary synthetic media, such as BG-11, Bold Basal medium, and Zarrouk media, ranges between 50% of the total production costs. It is crucial to continue exploring alternatives to the high-cost synthetic media to overcome this challenge (Mtaki, Kyewalyanga, and Mtolera 2023).

3.3.2 Organic Mediums

Previous research has indicated that different types of wastewater, such as dairy and piggery waste, municipal waste, and digested and undigested waste, can cultivate algal biomass and improve energy production (Shi et al. 2016). In order to enhance biomass production and approach the primary nutrients of the wastewater, a pretreatment method needs to be applied, such as hydrothermal, microwave, and UV irradiation. Furthermore, deep research is still needed to achieve a better understanding of improving these techniques for an industrial scale-up (Shi et al. 2016).

3.3.2.1 Wastewaters as Culture Medium

Since the depletion of natural resources is still increasing, replacing water use is essential nowadays. One of the recent solutions is using wastewater for microalgae production in a sustainable approach (Bani et al. 2021). The use of wastewaters leads to a heterotrophic cultivation, and freshwater and CO_2 utilization is removed since the nutrients come from this new kind of water. In fact, high chemical demand levels in water can be an indication of poor water quality, which can negatively impact the growth and health of microalgae. One problem in this kind of residues is that they may contain undesired microorganisms (Pawar 2016).

It has been studied that microalgae can remove inorganic compounds such as nitrogen, phosphorus, carbon dioxide, and heavy metals that are essential for their growth, creating a sustainable solution to the pollution of the rivers and lagoons to which these types of waters are discharged (Safi et al. 2014). However, the composition of the culture medium can significantly impact the metabolic pathway by altering the availability of nutrients, inducing stress responses, and providing alternative substrates or growth factors depending on the purpose given.

3.3.2.2 Agro-Industrial Residues as Culture Medium

One of the major problems causing environmental pollution is the organic matter coming from the agribusiness presented in high measurements of different components like ammonia, nitrates, phosphates, suspended solids, biochemical oxygen demand, and chemical oxygen demand; additionally, the waste also contains dissolved macromolecule components such as carbohydrates, fats, and proteins from the processing and manufacturing of food (Pereira et al. 2021).

It is estimated that 1.3 billion tons of food are wasted per year in the general process of the manufacturing of food from its production to consumption (Pereira et al. 2021). It is considered that the agro-industries are the biggest producers of waste globally. This waste is mainly generated from food production; the dairy industry is the principal sector contaminating the environment due to the disposal of the high amount of waste and the vast consumption of fresh water (Melo et al. 2018; Hamidian and Zamani 2022).

The application of agro-industrial residues as carbon sources can reduce the cost associated with the biomass production, creating a sustainable way to protect and restore the environment, leading to a biorefinery approach (Pereira et al. 2021).

3.3.2.3 Macroalgae as Culture Medium

As mentioned earlier, the third generation of the biorefinery includes all the aquatic biomass, such as macroalgae and microalgae, producing bio-compounds with high added value (Jambo et al. 2016). Utilizing macroalgae as a raw material source for energy production and chemical products for the food and pharmaceutical industries holds a significant promise (Ruiz et al. 2013).

Across all three types of macroalgae, red algae as *Rhodophyta*, green algae as *Chlorophyta*, and brown algae as *Phaeophyceae*, the polysaccharides in their cell walls offer a versatile source of active compounds (Lara et al. 2020). Another star property to consider about macroalgae is the carbohydrate content, which ranges from 30 to 70%. This value differs depending on the specific type of macroalgae (Ruiz et al. 2015).

Seaweeds are rich sources of macro and micronutrients such as nitrogen, potassium, other minerals, and trace elements. Brown algae, in particular, contain alginate, laminarin, and fucoidan as the major polysaccharides (Cervantes-Cisneros et al. 2017) and mannitol; and antioxidants like polyphenols and vitamin C; and minerals such as calcium, potassium, magnesium, sodium, sulfur, phosphorus, and others (Sudhakar et al. 2018).

However, *Sargassum sp.* is a brown algae species with a great potential for development. It is widely used as a raw material in various industries, including food, pharmacy, cosmetics, feed, fertilizer, textiles, paper, and more (González-Gloria et al. 2023). To obtain these components for a future application as a culture medium, it is necessary to apply a hydrothermal pretreatment process for the breakdown of the cell wall, extracting polysaccharides like fucoidan, alginate, and laminarin (Arguello-Esparza et al. 2019).

The conditions of the culture medium, including temperature, light, pH, and mostly nutrient availability, can impact not only the photosynthesis and productivity of microalgae cells, but also their cell metabolism and composition (Malik et al. 2018). A general review was made in Table 3.1 with different types of reactors, culture medium, and *Chlorella* species used for different applications.

3.4 BIOMASS HARVESTING

The biomass harvesting rate represents the concentration of algae developed in the culture medium. One of the main bottlenecks presented in the scaling up of microalgae industrialization is biomass recovery due to the high prices presented in the development of this technique since it varies between the 20 to 60% of the final costs of production (Li et al. 2017; Shen et al. 2015; Potocar et al. 2020).

Nowadays, there are five primary techniques for collecting microalgae cells: centrifugation, flocculation, flotation, filtration, and sedimentation. However, biomass harvesting is often time-consuming with sedimentation and flocculation methods, expensive due to high energy consumption and investment costs with the centrifugation method, and susceptible to clogging and fouling of membranes with filtration technique (Matho et al. 2019).

3.4.1 Centrifugation

The most employed method for harvesting microalgae biomass is the centrifugation technique. It is a soft process for *Chlorella vulgaris* because it does not harm its structure during the procedure. Studies have shown that this process can recover more than 95% and it is not time-consuming, making it highly efficient; the optimal conditions reported are 5,000 rpm within 15 min (Safi et al. 2014).

3.4.2 Flocculation

Different studies have reported that the cell size of *Chlorella vulgaris* is considered to be small ranging from 5 to 20 μm, making it a challenge when harvesting is considered (Shen et al. 2015). At the exponential phase of the microalgae growth, it is reported by Safi et al. (2014) that the algal cells contain a negative charge surface in which the traditional flocculation is conducted by adding to the culture medium cationic salts such as the report made by Zhang et al. (2022) and Zhu et al. (2018) in which inorganic metal salts like $AlCl_3$, $Al_2(SO_4)_3 \cdot K_2SO_4 \cdot 24H_2O$, $FeCl_3$, $Fe_2(SO_4)_3$, and $TiCl_4$ and organic polymers such as cationic poly-acrylamide, poly-dimethyl diallyl ammonium chloride, chitosan, and cationic starch were employed mainly to neutralize the negative charges if an auto-flocculation is not achieved.

This process is considered as an environmental hazard due to the difficulty of recollecting the metal ions; and the auto-flocculation occurs when the stationary phase is reached at the microalgae growth and the cells are no longer dispersed and the negative charge is decreased forming bumps between them (Shen et al. 2015).

3.4.3 Flotation

The harvesting technique of flotation involves using gas bubbles and the surface of cells to attach them to the bubbles, which removes the cells from the liquid phase. This method helps separate algal cells from liquid, as their low density makes sedimentation a challenging option. Flotation can be achieved through the continuous passage of air through a porous material or by dispersing air via a high-speed mechanical agitator, which produces bubbles of varying sizes ranging from 100 μm to 1,500 μm (Matho et al. 2019).

If the bubbles' size is similar to the algal cells' size, the recovery efficiency of flotation can be improved. Other gases like CO_2 or ozone can also be used in the process. Alternative flotation methods involve attaching cells to carrier particles instead of air bubbles. In the case of magnetic carrier particles, the cell–particle aggregates can then be separated through buoyancy or magnetic field gradient (Matho et al. 2019).

3.4.4 Filtration

Membrane filtration is a beneficial method for recollecting microalgae biomass as it provides several advantages, such as low energy consumption, high biomass retrieval, simple operation, scaling, and minimal chemical usage (Razak et al. 2020). This process entails the constant filtration of the broth containing the microalgae through a filter, where the algal cells gradually accumulate and increase in thickness until a certain point is reached (Safi et al. 2014). It can be a standalone process and in conjunction with a membrane photobioreactor (Discart et al. 2015).

Nevertheless, pressure-driven membrane technology is considerably restricted by membrane fouling caused by pore blockage, adsorption, cake layer formation, and gel layer formation (Razak et al. 2020). This limitation is caused by the high cost of the membranes' replacement and the pumping of the clog and foul (Gerulová et al. 2018). Additionally, they can be combined with another harvesting technique, such as flotation or flocculation, which enhances the efficiency of the process (Safi et al. 2014).

3.4.5 Sedimentation

The sedimentation technique effectively ensures low-effective operating costs and is easy to manage. The principle of the technique is based on the fact that algae cells are denser than the surrounding water and will settle under gravity if the medium is left undisturbed. Once the cells have settled, the clear supernatant is removed from the top, leaving a concentrated layer of algal cells at the bottom of the container, which can be further processed or used as desired (Patyna et al. 2018; Potocar et al. 2020). Studies have shown that the rate at which algal cells settle due to gravity averages between 0.1 and 2.6 cm/h. Furthermore, it is more effective for more giant cells like Spirulina (Patyna et al. 2018). However, it may not be suitable for all types of cultures, and alternative methods, such as centrifugation or filtration, may sometimes be required to complement the technique for better results.

Since *Chlorella vulgaris* has been reported to present a small size cell, the selection of the filtration technique needs to be more suitable for accomplishing the recollection of the biomass (Safi et al. 2014). Flocculation followed by flotation technique has been considered a combined method to be as effective as economical in large-scale algae cultivation (Zhang et al. 2022; Matho et al. 2019). Centrifugation is considered the most rapid and gentle method; the only disadvantage is the high cost making it a limitation to the industrial scale-up. (Castrillo et al. 2013). Overall, the different options of the harvesting method are determined by the size and density of the microalgae, as well as the characteristics of the desired products (Gerulová et al. 2018). Recent research needs to be made for cost-effective harvesting techniques (Patyna et al. 2018).

3.5 METHODS FOR EXTRACTION

In order to fully utilize the entire biomass of microalgae produced, gentle extraction methods must be employed. Traditional techniques, such as the physical–chemical methods for extraction and separation, prioritize the extraction of one specific product, often at the expense of damaging other components, and usually consume a significant amount of time to complete (Georgiopoulou et al. 2023; Kapoore et al. 2018).

For protein extraction from microalgae, the conventional solubilization in an alkaline extraction process is usually employed, but there are still various methods that can be combined or modified to optimize the protein extraction (Hildebrand et al. 2020).

When extracting total lipids from *Chlorella vulgaris*, researchers generally utilize the Bligh and Dyer method, which involves a mixture of chloroform, methanol, hexane, or petroleum ether. However, supercritical carbon dioxide (SC-CO_2) extraction has been suggested as a more environmentally friendly alternative (Safi et al. 2014). Nonetheless, some researchers have noted that SC-CO_2 extraction may not effectively extract other antioxidants and antimicrobial agents in *C. vulgaris* (Kitada et al. 2009). Various methods for lipid extraction from microalgae are available, such as organic solvent extraction, pressurized solvent extraction, and supercritical fluid extraction (Yang et al. 2015).

One of the significant energy-intensive steps within the overall process is drying microalgal biomass, which causes significant losses of valuable compounds. However, to create an economically viable microalgae biorefinery is imperative to extract intracellular compounds from different parts of the cells (Alavijeh et al. 2020). Nevertheless, this task is complicated by the rigid cell wall and cell membrane that safeguards the compounds. (Carullo et al. 2018). Table 3.2 shows different

TABLE 3.2
Different Extraction Methods Employed on *Chlorella vulgaris*.

Extraction method	Microalgae type	Component obtained	Recovery	Reference
Ultrasound-assisted single-solvent extraction	*Chlorella vulgaris*	Protein	79.1 ± 5.3% w/w	(Hildebrand et al. 2020)
Ultrasound-assisted enzymatic extraction	*Chlorella vulgaris*	Protein	58–82% w/w	(Hildebrand et al. 2020)
Microwave-assisted extraction	*Chlorella vulgaris*	Total carotenoid	23.09 mg gext^{-1}r	(Georgiopoulou et al. 2023)
Hydrothermal extraction	*Chlorella vulgaris*	Antioxidants and antimicrobials	42 mg gsample-1	(Kitada et al. 2009)
Microwave-assisted extraction	*Chlorella spp.*	Lipids— Fatty acid methyl esters (FAMEs)	26 mg g^{-1}	(Kapoore et al. 2018)
CO_2-expanded methanol	*Chlorella vulgaris*	Total lipid content	20.7 % w/w	(Yang et al. 2015)
Combined bead milling and enzymatic hydrolysis	*Chlorella vulgaris*	Total carbohydrates	74% w/w	(Alavijeh et al. 2020)
Pulsed electric fields	*Chlorella vulgaris*	Total carbohydrates	36% w/w	(Carullo et al. 2018)

extraction methods that can be employed for the different macro and micronutrients of *Chlorella vulgaris*.

3.6 FOUND IN THE MARKET WITH DIFFERENT APPLICATIONS

Due to their versatility, microalgae culture systems can engage in various processes, including producing biofuels, foods, feeds, and high-value bioactives (Álvarez-Díaz et al. 2014). There is a global estimation that more than 5,000 tons of dry microalgal biomass are produced and sold yearly, with an estimated value exceeding 1 billion euros on average (León-Vaz et al. 2019).

3.6.1 Biofuels

In the past decades, there has been a growing industrial interest in algal biotechnology, particularly in the production of biofuels. The significance of microalgal biofuels cannot be overstated as they play a crucial role in mitigating the energy crisis and promoting environmental protection, leading to a surge in interest in the microalgae high-value-added products (Li et al. 2017; Matho et al. 2019).

The extraction of lipids from microalgae biomass can be chemically processed to produce biodiesel, which has become increasingly popular due to its affordability. This bioproduct has led to an extensive research into finding low-cost carbon sources and ideal conditions for oil accumulation while also exploring the production of valuable co-products through a sustainable biorefinery approach (León-Vaz et al. 2019; Álvarez-Díaz et al. 2014).

3.6.2 Human Nutrition

The change toward a healthy diet is a trigger to stop the environmental contamination produced by the food system, substituting meat from animals with plant-based foods, such as fruits, vegetables, and legumes (Springmann et al. 2018; Caron et al. 2018).

Microalgae biomass is considered a natural resource containing rich amounts of bio-compounds due to their abundance of essential minerals, vitamins, antioxidants, and fatty acids; these and many other components make biomass a valuable source of feed supplements (Madeira et al. 2017).

The use of algal biomass has gained significant attention in the food industry as a renewable and promising resource for innovative food product development. Numerous studies have highlighted the potential of microalgae biomass as an ingredient in various food products, including noodles, bread, snacks, cookies, and yogurts (de Medeiros et al. 2020).

Chlorella is a notable example of a microalga that is highly valued as a nutrient-rich source and is widely marketed for use as a supplement. The consumption of microalgae has been shown to have numerous positive effects on consumer health, making it an increasingly popular choice (de Medeiros et al. 2020; Madeira et al. 2017).

3.6.3 Animal Feed

By 2050, the growing global population and rise in income levels are expected to result in a significant increase in demand for animal-derived products, particularly for the most commonly consumed meats worldwide. This surge in demand will have a substantial impact on the livestock industry, as corn and soybean—the two primary conventional feedstuffs for animals—will be in high demand. The overall demand for animal-derived products is set to double, with a particularly significant impact on livestock agriculture (FAO 2011; Madeira et al. 2017).

Numerous nutritional and toxicological assessments have supported using microalgal biomass as feed supplements. However, the nutritive value of microalgae for livestock production is subject to significant variation, primarily influenced by the microalgae species and its chemical composition, including protein, lipids, polysaccharides, vitamins, antioxidants, and minerals. Additionally, the effectiveness of microalgae as an ingredient depends on the animal's ability to adapt to its presence (Madeira et al. 2017).

Microalgae biomass is already used in various livestock industries employing *Chlorella vulgaris* as a source of live feed for the aquaculture industry, including crustaceans, zooplanktons, and mollusks like oysters, clams, and scallops (Mtaki et al. 2023).

3.6.4 Cosmetology, Nutraceutical, and Pharmaceutical

The microalgal industry is experiencing a period of rapid growth with a focus on nutraceuticals leading the way in terms of marketability, followed by food and feed applications (Ahmad et al. 2020). According to predictions, the market value of microalgal pigments is expected to reach USD 452.4 million by 2025, with a Compound Annual Growth Rate (CAGR) of 4%. Therefore, the market for microalgae products is expected to grow and be a significant business until that time (Sun et al. 2023).

The most common microalgae used in the extraction and application of cosmetics are related to the most traditional species in literature, such as *Dunaliella*, *Chlorella*, *Haematococcus*, and *Arthrospira* (Morocho-Jácome et al. 2020). Also, Morocho-Jácome et al. (2020) reported that the main pigments presented in the microalgae biomass include chlorophyll, β-carotene, astaxanthin, xanthophylls, and phycobiliproteins. These pigments have a high advantage over cosmetic products due to their multiple properties, such as antioxidant, immune-boosting, and anti-inflammatory activity; they are mainly used as thickeners, water retention agents, and antioxidants (Alam et al. 2020).

Spirulina and *Chlorella* are currently the most popular microalgae species in the global market, as they are increasingly being recognized and sold in food and health stores. Algae and their extracts have become crucial ingredients in the nutraceutical industry (Alam et al. 2020).

The interest in utilizing microalgae to produce antibiotics and other pharmacologically active compounds has increased. The pharmaceutical industry has already developed a diverse range of products derived from algae biomass, including antivirals, antimicrobials, antifungals, drugs, and therapeutic proteins. Microalgae are also considered to produce essential pharmaceutical products such as omega-3 fatty acids, EPA, DHA, beta-carotene, and astaxanthin (Sathasivam et al. 2019; Alam et al. 2020).

Microalgae are considered to be the primary natural source of carotenoids. Moreover, they are believed to possess significant potential as a source of bioactive compounds belonging to the pigments that are yet to be fully explored for their potential applications in the food, cosmetic, nutraceutical, and pharmaceutical industries (León-Vaz et al. 2019).

3.7 FUTURE PERSPECTIVES

Microalgae represent a promising source of renewable resources for the future since they can compete with nonrenewable resources as a sustainable solution for producing biofuels, chemicals, and food. However, despite the advantages of microalgae, there are still challenges to overcome for its large-scale implementation. The search for other culture media and reducing production costs are essential for the success of a biorefinery based on microalgae. Furthermore, the need for adequate industrial scaling up is a significant problem that must be solved to approach the potential of microalgae thoroughly. Despite these challenges, the prospects for microalgae are very positive. Using its resources can help meet the growing demand for food, chemicals, and biofuels sustainably and efficiently. Therefore, microalgae are emerging as an essential source of renewable resources for the future, and their development and use should continue to be a research priority.

3.8 CONCLUSION

Microalgae are very versatile and adaptable organisms that can grow in various ways according to the objectives of each researcher. These microorganisms can grow in different conditions of light, temperature, salinity, nutrients, and pH, allowing their cultivation in different culture media and their adaptation to different environments. In addition, microalgae can accumulate lipids, carbohydrates, proteins, and pigments, making them a valuable source of raw materials. *Chlorella vulgaris* has demonstrated throughout the years its different applications and adaptability of its components to different factors, offering major products for human and animal consumption, making it the best option for a future algae refinery.

ACKNOWLEDGMENTS

This research project was supported by the Consejo Estatal de Ciencia y Tecnología (COECYT, Coahuila, Mexico) with the "Fondo Destinado a Promover el Desarrollo de la Ciencia y la Tecnología en el Esatdo de Coahuila (FONCYT)" Project "Sustainable cultivation of Spirulina platensis on a large scale for use in livestock feed" – COAH-2022-C19-C123. The authors Miriam L. Rosales-Aguado, Samanta Machado-Cepeda, and Regina Barboza-Rodríguez thank the Mexican National

Council for Science and Technology (CONACYT) for their Master Fellowship support (grant number: 1229509, 1154370, and 1229508 respectively), and Alejandra Cabello also thanks the CONACYT for his Ph.D. fellowship support (Grant number: 711463).

CONTRIBUTIONS

Miriam L. Rosales-Aguado contributed toward conceptualization, writing – original draft, investigation, data analysis, and conceptualization. Samanta Machado-Cepeda contributed toward investigation, data analysis and revision of the manuscript. Regina Barboza-Rodríguez contributed toward investigation, data analysis, and the revision of the manuscript. Alejandra Cabello-Galindo contributed toward investigation, data analysis, and revision of the manuscript. Gilver Rosero-Chasoy contributed toward investigation, data analysis, and revision of the manuscript. Rosa M. Rodríguez-Jasso contributed toward visualization, supervision, funding acquisition, resources, project administration, and revision of the manuscript. Héctor A. Ruiz contributed toward conceptualization, visualization, supervision, funding acquisition, resources, project administration, and revision of the manuscript.

REFERENCES

Ahmad, Muhammad Talba, Mohamed Shariff, Fatimah Md. Yusoff, Yong Meng Goh, and Sanjoy Banerjee. 2020. "Applications of microalga *Chlorella vulgaris* in aquaculture." *Reviews in Aquaculture* 12 (1): 328–346. doi:10.1111/raq.12320.

Alam, Md. Asraful, Jing-Liang Xu, and Zhongming Wang, eds. 2020. *Microalgae Biotechnology for Food, Health and High Value Products*. Singapore: Springer. doi:10.1007/978-981-15-0169-2.

Alavijeh, Razieh Shafiei, Keikhosro Karimi, Rene H. Wijffels, Corjan van den Berg, and Michel Eppink. 2020. "Combined bead milling and enzymatic hydrolysis for efficient fractionation of lipids, proteins, and carbohydrates of *Chlorella vulgaris* microalgae." *Bioresource Technology* 309: 123321. doi:10.1016/j.biortech.2020.123321.

Ali, Sameh Samir, Savvas Giannis Mastropetros, Michael Schagerl, Myrsini Sakarika, Tamer Elsamahy, Mostafa El-Sheekh, Jianzhong Sun, and Michael Kornaros. 2022. "Recent advances in wastewater microalgae-based biofuels production: A state-of-the-art review." *Energy Reports* 8: 13253–13280 (Elsevier). doi:10.1016/J.EGYR.2022.09.143.

Álvarez-Díaz, P. D., J. Ruiz, Z. Arbib, J. Barragán, C. Garrido-Pérez, and J. A. Perales. 2014. "Factorial analysis of the biokinetic growth parameters and CO_2 fixation rate of *Chlorella vulgaris* and *Botryococcus Braunii* in wastewater and synthetic medium." *Desalination and Water Treatment* 52 (25–27): 4904–4914. doi:10.1080/19443994.2013.808590.

Arguello-Esparza, Dulce G., Héctor A. Ruiz, Cristobal N. Aguilar, Diana Jasso de Rodríguez, Bartolomeu W. S. Souza, and Rosa M. Rodríguez-Jasso. 2019. "Relevant marine biomass as feedstock for application in the food industry: An overview." In *Handbook of Research on Food and Science Technology*, edited by Mónica Lizeth Chávez-González, José Juan Buenrostro-Figueroa, and Cristóbal N. Aguilar, vol. 3. Apple Academic Press.

Assunção, Joana, and F. Xavier Malcata. 2020. "Enclosed 'non-conventional' photobioreactors for microalga production: A review." *Algal Research* 52: 102107. doi:10.1016/j.algal.2020.102107.

Bani, Alessia, Francisco Gabriel Acién Fernandez, Giuliana D'Imporzano, Katia Parati, and Fabrizio Adani. 2021. "Influence of photobioreactor set-up on the survival of microalgae inoculum." *Bioresource Technology* 320: 124408. doi:10.1016/j.biortech.2020.124408.

Begum, Hasina, Fatimah M. D. Yusoff, Sanjoy Banerjee, Helena Khatoon, and Mohamed Shariff. 2016. "Availability and utilization of pigments from microalgae." *Critical Reviews in Food Science and Nutrition* 56 (13): 2209–2222 (Taylor and Francis Inc.). doi:10.1080/10408398.2013.764841.

Burki, Talha. 2022. "Food security and nutrition in the world." *The Lancet Diabetes & Endocrinology* 10 (9): 622. doi:10.1016/S2213-8587(22)00220-0.

Caron, Patrick, Gabriel Ferrero y de Loma-Osorio, David Nabarro, Etienne Hainzelin, Marion Guillou, Inger Andersen, Tom Arnold, et al. 2018. "Food systems for sustainable development: Proposals for a profound four-part transformation." *Agronomy for Sustainable Development* 38 (4): 41. doi:10.1007/s13593-018-0519-1.

Carullo, Daniele, Biresaw Demelash Abera, Alessandro Alberto Casazza, Francesco Donsì, Patrizia Perego, Giovanna Ferrari, and Gianpiero Pataro. 2018. "Effect of pulsed electric fields and high pressure homogenization on the aqueous extraction of intracellular compounds from the microalgae *Chlorella vulgaris*." *Algal Research* 31: 60–69. doi:10.1016/j.algal.2018.01.017.

Castrillo, M., L. M. Lucas-Salas, C. Rodríguez-Gil, and D. Martínez. 2013. "High PH-induced flocculation–sedimentation and effect of supernatant reuse on growth rate and lipid productivity of scenedesmus obliquus and *Chlorella vulgaris*." *Bioresource Technology* 128: 324–329. doi:10.1016/j.biortech.2012.10.114.

Cervantes-Cisneros, Daniela E., Dulce Arguello-Esparza, Alejandra Cabello-Galindo, Brian Picazo, Cristóbal N. Aguilar, Héctor A. Ruiz, and Rosa M. Rodríguez-Jasso. 2017. "Hydrothermal processes for extraction of macroalgae high value-added compounds." In *Hydrothermal Processing in Biorefineries*, pp. 461–481. Cham: Springer International Publishing. doi:10.1007/978-3-319-56457-9_20.

Cherubini, Francesco. 2010. "The biorefinery concept: Using biomass instead of oil for producing energy and chemicals." *Energy Conversion and Management* 51 (7): 1412–1421. doi:10.1016/j.enconman.2010.01.015.

Dasan, Yaleeni Kanna, Man Kee Lam, Suzana Yusup, Jun Wei Lim, Pau Loke Show, Inn Shi Tan, and Keat Teong Lee. 2020. "Cultivation of *Chlorella vulgaris* using sequential-flow bubble column photobioreactor: A stress-inducing strategy for lipid accumulation and carbon dioxide fixation." *Journal of* CO_2 *Utilization* 41: 101226. doi:10.1016/j.jcou.2020.101226.

de Medeiros, Viviane Priscila Barros, Tatiana Colombo Pimentel, Roberta Conceição Ribeiro Varandas, Silvana Alves dos Santos, Geany Targino de Souza Pedrosa, Cristiane Francisca da Costa Sassi, Marta Maria da Conceição, and Marciane Magnani. 2020. "Exploiting the use of agro-industrial residues from fruit and vegetables as alternative microalgae culture medium." *Food Research International* 137: 109722. doi:10.1016/j.foodres.2020.109722.

Discart, V., M. R. Bilad, R. Moorkens, H. Arafat, and Ivo F. J. Vankelecom. 2015. "Decreasing membrane fouling during *Chlorella vulgaris* broth filtration via membrane development and coagulant assisted filtration." *Algal Research* 9: 55–64. doi:10.1016/j.algal.2015.02.029.

FAO. 2011. "World livestock 2011 – Livestock in food security." *Rome, Fao*. http://www.fao.org/3/i2373e/i2373e.pdf

Feng, Pingzhong, Zhongyang Deng, Zhengyu Hu, and Lu Fan. 2011. "Lipid accumulation and growth of chlorella zofingiensis in flat plate photobioreactors outdoors." *Bioresource Technology* 102 (22): 10577–10584. doi:10.1016/j.biortech.2011.08.109.

Freitas, B. C. B., M. G. Morais, and J. A. V. Costa. 2017. "Chlorella minutissima cultivation with CO_2 and pentoses: Effects on kinetic and nutritional parameters." *Bioresource Technology* 244): 338–344. doi:10.1016/j.biortech.2017.07.125.

Georgiopoulou, Ioulia, Soultana Tzima, Vasiliki Louli, and Kostis Magoulas. 2023. "Process optimization of microwave-assisted extraction of chlorophyll, carotenoid and phenolic compounds from *Chlorella vulgaris* and comparison with conventional and supercritical fluid extraction." *Applied Sciences* 13 (4): 2740. doi:10.3390/app13042740.

Gerulová, Kristína, Alica Bartošová, Lenka Blinová, Katarína Bártová, Mária Dománková, Zuzana Garaiová, and Marián Palcut. 2018. "Magnetic Fe3O4-polyethyleneimine nanocomposites for efficient harvesting of Chlorella Zofingiensis, *Chlorella vulgaris*, Chlorella Sorokiniana, Chlorella Ellipsoidea and Botryococcus Braunii." *Algal Research* 33: 165–172. doi:10.1016/j.algal.2018.05.003.

González-Gloria, K. D., Rosa M. Rodríguez-Jasso, Gilver Rosero-Chasoy, Shiva, Emily T. Kostas, E. Aparicio, Arturo Sanchez, Iosvany López-Sandin, and Héctor A. Ruiz. 2023. "Scale-up of hydrothermal processing: Liquid hot water and pilot-scale tubular steam explosion batch reactor for bioethanol production using macroalgae sargassum spp biomass." *Bioresource Technology* 369: 128448. doi:10.1016/j.biortech.2022.128448.

González-Gloria, K. D., Rosa M. Rodríguez-Jasso, Shiva, E. Aparicio, Mónica L. Chávez González, Emily T. Kostas, and Héctor A. Ruiz. 2021. "Macroalgal biomass in terms of third-generation biorefinery concept: Current status and techno-economic analysis – A review." *Bioresource Technology Reports* 16: 100863. doi:10.1016/j.biteb.2021.100863.

Hamidian, Najmeh, and Hajar Zamani. 2022. "Biomass production and nutritional properties of chlorella sorokiniana grown on dairy wastewater." *Journal of Water Process Engineering* 47: 102760. doi:10.1016/j.jwpe.2022.102760.

Han, Feifei, Weiliang Wang, Yuanguang Li, Guomin Shen, Minxi Wan, and Jun Wang. 2013. "Changes of biomass, lipid content and fatty acids composition under a light–dark cyclic culture of chlorella pyrenoidosa in response to different temperature." *Bioresource Technology* 132: 182–189. doi:10.1016/j.biortech.2012.12.175.

Hassan, Shady S., Gwilym A. Williams, and Amit K. Jaiswal. 2019. "Moving towards the second generation of lignocellulosic biorefineries in the EU: Drivers, challenges, and opportunities." *Renewable and Sustainable Energy Reviews* 101: 590–599. doi:10.1016/j.rser.2018.11.041.

Hildebrand, Gunda, Mahesha M. Poojary, Colm Ó'Donnell, Marianne N. Lund, Marco Garcia-Vaquero, and Brijesh K. Tiwari. 2020. "Ultrasound-assisted processing of *Chlorella vulgaris* for enhanced protein extraction." *Journal of Applied Phycology* 32 (3): 1709–1718. doi:10.1007/s10811-020-02105-4.

Huo, Shuhao, Zhongming Wang, Shunni Zhu, Qing Shu, Liandong Zhu, Lei Qin, Weizheng Zhou, et al. 2018. "Biomass accumulation of Chlorella Zofingiensis G1 cultures grown outdoors in photobioreactors." *Frontiers in Energy Research* 6. doi:10.3389/fenrg.2018.00049.

Ishika, Tasneema, Navid R. Moheimani, and Parisa A. Bahri. 2017. "Sustainable saline microalgae co-cultivation for biofuel production: A critical review." *Renewable and Sustainable Energy Reviews* 78. Pergamon: 356–368. doi:10.1016/J.RSER.2017.04.110.

Jambo, Siti Azmah, Rahmath Abdulla, Siti Hajar Mohd Azhar, Hartinie Marbawi, Jualang Azlan Gansau, and Pogaku Ravindra. 2016. "A review on third generation bioethanol feedstock." *Renewable and Sustainable Energy Reviews* 65: 756–769. doi:10.1016/j.rser.2016.07.064.

Kapoore, Rahul, Thomas Butler, Jagroop Pandhal, and Seetharaman Vaidyanathan. 2018. "Microwave-assisted extraction for microalgae: From biofuels to biorefinery." *Biology* 7 (1): 18. doi:10.3390/biology7010018.

Khanal, Uttam, Clevo Wilson, Sanzidur Rahman, Boon L. Lee, and Viet-Ngu Hoang. 2021. "Smallholder farmers' adaptation to climate change and its potential contribution to UN's sustainable development goals of zero hunger and no poverty." *Journal of Cleaner Production* 281: 124999. doi:10.1016/j.jclepro.2020.124999.

Kitada, Kiwa, Siti Machmudah, Mitsuru Sasaki, Motonobu Goto, Yuya Nakashima, Shoichiro Kumamoto, and Takashi Hasegawa. 2009. "Antioxidant and antibacterial activity of nutraceutical compounds from *Chlorella vulgaris* extracted in hydrothermal condition." *Separation Science and Technology* 44 (5): 1228–1239. doi:10.1080/01496390902729056.

Lara, Abraham, Rosa M. Rodríguez-Jasso, Araceli Loredo-Treviño, Cristóbal N. Aguilar, Anne S. Meyer, and Héctor A. Ruiz. 2020. "Enzymes in the third generation biorefinery for macroalgae biomass." In *Biomass, Biofuels, Biochemicals*, pp. 363–396. Amsterdam: Elsevier. doi:10.1016/B978-0-12-819820-9.00017-X.

Lara-Flores, Anely A., Rafael G. Araújo, Rosa M. Rodríguez-Jasso, Mario Aguedo, Cristóbal N. Aguilar, Heather L. Trajano, and Héctor A. Ruiz. 2018. "Bioeconomy and biorefinery: valorization of hemicellulose from lignocellulosic biomass and potential use of avocado residues as a promising resource of bioproducts." In, pp. 141–170. doi:10.1007/978-981-10-7431-8_8.

León-Vaz, Antonio, Rosa León, Encarnación Díaz-Santos, Javier Vigara, and Sara Raposo. 2019. "Using agro-industrial wastes for mixotrophic growth and lipids production by the green microalga chlorella sorokiniana." *New Biotechnology* 51: 31–38. doi:10.1016/j.nbt.2019.02.001.

Li, Yi, Yanting Xu, Tianling Zheng, and Hailei Wang. 2017. "Flocculation mechanism of the actinomycete streptomyces Sp. Hsn06 on *Chlorella vulgaris*." *Bioresource Technology* 239: 137–143. doi:10.1016/j.biortech.2017.05.028.

Lu, Qian, Wenguang Zhou, Min, Xiaochen Ma, Ceria Chandra, Yen T. T. Doan, Yiwei Ma, et al. 2015. "Growing chlorella Sp. On meat processing wastewater for nutrient removal and biomass production." *Bioresource Technology* 198: 189–197. doi:10.1016/j.biortech.2015.08.133.

Luangpipat, Tiyaporn, and Yusuf Chisti. 2017. "Biomass and Oil production by *Chlorella vulgaris* and four other microalgae—Effects of salinity and other factors." *Journal of Biotechnology* 257: 47–57. doi:10.1016/j.jbiotec.2016.11.029.

Madeira, Marta S., Carlos Cardoso, Paula A. Lopes, Diogo Coelho, Cláudia Afonso, Narcisa M. Bandarra, and José A. M. Prates. 2017. "Microalgae as feed ingredients for livestock production and meat quality: A review." *Livestock Science* 205: 111–121. doi:10.1016/j.livsci.2017.09.020.

Malik, Andi Adam, Khaeruddin, and Fitriani. 2018. "The effect of sargassum extract on culture medium to the growth of chaetoceros gracilis." *Aquacultura Indonesiana* 19 (1): 10. doi:10.21534/ai.v19i1.115.

Matho, Christoph, Karin Schwarzenberger, Kerstin Eckert, Behnam Keshavarzi, Thomas Walther, Juliane Steingroewer, and Felix Krujatz. 2019. "Bio-compatible flotation of *Chlorella vulgaris*: Study of zeta potential and flotation efficiency." *Algal Research* 44: 101705. doi:10.1016/j.algal.2019.101705.

Mehariya, Sanjeet, Martin Plöhn, Antonio Leon-Vaz, Alok Patel, and Christiane Funk. 2022. "Improving the content of high value compounds in nordic desmodesmus microalgal strains." *Bioresource Technology* 359: 127445 (Elsevier). doi:10.1016/J.BIORTECH.2022.127445.

Melo, Rebeca Gonçalves de, Alexsandra Frazão de Andrade, Raquel Pedrosa Bezerra, Dominick Spindola Correia, Vanessa Cristina de Souza, Ana Christina Brasileiro-Vidal, Daniela de Araújo Viana Marques, and Ana Lúcia Figueiredo Porto. 2018. "*Chlorella vulgaris* mixotrophic growth enhanced biomass productivity and reduced toxicity from agro-industrial by-products." *Chemosphere* 204: 344–350 (Pergamon). doi:10.1016/J.CHEMOSPHERE.2018.04.039.

Morocho-Jácome, Ana Lucía, Nadia Ruscinc, Renata Miliani Martinez, João Carlos Monteiro de Carvalho, Tânia Santos de Almeida, Catarina Rosado, João Guilherme Costa, Maria Valéria Robles Velasco, and André Rolim Baby. 2020. "(Bio)Technological aspects of microalgae pigments for cosmetics." *Applied Microbiology and Biotechnology*. Springer Science and Business Media Deutschland Gmbh. doi:10.1007/s00253-020-10936-x.

Mtaki, Kulwa, Margareth S. Kyewalyanga, and Matern S. P. Mtolera. 2023. "Replacing expensive synthetic media with banana stem compost extract medium for production of *Chlorella vulgaris*." *Applied Phycology* 4 (1): 34–43. doi:10.1080/26388081.2022.2140073.

Naik, S. N., Vaibhav V. Goud, Prasant K. Rout, and Ajay K. Dalai. 2010. "Production of first and second generation biofuels: A comprehensive review." *Renewable and Sustainable Energy Reviews* 14 (2): 578–597. doi:10.1016/j.rser.2009.10.003.

Nair, Ashvini, and Saikat Chakraborty. 2020. "Synergistic effects between autotrophy and heterotrophy in optimization of mixotrophic cultivation of chlorella sorokiniana in bubble-column photobioreactors." *Algal Research* 46: 101799. doi:10.1016/j.algal.2020.101799.

Patel, Anil Kumar, Jae Min Joun, Min Eui Hong, and Sang Jun Sim. 2019. "Effect of light conditions on mixotrophic cultivation of green microalgae." *Bioresource Technology* 282: 245–253. doi:10.1016/j.biortech.2019.03.024.

Patyna, Agnieszka, Małgorzata Płaczek, and Stanisław Witczak. 2018. "Study of *Chlorella vulgaris* sedimentation process." *MATEC Web of Conferences* 240: 05023. doi:10.1051/matecconf/201824005023.

Pawar, Sanjay. 2016. "Effectiveness mapping of open raceway pond and tubular photobioreactors for sustainable production of microalgae biofuel." *Renewable and Sustainable Energy Reviews* 62: 640–653. doi:10.1016/j.rser.2016.04.074.

Pereira, Izabel, Adriano Rangel, Bruna Chagas, Bruno de Moura, Stela Urbano, Roberto Sassi, Fabiana Camara, and Cíntia Castro. 2021. *Biotechnological Applications of Biomass* (Edited by Thalita Peixoto Basso, Thiago Olitta Basso, and Luiz Carlos Basso). London: IntechOpen. doi:10.5772/intechopen.89320.

Potocar, Tomas, Luan de Souza Leite, Luiz Antonio Daniel, Martin Pivokonsky, Dagmar Matoulkova, and Tomas Branyik. 2020. "Cooking oil-surfactant emulsion in water for harvesting *Chlorella vulgaris* by sedimentation or flotation." *Bioresource Technology* 311: 123508. doi:10.1016/j.biortech.2020.123508.

Razak, Nik Nurul Ain Nabilah, Ratri Rahmawati, Muhammad Roil Bilad, Amalia Enggar Pratiwi, Muthia Elma, Normi Izati Mat Nawi, Juhana Jaafar, and Man Kee Lam. 2020. "Finned spacer for enhancing the impact of air bubbles for membrane fouling control in *Chlorella vulgaris* filtration." *Bioresource Technology Reports* 11: 100429. doi:10.1016/j.biteb.2020.100429.

Romero-Villegas, Gabriel Ivan, Marco Fiamengo, Francisco Gabriel Acién-Fernández, and Emilio Molina-Grima. 2018. "Utilization of centrate for the outdoor production of marine microalgae at the pilot-scale in raceway photobioreactors." *Journal of Environmental Management* 228: 506–516. doi:10.1016/j.jenvman.2018.08.020.

Rosero-Chasoy, Gilver. 2022. *Intensification of a Cultivation System for Spirulina Platensis inside a Biorefinery Platform*. Saltillo: Autonomous University of Coahuila.

Ruiz, Héctor A., Juan C. Parajó, and José A. Teixeira. 2015. "Biorefinery strategies for macroalgae-based in bioethanol production." In *Energy science and technology: energy management. energy management*. vol. 12. Press LLC, USA.

Ruiz, Héctor A., Rosa M. Rodríguez-Jasso, Bruno D. Fernandes, António A. Vicente, and José A. Teixeira. 2013. "Hydrothermal processing, as an alternative for upgrading agriculture residues and marine biomass according to the biorefinery concept: A review." *Renewable and Sustainable Energy Reviews* 21: 35–51. doi:10.1016/j.rser.2012.11.069.

Safi, Carl, Bachar Zebib, Othmane Merah, Pierre-Yves Pontalier, and Carlos Vaca-Garcia. 2014. "Morphology, composition, production, processing and applications of *Chlorella vulgaris*: A review." *Renewable and Sustainable Energy Reviews* 35: 265–278. doi:10.1016/j.rser.2014.04.007.

Santos, C. A., B. Nobre, T. Lopes da Silva, H. M. Pinheiro, and A. Reis. 2014. "Dual-mode cultivation of chlorella protothecoides applying inter-reactors gas transfer improves microalgae biodiesel production." *Journal of Biotechnology* 184: 74–83. doi:10.1016/j.jbiotec.2014.05.012.

Sathasivam, Ramaraj, Ramalingam Radhakrishnan, Abeer Hashem, and Elsayed F. Abd_Allah. 2019. "Microalgae metabolites: A rich source for food and medicine." *Saudi Journal of Biological Sciences* 26 (4): 709–722. doi:10.1016/j.sjbs.2017.11.003.

Schwerhoff, Gregor, and Martin Stuermer. 2019. "Non-renewable resources, extraction technology, and endogenous growth." *Federal Reserve Bank of Dallas, Working Papers* 2015 (1506). doi:10.24149/wp1506r1.

Seo, Il-hwan, In-bok Lee, Hyun-seob Hwang, Se-woon Hong, Jessie P. Bitog, Kyeong-seok Kwon, Choul-gyun Lee, Z-hun Kim, and Joel L. Cuello. 2012. "Numerical investigation of a bubble-column photo-bioreactor design for microalgae cultivation." *Biosystems Engineering* 113 (3): 229–241. doi:10.1016/j.biosystemseng.2012.08.001.

Serra-Maia, Rui, Olivier Bernard, Ana Gonçalves, Sakina Bensalem, and Filipa Lopes. 2016. "Influence of temperature on chlorella vulgaris growth and mortality rates in a photobioreactor." *Algal Research* 18: 352–359. doi:10.1016/j.algal.2016.06.016.

Sharma, Pooja, Lohit Kumar Srinivas Gujjala, Sunita Varjani, and Sunil Kumar. 2022. "Emerging microalgae-based technologies in biorefinery and risk assessment issues: Bioeconomy for sustainable development." *Science of The Total Environment* 813: 152417 (Elsevier). doi:10.1016/J.SCITOTENV.2021.152417.

Shen, Y., Z. Fan, C. Chen, and X. Xu. 2015. "An auto-flocculation strategy for chlorella vulgaris." *Biotechnology Letters* 37 (1): 75–80. doi:10.1007/s10529-014-1655-6.

Shi, Jun, Pramod K. Pandey, Annaliese K. Franz, Huiping Deng, and Richard Jeannotte. 2016. "Chlorella vulgaris production enhancement with supplementation of synthetic medium in dairy manure wastewater." *AMB Express* 6 (1): 15. doi:10.1186/s13568-016-0184-1.

Singh, Anusuiya, Rosa M. Rodríguez Jasso, Karla D. Gonzalez-Gloria, Miriam Rosales, Ruth Belmares Cerda, Cristóbal N. Aguilar, Reeta Rani Singhania, and Héctor A. Ruiz. 2019. "The enzyme biorefinery platform for advanced biofuels production." *Bioresource Technology Reports* 7: 100257. doi:10.1016/j.biteb.2019.100257.

Sirohi, Ranjna, Ashutosh Kumar Pandey, Panneerselvam Ranganathan, Shikhangi Singh, Aswathy Udayan, Mukesh Kumar Awasthi, Anh Tuan Hoang, Chaitanya Reddy Chilakamarry, Sang Hyoun Kim, and Sang Jun Sim. 2022. "Design and applications of photobioreactors- a review." *Bioresource Technology* 349: 126858. doi:10.1016/j.biortech.2022.126858.

Springmann, Marco, Michael Clark, Daniel Mason-D'Croz, Keith Wiebe, Benjamin Leon Bodirsky, Luis Lassaletta, Wim de Vries, et al. 2018. "Options for keeping the food system within environmental limits." *Nature* 562 (7728): 519–525. doi:10.1038/s41586-018-0594-0.

Stirk, Wendy A., Péter Bálint, Gergely Maróti, Zoltán Varga, Zsuzsanna Lantos, Johannes van Staden, and Vince Ördög. 2021. "Comparison of monocultures and a mixed culture of three chlorellaceae strains to optimize biomass production and biochemical content in microalgae grown in a greenhouse." *Journal of Applied Phycology* 33 (5): 2755–2766. doi:10.1007/s10811-021-02515-y.

Sudhakar, K., R. Mamat, M. Samykano, W. H. Azmi, W. F. W. Ishak, and Talal Yusaf. 2018. "An overview of marine macroalgae as bioresource." *Renewable and Sustainable Energy Reviews* 91: 165–179. doi:10.1016/j.rser.2018.03.100.

Sun, Han, Yuxin Wang, Yongjin He, Bin Liu, Haijin Mou, Feng Chen, and Shufang Yang. 2023. "Microalgae-derived pigments for the food industry." *Marine Drugs* 21 (2): 82. doi:10.3390/md21020082.

Tang, Doris Ying, Kuan Shiong Khoo, Kit Wayne Chew, Yang Tao, Shih Hsin Ho, and Pau Loke Show. 2020. "Potential utilization of bioproducts from microalgae for the quality enhancement of natural products." *Bioresource Technology* 304: 122997 (Elsevier). doi:10.1016/J.BIORTECH.2020.122997.

Thornton, Philip, Dhanush Dinesh, Laura Cramer, Ana Maria Loboguerrero, and Bruce Campbell. 2018. "Agriculture in a changing climate: Keeping our cool in the face of the hothouse." *Outlook on Agriculture* 47 (4): 283–290. doi:10.1177/0030727018815332.

Wan Mahari, Wan Adibah, Wan Aizuddin Wan Razali, Hidayah Manan, Mursal Abdulkadir Hersi, Sairatul Dahlianis Ishak, Wee Cheah, Derek Juinn Chieh Chan, Christian Sonne, Pau Loke Show, and Su Shiung Lam. 2022. "Recent advances on microalgae cultivation for simultaneous biomass production and removal of wastewater pollutants to achieve circular economy." *Bioresource Technology* 364: 128085. doi:10.1016/j.biortech.2022.128085.

Wells, Mark L., Philippe Potin, James S. Craigie, John A. Raven, Sabeeha S. Merchant, Katherine E. Helliwell, Alison G. Smith, Mary Ellen Camire, and Susan H. Brawley. 2017. "Algae as nutritional and functional food sources: revisiting our understanding." *Journal of Applied Phycology* 29 (2): 949–982. doi:10.1007/s10811-016-0974-5.

Yang, Yi-Hung, Worasaung Klinthong, and Chung-Sung Tan. 2015. "Optimization of continuous lipid extraction from chlorella vulgaris by CO_2 -expanded methanol for biodiesel production." *Bioresource Technology* 198: 550–556. doi:10.1016/j.biortech.2015.09.076.

Yun, Chol-Jin, Kum-Ok Hwang, Song-Su Han, and Hyong-Guan Ri. 2019. "The effect of salinity stress on the biofuel production potential of freshwater microalgae chlorella vulgaris YH703." *Biomass and Bioenergy* 127: 105277. doi:10.1016/j.biombioe.2019.105277.

Zhang, Ping, Sihan Zhu, Chao Xiong, Bin Yan, Zhikang Wang, Kai Li, Irumva Olivier, and Han Wang. 2022. "Flocculation of *Chlorella vulgaris*–Induced algal blooms: critical conditions and mechanisms." *Environmental Science and Pollution Research* 29 (52): 78809–78820. doi:10.1007/s11356-022-21383-8.

Zhu, Liandong, Zhaohua Li, and Erkki Hiltunen. 2018. "Microalgae *Chlorella vulgaris* biomass harvesting by natural flocculant: Effects on biomass sedimentation, spent medium recycling and lipid extraction." *Biotechnology for Biofuels* 11 (1): 183. doi:10.1186/s13068-018-1183-z.

Zuccaro, Gaetano, Abu Yousuf, Antonino Pollio, and Jean-Philippe Steyer. 2020. "Microalgae cultivation systems." In *Microalgae Cultivation for Biofuels Production*, pp. 11–29. Amsterdam: Elsevier. doi:10.1016/B978-0-12-817536-1.00002-3.

4 Algae Based Nutrient Recovery from Different Waste Streams

Meenakshi Fartyal and Chitra Jain

4.1 INTRODUCTION

There are currently two crucial concerns for worldwide sustainable growth, and they are the energy crisis and the deterioration of the environment (OECD 2008). Fossil fuels, which are used extensively in both advanced and emerging countries, are the primary source of potential energy (Singh et al. 2014; Zou et al. 2016). Global warming has been exacerbated by our reliance on conventional fuels (Srivastava et al. 2015; Vaish et al. 2016). Consistent demand growth has accelerated the depletion of renewable energy sources (Singh et al., 2011; Oyedepo 2012). Therefore, for the sake of all living things, the globe is searching for a readily available, carbon–light energy system (Wang et al. 2019; Vaish et al. 2017). As a result of recent developments in genetics, biotechnology, engineering, and process chemistry, a new manufacturing idea known as "biorefinery" has emerged. The emerging evidence of the benefits of biorefinery in terms of material savings and carbon reductions has piqued the public's curiosity (Ahuja and Tatsutani 2009; Mitchell and Morgan 2015), and has led to commensurate focus on biorefinery. Recovering a variety of nitrogen, phosphorus, or potassium products of agricultural relevance has been a primary focus in the design and operation of biorefineries, alongside the production of biofuels. In some ways, a biorefinery can be thought of as being related to a petroleum refinery industry in that it is integrated facilities that use extraction/conversion technologies to produce a wide range of platform chemicals, fuels, and energy (Naik et al. 2010; Blades et al. 2017; Vaish et al. 2019).

In order to convert algal biomass into numerous higher valued products with little waste creation, a new, cost-effective, and sustainable idea known as the algal biorefinery is evolving. Recovering a single product does not make economic sense. In order to bring several algae-based goods to market, biomass must be biorefined selectively and processed further into a variety of products (Ruiz et al. 2016). Cascading biorefineries gradually extract several major and intermediate products from biological feedstock with the goal of generating as little trash as possible (Gifuni et al. 2019; Haider et al. 2022).

Nutrient recovery is one resourceful approach that is currently fashionable in the wastewater sector and highlighting these advantages. Recovering nutrients like nitrogen and phosphorus from wastewater streams that would otherwise be flushed down

DOI: 10.1201/9781003455950-4

the drain is called nutrient recovery, and the by-product is a fertilizer that is good for the environment and can be used in agriculture. By eliminating these nutrients, this procedure aids in the cleaning of the effluent and ultimately converts it into a useful, recyclable feedstock. Nutrients, especially phosphorus, found in wastewater can be hazardous to both infrastructure and the environment, leading to issues like eutrophications in aquatic systems and struvite build in mechanical system. Wastewater treatment plants can deal with these problems, boost water quality, and conform to strict phosphorus discharge regulations by employing nitrogen recovery. The recovery method provides an opportunity for local governments to earn money while simultaneously meeting the needs of agricultural industries for purified, usable phosphorus. Furthermore, it enables wastewater organizations to serve as resource recovery agents in addition to treatment facilities, changing the way people view conventional wastewater treatment (Haddaway 2015).

The cultivation of cyanobacteria for the purpose of resource recovery from wastewater is gaining popularity around the world and could be a useful contribution to the circular bioeconomy. In the 1950s, algae were being employed to give O_2 for the breakdown of biological waste, and William J. Ostwald initially presented this idea. Algal nutrient removal helps with wastewater treatment because it reduces eutrophication, stores carbon, and generates useful biomass. It is getting increasingly difficult to manage the amount of 61–2,310 million m^3 of wastewater produced daily over the world (Renuka et al. 2021); the alternative is to use it as a resource to efficiently manufacture algal biomass. Wastewater that is discharged directly into bodies of water without being treated has the potential to damage aquatic ecosystems and endanger aquatic life. Data suggests that the percentage of treated wastewater is significantly lower than 40%. In industrialized nations, the figure is closer to 70%, whereas in less developed nations, it is closer to 8% (Batool and Shahzad 2021). Oedogonium, Spirogyra, Klebsormidium, and Cladophora are only few of the filamentous cyanobacterial species that have been documented to cleanse wastewater. Species of chlorella have been demonstrated to remove NH4-N at 82% rate and P at 70% rate. The effectiveness of *C. vulgaris* at removing COD was 80% (Lin et al. 2019). Continual shifts in the compositions of the wastewater due to seasonal variation impact the development of the algal strain used in wastewater treatment. Both N and P are necessary macronutrients for plant and algae development. Nitrogen contributes to the synthesis of nucleic acids and proteins and amounts for 7%to 20% of the dry cell weight, making phosphorus an essential component of the energy-storage molecule ATP (Pancha et al. 2015). Scarcity of macronutrients, for instance, prompts the body to devote more energy to producing store lipids. Increased fat and starch buildup occurred during nitrogen deficiency, and protein synthesis and phycobilin synthesis were bolstered by an additional nitrogen source (Haider et al. 2022).

4.2 ALGAE AND THEIR ROLE IN BIOTECHNOLOGY

Aquatic creatures that do photosynthesis are known collectively as algae. Microalgae are single-celled organisms whereas macroalgae are communities of thousands of cells. Algae, in contrast to higher plants, lack all three of these structural components (Bajpai 2019). Microalgae, or unicellular algae, are eukaryotic organisms that are only a few micrometers to a few hundred micrometers in size. Rapid multiplication and food uptake allow them to thrive in autotrophic or mixotrophic environments, where they can expand their numbers every few hours (Su 2021; Aditya et al. 2022; Nguyen et al. 2022). Macroalgae can grow to be many meters in length and reproduce rapidly in watery environments. Their thallus (leaf-like bodies) are buoyant thanks to enclosed gas-filled structures (foats), and they cling to surfaces via specialized basal structures (holdfasts) (Suganya et al. 2016). Microalgae and macroalgae are able to naturally accumulate nutrients having nitrogen and phosphorous from various water bodies to produce a wide variety of bioactive components such as "pigments, carbohydrates, proteins, and lipids" with a variety of commercial applications due to their diverse physiological and biochemical characteristics (Suganya et al. 2016; Khoo et al. 2019; Liu et al. 2020; Nguyen et al. 2022).

Among the cyanobacteria and algae, you'll find anything from tiny unicellular species to massive multicellular marine and freshwater animals measuring more than a meter in length. While both plants and algae perform photosynthesis, algae have a less complex cellular structure. Algae don't contain any of the typical plant parts, such as a vascular system or a root system. Without a lot of cell differentiation, they can take the shape of "unicellular, multicellular, simple or branched filamentous, leafy or blade forms" (Barsanti and Gualtieri 2006). Algae and cyanobacteria are capable of colonizing any ecosystem. In the euphotic zone of the ocean, lakes, and other bodies of water, they can survive as planktonic organisms. They also colonize and reside attached on submerged hard surfaces like sediments, stones, plants, and so on. "Microalgae" refers to the microscopic species. "Macroalgae" refers to a group of large, multicellular organisms. In the field of applied phycology, the term "microalgae" has a broad meaning. Both bacterial cyanobacteria and eukaryotic algae can be found in this group (Masojídek and Prášil 2010).

Several biotechnological processes depend on algae. Microalgae are mostly grown commercially for use in nutritional supplements, cosmetics, and aquaculture. A variety of beneficial substances, including "proteins, amino acids, vital unsaturated fatty acids, and vitamins", are concentrated in microalgae. "Chlorella, Arthrospira (Spirulina), Dunaliella, Nannochloropsis, and Haematococcus" are the most common genera used in industrial production (Mimouni et al. 2012). It has been discovered that microalgae are a rich resource for lipids which could be used as a feedstock in the manufacturing of biofuels. Unsaturated fatty acids (such eicosapentanoic acid and docosahexanoic acid) as well as neutral lipids like triacylglycerides can be produced by microalgae (Markou and Nerantzis 2013). After the lipids are converted to fatty acid methyl esters, they can be used as a viable feedstock for biofuel production. Microalgal biotechnology industry has recently become interested in the possibility of employing algae for bioremediation. Different pollutants can be eliminated from wastewater by the employment of microalgae. Together with the study of nitrogen and phosphorus removal in agricultural, home, or other wastewater, studies have focused on reducing chemical and biological oxygen demands (Arcila and Buitrón 2016). Algae are a valuable biosorbent for the removal of heavy metals because of their negatively charged cell surfaces and high-cell-surface-to-volume ratios (Roberts et al. 2015; Li et al. 2015). Algae have been found to grow in wastewater and lower nutrient concentrations in several lab-scale investigations, as was described earlier (Doria et al. 2012). Microalgae can be grown for this purpose either in suspension or adhered to a solid surface. Algae culture in wastewater treatment ponds and natural connected algal-based systems have been considerably studied in recent years (Adey et al. 2011; Park et al. 2011). Biofuel production and nutrient recycling have both been investigated (Sukačová and Červený2017).

Algae mechanisms to take in nutrients- Wastewater phycoremediation is based on the metabolic processes of algae. Algae absorb nutrients and convert them into structural building blocks like nucleic acids and proteins in order to increase their biomass (Cai et al. 2013; Whitton et al. 2015). Nitrogen can be found in wastewater in the form of ammonium and nitrate as well as organic nitrogen. Algae prefer to take in nitrogen in the form of ammonium, and nitrate is converted back to ammonium and used to make proteins within the algae cell (Beuckels et al. 2015). Ribosomal RNA production requires inorganic phosphorus. When external phosphorus availability is low, microalgae can still use the additional phosphorus they took in during a luxury uptake and stored as polyphosphate (Wu et al. 2012; Lavrinovičs and Juhna 2017).

In order to grow, algae must assimilate nitrogen and phosphorus from their growing medium. Organic nitrogen, which includes peptides, proteins, enzymes, chlorophylls, energy transfer molecules like adenosine diphosphate (ADP) and adenosine triphosphate (ATP), and genetic materials like RNA and DNA is produced through the process of nitrogen assimilation (Barsanti and Gualtieri 2014). Nitrate and nitrite are changed into ammonium ions with the help of nitrate reductases and nitrite reductase during the assimilation process (Gonçalves et al. 2017). The amino acid glutamine (intracellular) is synthesized when ammonium is incorporated into it via the action of glutamate (Glu) and adenosine triphosphate (ATP) (Cai et al. 2013; Su 2021). Algae use a method of nitrogen uptake that is distinct from that used by bacteria. There are two possible phases of bacterial nitrogen elimination: nitrification and denitrification. Alternatively, nitrogen can be removed in one stage via anaerobic ammonium oxidation (ANAMMOX) (Mishra et al. 2022; Ren et al. 2020). In addition to nitrogen, phosphorus is essential for the metabolism and development of microalgae. Intracellular organic molecules (such as nucleic acids, lipids, and proteins) are phosphorylated from inorganic phosphorous (H_2PO_4 and HPO_4) (Gonçalves et al. 2017; Mohsenpour et al. 2021). Microalgae use a variety of phosphate transporters in their plasma membranes to take up inorganic P for phosphorous transformation within their cells. Under light conditions, a series of chemical reactions takes place, including the generation of ATP from ADP and the synthesis of polyphosphate (both acid-soluble and acid-insoluble polyphosphate) by polyphosphate kinase (i.e., photosynthesis). Algal nutrient absorption patterns may change depending on the availability of specific nutrients (Su 2021). For instance, slow-growing macroalgae accumulate nitrogen and phosphorous during nutrient-having seasons so that they can continue developing during nutrient-deficient periods (i.e., higher uptake) (Martínez et al. 2012; Cole et al. 2016). These slow-growing algae reduce their fertilizer need by slowing their growth rate and maximize their C uptake ability, benefiting from the abundant sunshine, when resources are scarce (Beuckels et al. 2015; Nguyen et al. 2022).

4.3 NUTRIENTS FROM WASTEWATER STREAMS

Where wastewater was produced has a significant impact on the N:P ratio. N:P ratios of above 40 can be found in wastewaters including animal manure and human excreta (Kumar et al. 2010; Tuantet et al. 2014). However, municipal wastewaters contribute very little compared to industrial wastewaters, and their N:P ratios are normally around 10 (Henze and Comeau 2008). Municipal wastewater typically has a COD and BOD of around 300–500 mg L^{-1} and 200–300 MG L^{-1}, respectively. Eutrophication can be avoided, and new nutrient sources can be made available, if nutrients are recovered from wastewaters (Figure 4.1) (Muthoni et al. 2022).

The efficiency of nutrient recovery using microalgae utilizing current technology is constrained by a number of biological and technological factors, including but not limited to light, fertilizer supply, carbon dioxide supply, temperature, pH, and solar radiation. Technologies are being tweaked and enhanced to increase biomass production capacity while recovering all major nutrients. Tube reactors are commonly used for cultivating sensitive strains of microalgae. Rapidly growing microalgae can use atmospheric nitrogen and phosphorus to produce algal biomass, which has several potential applications including but not limited to biofuel, carbohydrates, animal feed, polymers, and agricultural fertilizer when exposed to light and carbon dioxide (Muthoni et al. 2022).

Nutrient uptake by microalgae is a key issue in the nutrient removal process because of the impending necessity to recover nutrients. Multiple applications exist

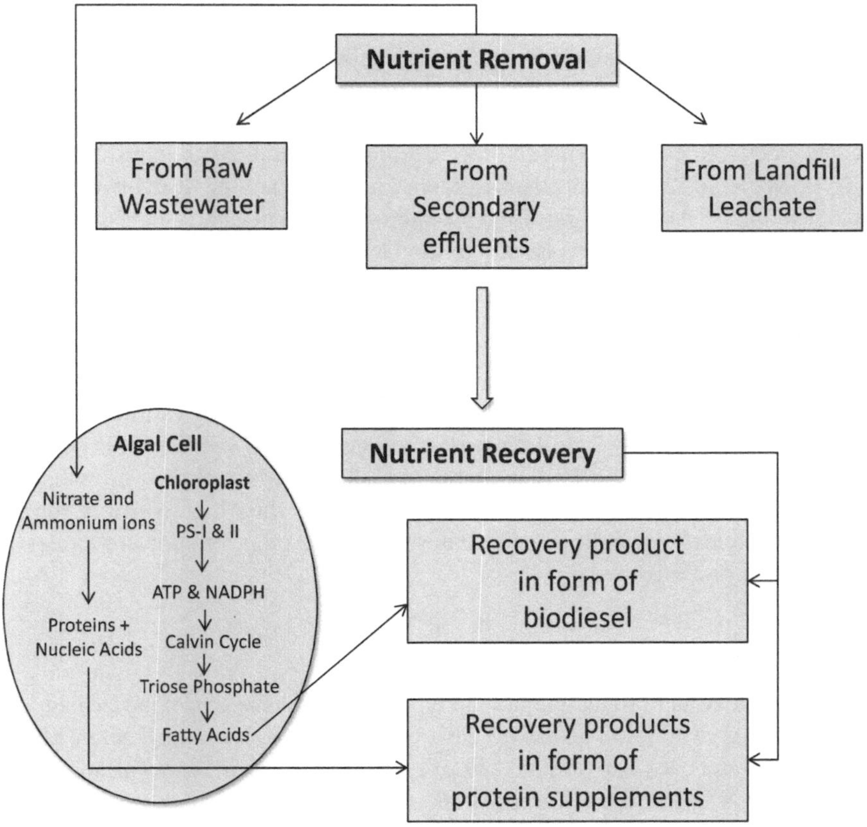

FIGURE 4.1 Nutrient recovery technology.

for the algal biomass that grows in sewage (Pittman et al. 2011). The potential for recycling the nutrients is diminished, however, if they are contaminated with heavy metals, micropollutants, or pathogens. The algal biomass could be used as a biofertilizer. It has been documented that blue-green algae can be used as a biofertilizer and soil conditioner in rice farming. As a form of slow-release fertilizer, the algal biomass recovered from treating cow feces has been put to use by Mulbry et al. (2005). They used commercial potting soil altered with either ATS biomass or a commercial fertilizer that was nearly equivalent and found no difference in plant growth between the two treatments. Metals in wastewater from a coal-fired power plant can be sequestered by algae growing in bioremediation ponds, according to research published by Roberts et al. in 2015. The filamentous alga Oedogonium makes up the algal biomass that can be used to make algal biochar for soil improvement. The addition of this biochar to poor soil increased fertilizer efficiency by 35–40%, leading to increased radish yields (Roberts et al. 2015).

Although biochar is typically used to enhance soil by replenishing the carbon pool and providing critical trace elements, we predict that algal biomass high in phosphorus can be successfully converted to biochar for supplementing soils with phosphorus. Because of having higher proteins, algae can be used as a supplementary feed for animals (Spolaore et al. 2006). However, research on the feasibility of feeding animals using algae recovered from wastewater has not yet been conducted. In order for the algal biomass to be used as animal feed, it would need to be free of infections and other potentially dangerous compounds. Nutrient recovery utilizing algal biomass from wastewater treatment is an area that still needs more research and development. A wide range of algae species have been used to remove nutrients in controlled laboratory growing settings (Chinnasamy et al. 2010; Johnson and Wen 2010; Boelee et al. 2011). Nonetheless, there are only a few number of widespread implementations. Ability to release Phosphorous, having algal biomass has been studied but only on a small scale due to the lack of large-scale systems (Sukačová and Červený 2017).

Research into microalgae cultivation in wastewaters has advanced dramatically during the past few decades. The use of microalgae for removing of N, P, and other components from wastewaters has been studied, and the findings have been encouraging. In 2008, the Clean Watersheds Needs Survey estimated that the United States produced 12 million t dl of wastewater (USEPA 2008). Different types of wastewater have different chemical makeups. Microalgae's growth, lipid composition, and yield are all significantly affected by their diet. Nutrient components in municipal, agricultural, and industrial wastewater streams are explained in subsequent sections.

4.3.1 Municipal Wastewater

Municipal wastewater production has increased as cities have grown in size and population. If a city of 500,000 people used 0.2 t dl per capital, it would generate 85,000 t dl of wastewater each day (Pescod 1992). Table 4.1 displays the concentrations of N andP in various wastewaters. The amounts of nitrogen and phosphorus in municipal wastewater are lower than those found in animal waste water. Raw sewage from a city, however, frequently contains significant levels of heavy metals including lead, zinc, and copper. Municipal wastewater treatment normally consists of three stages: primary, secondary, and advanced. Primary treatment involves the use of physical or chemical processes to distinguish between buoyant and non-buoyant substances. Secondary treatment involves the elimination of dissolved organics and colloidal pollutants using biological or chemical procedures. Dissolved inorganic components like nitrogen and phosphorus are removed via unit activities in the advanced treatment process viz. post-aeration, filtration, carbon adsorption, and membrane separation. The most research into microalgae has focused on their potential to reduce nitrogen and phosphorus in municipal wastewaters (Bhatnagar et al. 20 10; Ruiz-Marin et al. 20 10; Chi et al. 20 11; Li et al. 2011).

4.3.2 Agricultural Wastewater

Wastewater from animal production is a major contributor to agricultural wastewater. Current animal feeding facilities in the United States create more than 450 million metric tons of manure and manure-contaminated runoff water each year. Over the past decade, there has been a movement from small and medium-sized livestock enterprises to larger ones. Because of this, nutrients are increasingly concentrated in places with a lot of cattle (Kellogg et al. 2000). Table 4.1

TABLE 4.1
Total Nitrogen (TN) and Total Phosphorus (TP) Content of Different Waste Streams.

Wastewater category	Description	TN (mg L1)	TP (mg L1)	N/P	Reference(s)
Municipal	Sewage	15–90	5–20	3.3	(Burks and Minnis 1994)
wastewater	Dairy	185–2636	30–727	3.6–7.2	(Barker et al. 2001;
Animal	Poultry	802–1825	50–446	4–16	Bradford et al. 2008)
wastewater	Swine	1110[a]–3213	310–987	3.0–7.8	(Barker et al. 2001;
Industrial	Beef feedlot	63–4165	14–1195	2.0–4.5	Yetilmezsoy and Suleyman
wastewater	Textile	21–57[a]	1.0–9.7[b]	2.0–4.1	2008)
	Winery	110[a]	52	2.1	(Barker et al. 2001;
	Tannery	273[a]	21[b]	13.0	Millmier et al. 2000).
	Paper mill	1.1–10.9	0.6–5.8	3.0–4.3	(Barker et al. 2001;
	Olive mill	532	182	2.9	Yetilmezsoy and Suleyman 2008)
					(Chinnasamy et al. 2010; Şen and Demirer 2003)
					(Mosse et al. 2011)
					(Durai and Rajasimman, 2010)
					(Slade et al., 2004)

Note: a, Total Kjeldahl nitrogen (TKN); b, total orthophosphates (PO4 3–P).
Source: Cai et al. (2013)

shows that nitrogen and phosphorus levels tend to be high in effluents from livestock farms. Half of the nitrogen in animal excrement is inorganic nitrogen, and half is in the form of ammonium. The nutrient content of animal wastewater will vary greatly depending on factors including the animals' diet, age, consumption, production, management, and location. Wastewater from dairy, swine, and beef feedlots often has a N/P ratio of 2–8. Manure from animals is commonly applied to fields as fertilizer. Crops can't fully repair the nutrients in manure due to varying N/P ratio needs and nutrient availability. Soil eutrophication occurs when an abundance of nutrients causes an increase in runoff and subsequent nutrient losses. Herbicides, fungicides, insecticides, and nitrates and phosphates are all possible components of agricultural runoff in addition to manure. Particles, dissolved ions and chemicals, and living microorganisms may all be carried into receiving rivers by agricultural runoff, drastically decreasing water quality.

4.3.3 Industrial Wastewater

Between 1987 and 2003, it was estimated that global industry consumed an average of 665 billion metric tons (t) of water per year. It has been estimated that the United States accounts for almost one-third of global water use (Newman 2011). In general, industrial wastewaters have a higher concentration of heavy metal contaminants and a lower concentration of nitrogen or phosphorus than municipal wastewaters (Ahluwalia and Goyal 2007). High metal removal efficiency requires the careful selection of microalgae strains with a high metal sorption capacity. The potential of only a few algae species to sorb metals has been investigated thus far. Several studies have examined the efficacy of using carpet factory wastewater, which contains high levels of nitrogen and phosphorus, as an algae growth medium (Chinnasamy et al. 2010; Cai et al. 2013).

4.4 TECHNOLOGIES TO RECOVER NUTRIENTS FROM WASTE STREAMS

Physical, biological, and chemical procedures can all be used at different stages of wastewater treatment, from basic to secondary to tertiary. Sedimentation, filtration, and evaporation are all examples of physical processes. Physical and chemical processes include things like coagulation and flocculation, adsorption, ion exchange, precipitation, membranes, and oxidation reduction. There are two types of biological processes: those that require oxygen and those that don't. The first step is to get rid of all the heavy stuff. Nitrates, phosphates, and trace organic compounds are removed in the final step of treatment, after the dissolved organic matter has been consumed and key nutrients oxidized (Molazadeh et al. 2019). The N:P ratio of wastewater varies greatly depending on where it was generated. N:P ratios of above 40 can be found in wastewaters including animal manure and human excreta (Kumar et al. 2010; Tuantet et al. 2014). However, municipal wastewaters contribute very little compared to industrial wastewaters, and their N:P ratios are normally around 10 (Henze, 2008). Municipal wastewater typically has a COD and BOD of around 300–500 mg L^{-1} and

200–300 mg L^{-1}, respectively. Eutrophication can be avoided, and new sources of nutrients can be made available, if we recover nutrients from wastewaters. Biological nutrient removal (BNR) is used in conventional water treatment to reduce phosphorus levels to 1–3 mg L^{-1} and nitrogen levels to 8–10 mg L^{-1}, respectively. Improved nutrient removal prior to effluent discharge is necessary for wastewater plants to maintain regulatory compliance (Nutrient Pollution in Wastewater 2016; Muthoni et al. 2022).

4.4.1 Algae-Based Technologies

Easily settled materials are removed during the first wastewater treatment process, and organic pollutants are oxidized during the secondary wastewater treatment process, leaving a clear apparent effluent that can be released into water bodies. However, eutrophication and other long-term issues can be caused by excessive inorganic nitrogen and phosphorus in this effluent. The combination of microalgae's ability to produce potentially valuable biomass and its use in tertiary bio-treatment is intriguing. Due to their potential applications in the renewable energy and nutraceutical industries, microalgae are currently receiving widespread attention across the world. They are cheap to produce and can be used to make biofuels or added to food indefinitely. Recovering nutrients from wastewaters via biological processes is sustainable and kind to the environment. The algae generate oxygen as a by-product of the wastewater treatment process, which the bacteria in the water utilize to generate carbon dioxide for the algae, as shown in Figure 4.2 (Muthoni et al. 2022).

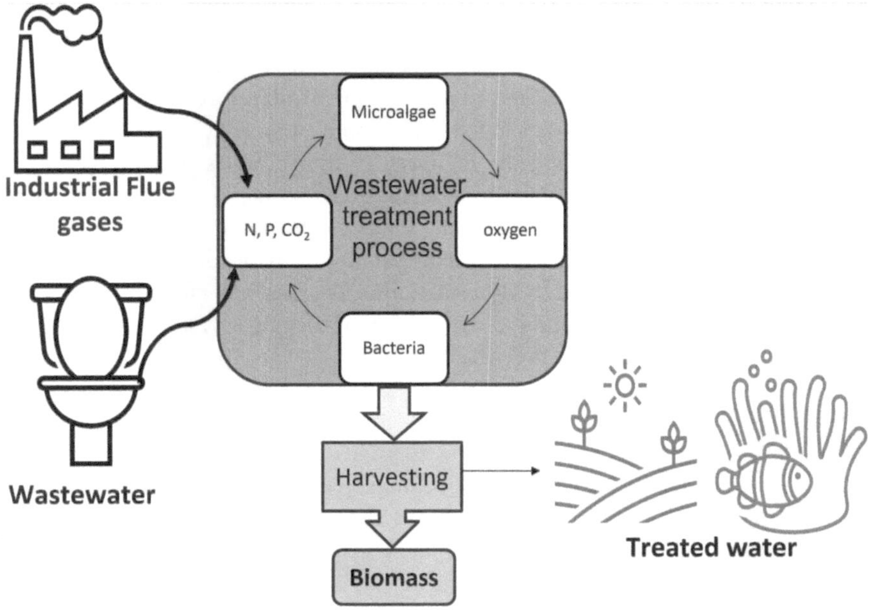

FIGURE 4.2 Microalgae–bacteria consortium.

The three main cultivation methods for wastewater treatment using algae are:

1. Open systems
2. Closed systems
3. Immobilized systems

Because of its cheaper running expenses and user-friendliness, the open system is preferred. It can be done on a vast scale in either natural or man-made ponds. Common designs for open ponds in use include raceways, unstirred ponds, and circular ponds.

The two most common forms of closed photobioreactors are covered raceways and tubular reactors. There are primarily four types of closed systems: plate reactors, column reactors, annular reactors, and tubular reactors. They allow more light to enter than open ponds, leading to greater biomass yield with less time spent in storage (Harshad 2015). However, the biggest drawbacks of this technology are the high maintenance expenses and increased energy requirements.

To create an immobilized system, algae cells are encased in a solid medium and rendered unable to move on their own. Beads are the standard form for the algae media. Evidence from the multiple documented laboratory systems shows that the imprisoned algae are more effective than other methods at removing nitrogen from wastewater (Harshad, 2015; Muthoni et al. 2022).

The focus of sewage treatment has shifted from nutrient removal to nutrient recovery over the past two decades. The urgent need to preserve natural resources, reduce energy consumption during fertilizer production, and lessen the negative impact that STPs have on the environment are driving this change (Verstraete et al. 2009; Batstone et al. 2015). There are two main ways in which this trend has progressed so far. One strategy involves the incorporation of specialized nutrient recovery techniques into the conventional sewage treatment infrastructure. On the other hand, novel technologies have been developed for sewage treatment, which reduce dissipative losses while maximizing the recovery of the sewage's inherent resources (Batstone et al. 2015). This chapter summarizes current perspectives and applications of a new algal-based technology that enables efficient sewage treatment and nitrogen recovery for use as fertilizers. Material and energy needs, as well as hurdles that must be addressed, are briefly described after the nitrogen and phosphorus fertilizer yield recovery rates in a pilot-scale version of this algal system are compared to those reported for the current processes.

4.4.1.1 Processes for N-recovery

Most STPs have a specialized tertiary stage for removing N, and this is often achieved through nitrification and denitrification. Many other process designs have been used for N-removal, including the modified Ludzak–Ettinger process, the four-stage Bardenpho method, and sequencing batch reactors.

4.4.1.2 Processes for P-recovery

Precipitation with iron, aluminum, magnesium, or calcium salts has long been the standard method used by STPs for removing phosphorus from wastewater. For instance, the usual chemical dosage for iron-based precipitation is 5.23 g $FeCl_3$ g^{-1}

PO43-P. Phosphates can also be concentrated in P-accumulating organisms using the anaerobic–anoxic–aerobic process, a modified Bardenpho technique, or the University of Cape Town procedure.

4.4.1.3 Reinvented Systems for Sewage Treatment and Nutrient Recovery

In recent years, innovative strategies have been presented for treating sewage while minimizing energy consumption and maximizing resource recovery. Two methods within this trend, published by Verstraete et al. (2009), have been singled out by Batstone et al. (2015) as significant departures from the standard practice. "Up-concentration" of screened sewage for treatment by anaerobic digestion and recovery of nutrients from the digestate is the method proposed by Verstraete et al. (2009).

4.4.1.4 Algal Pathway for Sewage Treatment and Nutrient Recovery

When compared to the algal pathway, the currently employed bacterial methods for sewage treatment, such as activated sludge and nitrification–denitrification, have some limitations. The fundamental reason is that sewage has lower stoichiometric C:N:P ratios than heterotrophic biomass does. Due to a lack of nitrogen and phosphorus removal, the activated sludge process relies on carbon-limited heterotrophic carbon metabolism.

4.4.1.5 Perspective: Status and Prospects for Nutrient Recovery from Sewage

Most municipalities use energy-intensive active sludge-based sewage treatment systems, and the few viable options for nutrient recovery from sewage are usually only used as additional procedures. The STaRR system is an alternative that enables a single-step sewage treatment procedure that is both energy-efficient and integrated with N- and P-recovery operations (Munasinghe-Arachchige et al. 2020).

4.5 MECHANISM OF NUTRIENT RECOVERY

For the optimization of microalgal nutrient removal from the aforementioned wastewater streams, it is important to have a firm grasp on how different algal species absorb nutrients. It is possible to make a rough classification of the algal cell's metabolic processes according to the elements involved. Algal development requires a number of ionic components in addition to the four basic nutrients (C, N, P, and S). These include Na, K, Fe, Mg, Ca, and trace elements. Nitrogen and phosphorus are the two nutrients typically highlighted. An increase in these nutrients due to human waste generation raises the likelihood of nutrient runoff, which can promote eutrophication.

4.5.1 Carbon

Carbon dioxide in the environment and industrial exhaust gases can be converted into usable forms of carbon through the photosynthetic activity of autotrophic microalgae. There are microalgae that are heterotrophs, meaning they get their carbon from sources other than photosynthesis, and there are those that are autotrophs or which may be both. Cells can utilize soluble carbonates for expansion either by directly ingesting them or by releasing carbon dioxide from carbonate via carboanhydrase activity. Algae as a carbon dioxide sink for flue gases is an additional research field that, if fruitful, might benefit the environment and biofuel production.

4.5.2 Nitrogen

Nitrogen is an essential element for all living things to thrive. Proteins, peptides, enzymes, chlorophylls, nucleic acid bases, ribonucleic acid bases, and adenosine triphosphate (ATP) are all examples of organic nitrogen-containing biomolecules (Barsanti and Gualtieri 2014). Inorganic nitrogen compounds such as nitrate (NO_3), nitrite (NO_2), nitric acid (HNO_3), ammonium (NH_4), ammonia (NH_3), and nitrogen gas (N_2) erve as precursors to their organic counterparts. Assimilation is the process by which microalgae convert inorganic nitrogen to its organic form. Furthermore, through a process called fixation, atmospheric nitrogen can be converted into ammonia by cyanobacteria. All eukaryotic algae conduct assimilation, which is only possible with "inorganic nitrogen in the forms of nitrate, nitrite, or ammonium. In a process, inorganic nitrogen is transported across the plasma membrane, oxidized nitrogen is reduced, and ammonium is incorporated into amino acids". Reductases that help reduce nitrate and nitrite are called nitrate reductase and nitrite reductase. The enzyme nitrate reductase transfers two electrons from reduced nicotinamide adenine dinucleotide (NADH) to nitrate, producing nitrite as a by-product. In the process of converting nitrite to ammonium, a total of six electrons are transferred from nitrite reductase to ferredoxin (Fd). All inorganic nitrogen in the intracellular fluid is first transformed to ammonium and then utilized in the synthesis of amino acids. Finally, glutamine synthase uses glutamate (Glu) and adenosine triphosphate (ATP) to incorporate ammonium into the amino acid glutamine. Assimilation of ammonium does not require a redox reaction, making it the preferred source of nitrogen. Algae prefer ammonium to nitrate, and nitrate is not consumed until almost all ammonium has been used up, according to the research (Maestrini et al. 1986). As a result, high-ammonium wastewaters can be used to efficiently cultivate microalgae at a quick rate. Nitrate predominates in oxidized aquatic settings because it is more strongly oxidized and thermodynamically stable than ammonium. In contrast, nitrate can be an important nitrogen source for microalgae because it activates nitrate reductase. However, ammonium in excess can inhibit growth (Morris and Syrett 1963). Algae species range in their ammonium tolerance from 25 mmol NH4 -N L1 to 1000 mmol NH4 -N L1 (Collos and Berges 2004). Using the plant enzyme glutamine synthetase, which has a high affinity for ammonium, is one strategy for growing microalgae under high ammonium concentrations. It has been shown that when *Chlorella vulgaris* is grown in natural wastewater with the addition of glutamic acid, ammonium reduction per cell increases by 70% (Khan and Yoshida 2008). Cell metabolism isn't the only way ammonium leaves the body; ammonia stripping allows for the considerable volatilization of ammonia at elevated pH and temperature. Ammonia stripping was found to be the most significant process in high growth rate algal ponds by García et al. (2000). Ammonia emission was shown to be enhanced even AT pH levels below 9 when high-rate algal ponds were exposed to warm weather.

4.5.3 Phosphorus

Algal "nucleic acids, lipids, proteins, and the intermediates of their carbohydrate metabolism" all contain phosphorus, making it an essential element in their energy metabolism. Algal cell development and metabolism rely heavily on inorganic phosphates. Phosphorus is integrated into organic molecules during algal metabolism via phosphorylation, with much of this process including the synthesis of ATP from adenosine diphosphate (ADP) in conjunction with some sort of energy input (Martínez et al. 1999). Phosphorus is most commonly found in the forms of H_2PO_4 and HPO_{4-}^{2}. Oxidation of respiratory substrates, the mitochondrial electron transport chain, and, in the case of photosynthesis, light are all potential sources of energy input. Algal phosphates are transported across the plasma membrane by a process that requires energy. Microalgae can use inorganic phosphorus sources, while some can also utilize organic phosphorus esters for growth (Kuenzler 1965). Many examples of eutrophication are induced by excess phosphorus, which can come from wastewater discharge (Correll 1998), despite the fact that orthophosphate is usually acknowledged as the limiting nutrient in freshwater systems. Phosphorus removal in wastewater is regulated not just by the uptake into the cell, but also by external variables like pH and dissolved oxygen, much like nitrogen removal. Since gaseous phosphorus does not exist, increased pH and high levels of dissolved oxygen will cause phosphate to precipitate out of the medium.

4.5.4 Other Nutrients

Other micronutrients, such as silicon and iron, might alter the abundance of phytoplankton communities (Hecky and Kilham 1988), while nitrogen and phosphorus are the two main nutrients of concern in eutrophication, being limiting variables in most growth scenarios (Anderson et al. 2002). Many of the micronutrients, however, are harmful to the vast majority of algal species when present in sufficient quantities. Some of them also combine with other necessary components to diminish their availability by precipitating with them. It has been shown, however, that certain strains of algae are exceptionally tolerant of heavy metals and have the ability to absorb metals (Mehta and Gaur 2005; Cai et al. 2013).

Increases in fertilizer pricing and tighter discharge limits on these nutrients have spurred rapid innovation in technologies to recover nitrogen, phosphorous, and potassium from waste streams in the past decade. This study presents a critical state-of-the-art analysis of relevant technologies, analyzing the current and future potential of each technology for nutrient recovery and outlining the pathways and obstacles that must be overcome to fully implement these technologies. Nutrient accumulation comes first, then nutrient release, and finally nutrient extraction, when considering the many technologies available. Plants, microbes (algae, prokaryotic), and physicochemical mechanisms such as chemical precipitation, membrane separation, sorption, and binding with magnetic particles can all accumulate nutrients. Biochemical processes (such as anaerobic digestion and bioleaching) and thermal processing can both be used to recover nutrients. The crystallization, gas-permeable membrane, liquid–gas stripping, and electrodialysis processes are all viable options for nutrient extraction. Different types of waste streams, the recovered product, and the technological maturity level of these methods were all considered in the analysis. Concentrated nutrient recovery (through methods like the inorganic precipitate struvite) is seen favorably since it provides more flexibility in terms of future reuse, lowers the danger of pathogens, and facilitates transport. Overall, the efficiency of nutrient recovery might be enhanced by further developing methods for nitrogen and potassium recovery and integrating accumulation-release-extraction technologies. Newer and better technologies need to be put to use, demonstrated, and validated so that they can go past the experimental stage. Finally, the recovered nutritional products should be studied and developed for use in agriculture. Redirecting nutrients from trash into recovered nutrient products provides a long-term sustainable source of nutrients and helps cushion future nutrient price hikes, hence reducing pollution in waterways and the air.

The most exhaustive case is recovering nutrients from treated wastewater. In this case, microalgae-based technologies would take the place of traditional activated sludge systems, making the process more cost-effective and energy-efficient. One kilogram of microalgae biomass can be harvested from 1 m^3 of treated wastewater. The final product water must meet EU standards; therefore, its nitrogen concentration must be less than 10 mg L^{-1}, and its phosphorus concentration must be less than 2 mg L^{-1}. These factors must be taken into account during the planning and execution of the entire process. Some works have shown that this is a promising and doable approach (Cabanelas et al. 2013a; Craggs et al. 2014). The content of nitrogen and phosphorus in the effluent from traditional wastewater treatment methods can be lowered through the recovery of nutrients from wastewaters after secondary treatment, which is offered as an alternative to tertiary treatment (Cabanelas et al. 2013b; Acién Fernández et al. 2018).

4.6 CHALLENGES AND LIMITATIONS

The key difficulty here is that the microalgae biomass production potential is capped by the low concentration of nutrients in treated wastewater. Membranes have been offered as a means of separating the hydraulic retention period from the cellular retention time. By employing membranes, the capacity of wastewater treatment plants can be greatly increased while still retaining the optimal cellular retention duration of three to four days (Marbelia et al. 2014; Gómez-Serrano et al. 2015). It is important to consider how to extract nutrients from the centrate produced by anaerobic digestion of activated sludge. The removal of the 1,000 mgN L^{-1} and 30 mgP L^{-1} found in centrate from anaerobic digestion is expensive and energy intensive for wastewater treatment plants. As much as 2% of total wastewater flow is centrate, which is recirculated to the process and adds to the expense and energy consumption of wastewater treatment systems. By eliminating the need to recirculate and treat this centrate, a medium-sized conventional plant that processes 50,000 m^3/day of wastewater can save as much as 12,000 kWh/day in energy costs (FCC Aqualia, personal communication). This centrate can be used for nutrient recovery by microalgae; however, it must be fed to the photobioreactor with caution to prevent an excess of turbidity and ammonium. Common practice calls for diluting the centrate before adding it to the photobioreactor; however, this step can be made more environmentally friendly by recycling water from the effluent (Morales-Amaral et al. 2015). Centrate, a by-product of traditional wastewater treatment facilities, has been recommended as a viable single nutrient source for the cultivation of marine strains due to its high nutritional concentration (Ledda et al. 2016; Sepúlveda et al. 2015; Acién Fernández et al. 2018).

4.7 CONCLUSIONS

For future agricultural and environmental sustainability, nutrient recovery, particularly phosphorus, appears to be essential. Algae have been shown to be quite effective at cleaning up wastewater, according to a number of studies. Algae's significant capability for nutrient recycling has also been established in less investigations. The challenge of developing algal biotechnologies for nutrient recycling requires a multi-pronged approach. The primary focus is on improving the efficiency of nutrient sequestration by making use of the existing technologies. The conventional application of HRAP for wastewater treatment is particularly deserving of these efforts. Getting lab-grown techniques ready for industrial application is crucial. Here, we prioritize species with a high capacity for nutrient removal and choose growing strategies accordingly. Rarely is it reported that microalgae were grown in closed photobioreactors in wastewater on a significant scale. However, advancements in this field may be possible with the help of new harvesting methods and the optimization of energy inputs during production. Utilization of nutrient-rich algal biomass recovered during wastewater treatment is the next step in the research process. Among the many complicated challenges is the incorporation of various biomasses into soil, with studies of nutrient release and plant uptake. An efficient fix has the potential to complete the food chain. However, the use of microalgae for nutrient removal in water management has received very little attention despite their considerable potential.

ACKNOWLEDGMENTS

Dr. Meenakshi is grateful to Head, Kanoria PG Mahila Mahavidhyalaya, and Dr. Chitra Jain is grateful to Biomitra Life Sciences Pvt. Ltd. for providing opportunity to search more literature and writing this book chapter. Authors did not receive any specific grant from funding agencies in the public, commercial, or not-for-profit sectors.

CONTRIBUTIONS

Meenakshi Fartyal contributed toward the conceptualization, investigation, resources, and writing the original draft, and **Chitra Jain** contributed to review, editing, and finalization.

REFERENCES

Acién Fernández, Francisco Gabriel, Cintia Gómez-Serrano, and José María Fernández-Sevilla. 2018. "Recovery of Nutrients From Wastewaters Using Microalgae." *Frontiers in Sustainable Food Systems* 2: 59. https://doi.org/10.3389/fsufs.2018.00059.

Adey, Walter H., Patrick C. Kangas, and Walter Mulbry. 2011. "Algal Turf Scrubbing: Cleaning Surface Waters with Solar Energy While Producing a Biofuel." *BioScience* 61 (6): 434–41. https://doi.org/10.1525/bio.2011.61.6.5.

Aditya, Lisa, Indra Mahlia T. M., Luong N. Nguyen, Hang P. Vu, and Long D. Nghiem. 2022. "Microalgae-Bacteria Consortium for Wastewater Treatment and Biomass Production." *Science of The Total Environment* 838: 155871. https://doi.org/10.1016/j.scitotenv.2022.155871.

Ahluwalia, Sarabjeet Singh, and Dinesh Goyal. 2007. "Microbial and Plant Derived Biomass for Removal of Heavy Metals from Wastewater." *Bioresource Technology* 98 (12): 2243–57. https://doi.org/10.1016/j.biortech.2005.12.006.

Ahuja, Dilip,and Tatsutani Marika. 2009. "Sustainable Energy for Developing Countries". *SAPIENS* 2:1–16.

Anderson, Donald M., Patricia M. Glibert, and JoAnn M. Burkholder. 2002. "Harmful Algal Blooms and Eutrophication: Nutrient Sources, Composition, and Consequences." *Estuaries* 25, 4: 704–26. www.jstor.org/stable/1353028.

Arcila, Juan S., and Germán Buitrón. 2016. "Microalgae-Bacteria Aggregates: Effect of the Hydraulic Retention Time on the Municipal Wastewater Treatment, Biomass Settleability and Methane Potential: Microalgae-Bacteria Aggregates for Wastewater Treatment." *Journal of Chemical Technology & Biotechnology* 91 (11): 2862–70. https://doi.org/10.1002/jctb.4901.

Bajpai, Pratima. 2019. *Third Generation Biofuels. SpringerBriefs in Energy*. Singapore: Springer Singapore. https://doi.org/10.1007/978-981-13-2378-2.

Barker, J. C., Zublena, J. P., and Walls, F. R. 2001. "Livestock and poultry manure characteristics." *Biotechnology and Bioengineering* 98: 764–770.

Barsanti, Laura, and Paolo Gualtieri. 2014. *Algae: Anatomy, Biochemistry, and Biotechnology*. Second edition. Boca Raton [Florida]: CRC Press/Taylor & Francis Group. https://doi.org/10.1201/b16544

Batool, Masooma, and Laila Shahzad. 2021. "An Analytical Study on Municipal Wastewater to Energy Generation, Current Trends, and Future Prospects in South Asian Developing Countries (an Update on Pakistan Scenario)." *Environmental Science and Pollution Research* 28 (25): 32075-94. https://doi.org/10.1007/s11356-021-14029-8.

Batstone, D. J., T. Hülsen, C. M. Mehta, and J. Keller. 2015. "Platforms for Energy and Nutrient Recovery from Domestic Wastewater: A Review." *Chemosphere* 140: 2–11. https://doi.org/10.1016/j.chemosphere.2014.10.021.

Beuckels, Annelies, Erik Smolders, and Koenraad Muylaert. 2015. "Nitrogen Availability Influences Phosphorus Removal in Microalgae-Based Wastewater Treatment." *Water Research* 77: 98–106. https://doi.org/10.1016/j.watres.2015.03.018.

Bhatnagar, Ashish, Monica Bhatnagar, Senthil Chinnasamy, and K. C. Das. 2010. "Chlorella Minutissima—A Promising Fuel Alga for Cultivation in Municipal Wastewaters." *Applied Biochemistry and Biotechnology* 161 (1–8): 523–536. https://doi.org/10.1007/s12010-009-8771-0.

Blades, Luke, Kevin Morgan, Roy Douglas, Stephen Glover, Mattia De Rosa, Thomas Cromie, and Beatrice Smyth. 2017. "Circular Biogas-Based Economy in a Rural Agricultural Setting." *Energy Procedia* 123: 89–96. https://doi.org/10.1016/j.egypro.2017.07.255.

Boelee, N. C, Temmink H., Janssen M., Buisman C. J. N., and Wijffels R. H. 2011. "Nitrogen and Phosphorus Removal from Municipal Wastewater Effluent Using Microalgal Biofilms." *Water Research* 45 (18): 5925–33. https://doi.org/10.1016/j.watres.2011.08.044.

Bradford, Scott A., Eran Segal, Wei Zheng, Qiquan Wang, and Stephen R. Hutchins. 2008. "Reuse of Concentrated Animal Feeding Operation Wastewater on Agricultural Lands." *Journal of Environmental Quality* 37 (S5): S97–S115. https://doi.org/10.2134/jeq2007.0393.

Burks, B. D., and Minnis M. M. 1994. "Onsite Wastewater Treatment Systems: Madison." Wisconsin, Hogarth House, Ltd.

Cabanelas, Iago Teles Dominguez, Jésus Ruiz, Zouhayr Arbib, Fábio Alexandre Chinalia, Carmen Garrido-Pérez, Frank Rogalla, Iracema Andrade Nascimento, and José A. Perales. 2013a. "Comparing the Use of Different Domestic Wastewaters for Coupling Microalgal Production and Nutrient Removal." *Bioresource Technology* 13: 429–36. https://doi.org/10.1016/j.biortech.2012.12.152.

Cabanelas, Iago Teles Dominguez, Zouhayr Arbib, Fábio A. Chinalia, Carolina Oliveira Souza, José A. Perales, Paulo Fernando Almeida, Janice Izabel Druzian, and Iracema Andrade Nascimento. 2013b. "From Waste to Energy: Microalgae Production in Wastewater and Glycerol." *Applied Energy* 109: 283–90. https://doi.org/10.1016/j.apenergy.2013.04.023.

Cai, Ting, Stephen Y Park, and Yebo Li. 2013. "Nutrient Recovery from Wastewater Streams by Microalgae: Status and Prospects." *Renewable and Sustainable Energy Reviews* 19: 360–69. https://doi.org/10.1016/j.rser.2012.11.030.

Chi, Zhanyou, Yubin Zheng, Anping Jiang, and Shulin Chen. 2011. "Lipid Production by Culturing Oleaginous Yeast and Algae with Food Waste and Municipal Wastewater in an Integrated Process." *Applied Biochemistry and Biotechnology* 165 (2): 442–53. https://doi.org/10.1007/s12010-011-9263-6.

Chinnasamy, Senthil, Ashish Bhatnagar, Ryan W. Hunt, and Das K. C. 2010. "Microalgae Cultivation in a Wastewater Dominated by Carpet Mill Effluents for Biofuel Applications." *Bioresource Technology* 101 (9): 3097–3105. https://doi.org/10.1016/j.biortech.2009.12.026.

Cole, Andrew J., Nicolas Neveux, Anna Whelan, Jeff Morton, Mark Vis, Rocky De Nys, and Nicholas A. Paul. 2016. "Adding Value to the Treatment of Municipal Wastewater through the Intensive Production of Freshwater Macroalgae." *Algal Research* 20: 100–109. https://doi.org/10.1016/j.algal.2016.09.026.

Collos, Y., and Berges J. A. 2004. Nitrogen Metabolism in Phytoplankton. In C. M. Duarte [ed.], Encyclopedia of Life Support Systems, EOLSS Publishers (UNESCO). http://www.eolss.net

Correll, David L. 1998. "The Role of Phosphorus in the Eutrophication of Receiving Waters: A Review." *Journal of Environmental Quality* 27 (2): 261–66. https://doi.org/10.2134/jeq1998.00472425002700020004x.

Craggs, R., Park J., Heubeck S., and Sutherland D. 2014. "High Rate Algal Pond Systems for Low-Energy Wastewater Treatment, Nutrient Recovery and Energy Production." *New Zealand Journal of Botany* 52 (1): 60–73. https://doi.org/10.1080/0028825X.2013.861855.

Doria, E., Longoni P., Scibilia L., Iazzi N., Cella R., and Nielsen E. 2012. "Isolation and Characterization of a Scenedesmus Acutus Strain to Be Used for Bioremediation of Urban Wastewater." *Journal of Applied Phycology* 24 (3): 375–83. https://doi.org/10.1007/s10811-011-9759-z.

Durai, G., and Rajasimman M. 2010. "Biological Treatment of Tannery Wastewater—A Review." *Journal of Environmental Science and Technology* 4 (1): 1–17. https://doi.org/10.3923/jest.2011.1.17.

García, J., Mujeriego R., and Hernández-Mariné M. 2000. "High Rate Algal Pond Operating Strategies for Urban Wastewater Nitrogen Removal." *Journal of Applied Phycology* 12 (3–5): 331–39. https://doi.org/10.1023/A:1008146421368.

Gifuni, Imma, Antonino Pollio, Carl Safi, Antonio Marzocchella, and Giuseppe Olivieri. 2019. "Current Bottlenecks and Challenges of the Microalgal Biorefinery." *Trends in Biotechnology* 37 (3): 242–52. https://doi.org/10.1016/j.tibtech.2018.09.006.

Gómez-Serrano, C., Morales-Amaral M. M., Acién F. G., Escudero R., Fernández-Sevilla J. M., and Molina-Grima E. 2015. "Utilization of Secondary-Treated Wastewater for the Production of Freshwater Microalgae." *Applied Microbiology and Biotechnology* 99 (16): 6931–44. https://doi.org/10.1007/s00253-015-6694-y.

Gonçalves, Ana L., José C. M. Pires, and Manuel Simões. 2017. "A Review on the Use of Microalgal Consortia for Wastewater Treatment." *Algal Research* 24: 403–15. https://doi.org/10.1016/j.algal.2016.11.008.

Haddaway Art. 2015. *Nutrient Recovery Technology Transforms World's Largest Wastewater Treatment Plant*. Waste Water, Water World.

Haider, Muhammad Nabeel, Chen-Guang Liu, Tanveer A. Tabish, Deepanraj Balakrishnan, Pau-Loke Show, Shaza Yehya Abdulhamed Qattan, Munazza Gull, and Muhammad Aamer Mehmood. 2022. "Resource Recovery of the Wastewater-Derived Nutrients into Algal Biomass Followed by Its Cascading Processing to Multiple Products in a Circular Bioeconomy Paradigm." *Fermentation* 8 (11): 650. https://doi.org/10.3390/fermentation8110650.

Harshad, Rathod. 2015. "Algae Based Waste Water Treatment." Environmental Engineering Group Department of Civil Engineering."*Indian Institute of Technology Roorkee, Roorkee, Uttarakhand, India*. https://doi.org/10.13140/2.1.1241.8885.

Hecky, R. E., and Kilham P. 1988. "Nutrient Limitation of Phytoplankton in Freshwater and Marine Environments: A Review of Recent Evidence on the Effects of Enrichment1: Nutrient Enrichment." *Limnology and Oceanography* 33 (4part2): 796–822. https://doi.org/10.4319/lo.1988.33.4part2.0796.

Henze, M., ed. 2008. *Biological Wastewater Treatment: Principles, Modelling and Design*. London: IWA.

Johnson, Michael B., and Zhiyou Wen. 2010. "Development of an Attached Microalgal Growth System for Biofuel Production." *Applied Microbiology and Biotechnology* 85 (3): 525–34. https://doi.org/10.1007/s00253-009-2133-2.

Kellogg, Robert L., Charles H. Lander, David C. Moffitt, and Noel Gollehon. 2000. "Manure Nutrients Relative to The Capacity Of Cropland And Pastureland To Assimilate Nutrients: Spatial and Temporal Trends for the United States." *Proceedings of the Water Environment Federation* 2000 (16): 18–157. https://doi.org/10.2175/193864700784994812.

Khan, Masil, and Naoto Yoshida. 2008. "Effect of L-Glutamic Acid on the Growth and Ammonium Removal from Ammonium Solution and Natural Wastewater by Chlorella Vulgaris NTM06." *Bioresource Technology* 99 (3): 575–82. https://doi.org/10.1016/j.biortech.2006.12.031.

Khoo, Choon Gek, Yaleeni Kanna Dasan, Man Kee Lam, and Keat Teong Lee. 2019. "Algae Biorefinery: Review on a Broad Spectrum of Downstream Processes and Products." *Bioresource Technology* 292: 121964. https://doi.org/10.1016/j.biortech.2019.121964.

Kuenzler, Edward J. 1965. "Glucose-6-Phosphate Utilization By Marine Algae." *Journal of Phycology* 1 (4): 156–64. https://doi.org/10.1111/j.1529-8817.1965.tb04577.x.

Kumar, Martin S., Zhihong H. Miao, and Sandy K. Wyatt. 2010. "Influence of Nutrient Loads, Feeding Frequency and Inoculum Source on Growth of Chlorella Vulgaris in Digested Piggery Effluent Culture Medium." *Bioresource Technology* 101 (15): 6012–18. https://doi.org/10.1016/j.biortech.2010.02.080.

Lavrinovičs, Aigars, and Tālis Juhna. 2017. "Review on Challenges and Limitations for Algae-Based Wastewater Treatment." *Construction Science* 20 (1). https://doi.org/10.2478/cons-2017-0003.

Ledda, Claudio, Andrea Schievano, Barbara Scaglia, Mara Rossoni, Francisco Gabriel Acién Fernández, and Fabrizio Adani. 2016. "Integration of Microalgae Production with Anaerobic Digestion of Dairy Cattle Manure: An Overall Mass and Energy Balance of the Process." *Journal of Cleaner Production* 112: 103–12. https://doi.org/10.1016/j.jclepro.2015.07.151.

Li, Tong, Gengyi Lin, Björn Podola, and Michael Melkonian. 2015. "Continuous Removal of Zinc from Wastewater and Mine Dump Leachate by a Microalgal Biofilm PSBR." *Journal of Hazardous Materials* 297: 112–18. https://doi.org/10.1016/j.jhazmat.2015.04.080.

Li, Yecong, Yi-Feng Chen, Paul Chen, Min Min, Wenguang Zhou, Blanca Martinez, Jun Zhu, and Roger Ruan. 2011. "Characterization of a Microalga Chlorella Sp. Well Adapted to Highly Concentrated Municipal Wastewater for Nutrient Removal and Biodiesel Production." *Bioresource Technology* 102 (8): 5138–44. https://doi.org/10.1016/j.biortech.2011.01.091.

Liu, Jiajun, Bill Pemberton, Justin Lewis, Peter J. Scales, and Gregory J. O. Martin. 2020. "Wastewater Treatment Using Filamentous Algae – A Review." *Bioresource Technology* 298: 122556. https://doi.org/10.1016/j.biortech.2019.122556.

Maestrini, Serge Y., Jean-Michel Robert, John W. Leftley, and Yves Collos. 1986. "Ammonium Thresholds for Simultaneous Uptake of Ammonium and Nitrate by Oyster-Pond Algae." *Journal of Experimental Marine Biology and Ecology* 102 (1): 75–98. https://doi.org/10.1016/0022-0981(86)90127-9.

Marbelia, L., Bilad M. R., Passaris I., Discart V., Vandamme D., Beuckels A., Muylaert K., and Ivo F. J. Vankelecom. 2014. "Membrane Photobioreactors for Integrated Microalgae Cultivation and Nutrient Remediation of Membrane Bioreactors Effluent." *Bioresource Technology* 163: 228–35. https://doi.org/10.1016/j.biortech.2014.04.012.

Markou, Giorgos, and Elias Nerantzis. 2013. "Microalgae for High-Value Compounds and Biofuels Production: A Review with Focus on Cultivation under Stress Conditions." *Biotechnology Advances* 31 (8): 1532–42. https://doi.org/10.1016/j.biotechadv.2013.07.011.

Martínez, Brezo, Lorena Sordo Pato, and Jose Manuel Rico. 2012. "Nutrient Uptake and Growth Responses of Three Intertidal Macroalgae with Perennial, Opportunistic and Summer-Annual Strategies." *Aquatic Botany* 96 (1): 14–22. https://doi.org/10.1016/j.aquabot.2011.09.004.

Martínez, M. E., Jiménez J. M., and El Yousfi F. 1999. "Influence of Phosphorus Concentration and Temperature on Growth and Phosphorus Uptake by the Microalga Scenedesmus Obliquus." *Bioresource Technology* 67 (3): 233–40. https://doi.org/10.1016/S0960-8524(98)00120-5.

Masojídek, Jiří, and Ondřej Prášil. 2010. "The Development of Microalgal Biotechnology in the Czech Republic." *Journal of Industrial Microbiology & Biotechnology* 37 (12): 1307–17. https://doi.org/10.1007/s10295-010-0802-x.

Mehta, S. K., and Gaur J. P. 2005. "Use of Algae for Removing Heavy Metal Ions From Wastewater: Progress and Prospects." *Critical Reviews in Biotechnology* 25 (3): 113–52. https://doi.org/10.1080/07388550500248571.

Millmier, A., Lorimor J., Hurburgh Jr. C., Fulhage C., Hattey J., and Zhang H. 2000. "Near-Infrared Sensing Of Manure Nutrients." *Transactions of the ASAE* 43 (4): 903–8. https://doi.org/10.13031/2013.2986.

Mimouni, Virginie, Lionel Ulmann, Virginie Pasquet, Marie Mathieu, Laurent Picot, Gael Bougaran, Jean-Paul Cadoret, Annick Morant-Manceau, and Benoit Schoefs. 2012. "The Potential of Microalgae for the Production of Bioactive Molecules of Pharmaceutical Interest." *Current Pharmaceutical Biotechnology* 13 (15): 2733–50. https://doi.org/10.2174/138920112804724828.

Mishra, Saurabh, Virender Singh, Liu Cheng, Abid Hussain, and Banu Ormeci. 2022. "Nitrogen Removal from Wastewater: A Comprehensive Review of Biological Nitrogen Removal Processes, Critical Operation Parameters and Bioreactor Design." *Journal of Environmental Chemical Engineering* 10 (3): 107387. https://doi.org/10.1016/j.jece.2022.107387.

Mitchell, Peter, and Julian Morgan. 2015. *Employment and the Circular Economy Job Creation in a More Resource Efficient Britain.* Green Alliance. https://doi.org/10.13140/RG.2.1.1026.5049.

Mohsenpour, Seyedeh Fatemeh, Sebastian Hennige, Nicholas Willoughby, Adebayo Adeloye, and Tony Gutierrez. 2021. "Integrating Micro-Algae into Wastewater Treatment: A Review." *Science of The Total Environment* 752: 142168. https://doi.org/10.1016/j.scitotenv.2020.142168.

Molazadeh, Marziyeh, Hossein Ahmadzadeh, Hamid R. Pourianfar, Stephen Lyon, and Pabulo Henrique Rampelotto. 2019. "The Use of Microalgae for Coupling Wastewater Treatment With CO_2 Biofixation." *Frontiers in Bioengineering and Biotechnology* 7: 42. https://doi.org/10.3389/fbioe.2019.00042.

Morales-Amaral, Maria Del Mar, Cintia Gómez-Serrano, F. Gabriel Acién, José M. Fernández-Sevilla, and Molina-Grima E. 2015. "Outdoor Production of Scenedesmus Sp. in Thin-Layer and Raceway Reactors Using Centrate from Anaerobic Digestion as the Sole Nutrient Source." *Algal Research* 12: 99–108. https://doi.org/10.1016/j.algal.2015.08.020.

Morris, I., and Syrett P. J. 1963. "The Development of Nitrate Reductase in Chlorella and Its Repression by Ammonium." *Archiv Für Mikrobiologie* 47 (1): 32–41. https://doi.org/10.1007/BF00408287.

Mosse, K. P. M., Patti A. F., Christen E. W., and Cavagnaro T. R. 2011. "Review: Winery Wastewater Quality and Treatment Options in Australia: Winery Wastewater Review." *Australian Journal of Grape and Wine Research* 17 (2): 111–22. https://doi.org/10.1111/j.1755-0238.2011.00132.x.

Mulbry, Walter, Elizabeth Kebede Westhead, Carolina Pizarro, and Lawrence Sikora. 2005. "Recycling of Manure Nutrients: Use of Algal Biomass from Dairy Manure Treatment as a Slow Release Fertilizer." *Bioresource Technology* 96 (4): 451–458. https://doi.org/10.1016/j.biortech.2004.05.026.

Munasinghe-Arachchige, S. P., Abeysiriwardana-Arachchige I. S. A, Delanka-Pedige H. M. K., and Nirmalakhandan N. 2020. "Algal Pathway for Nutrient Recovery from Urban Sewage." *Algal Research* 51: 102023. https://doi.org/10.1016/j.algal.2020.102023.

Muthoni, Fridah Valenciah, María Veiga, and Christian Kennes. 2022. "Bioprocesses and Technologies for Nutrient Recovery From Waste Water: Microalgae and Jellyfish."*Earth Sciences*. https://doi.org/10.20944/preprints202201.0201.v1.

Naik, S. N., Vaibhav V. Goud, Prasant K. Rout, and Ajay K. Dalai. 2010. "Production of First and Second Generation Biofuels: A Comprehensive Review." *Renewable and Sustainable Energy Reviews* 14 (2): 578–597. https://doi.org/10.1016/j.rser.2009.10.003.

Newman, Mark. 2011. *Industrial Water Use* (Worldmapper map no. 325. Sheffield.

Nguyen, Luong N., Lisa Aditya, Hang P. Vu, Abu Hasan Johir, Lucia Bennar, Peter Ralph, Ngoc B. Hoang, Jakub Zdarta, and Long D. Nghiem. 2022. "Nutrient Removal by Algae-Based Wastewater Treatment." *Current Pollution Reports* 8 (4): 369–83. https://doi.org/10.1007/s40726-022-00230-x.

Nutrient Pollution in Wastewater: An Emerging Global Problem—Analyte Guru—Post. 2016. Analyte Guru. https://www.analyteguru.com/t5/forums/searchpage/tab/message?advanced=false&allow_punctuation=false&q=An%20Emerging%20Global%20Problem%20

OECD. 2008. *Annual Report on Sustainable Development Work in the OECD*. Paris: OECD. www.oecd.org/publishing/corrigenda.

Oyedepo, Sunday Olayinka. 2012. "Energy and Sustainable Development in Nigeria: The Way Forward." *Energy, Sustainability and Society* 2 (1): 15. https://doi.org/10.1186/2192-0567-2-15.

Pancha, Imran, Kaumeel Chokshi, and Sandhya Mishra. 2015. "Enhanced Biofuel Production Potential with Nutritional Stress Amelioration through Optimization of Carbon Source and Light Intensity in Scenedesmus Sp. CCNM 1077." *Bioresource Technology* 179: 565–72. https://doi.org/10.1016/j.biortech.2014.12.079.

Park, J. B. K., Craggs R. J., and Shilton A. N. 2011. "Wastewater Treatment High Rate Algal Ponds for Biofuel Production." *Bioresource Technology* 102 (1): 35–42. https://doi.org/10.1016/j.biortech.2010.06.158.

Pescod, M. B. 1992. "Wastewater Treatment and Use in Agriculture."*FAO Irrigation and Drainage Paper 47*. Rome: Food and Agriculture Organization of the United Nations.

Pittman, Jon K., Andrew P. Dean, and Olumayowa Osundeko. 2011. "The Potential of Sustainable Algal Biofuel Production Using Wastewater Resources." *Bioresource Technology* 102 (1): 17–25. https://doi.org/10.1016/j.biortech.2010.06.035.

Ren, Yi, Huu Hao Ngo, Wenshan Guo, Dongbo Wang, Lai Peng, Bing-Jie Ni, Wei, and Yiwen Liu. 2020. "New Perspectives on Microbial Communities and Biological Nitrogen Removal Processes in Wastewater Treatment Systems." *Bioresource Technology* 297: 12249 1. https://doi.org/10.1016/j.biortech.2019.122491.

Renuka, Nirmal, Sachitra Kumar Ratha, Farzana Kader, Ismail Rawat, and Faizal Bux. 2021. "Insights into the Potential Impact of Algae-Mediated Wastewater Beneficiation for the Circular Bioeconomy: A Global Perspective." *Journal of Environmental Management* 297: 113257. https://doi.org/10.1016/j.jenvman.2021.113257.

Roberts, David A., Nicholas A. Paul, Andrew J. Cole, and Rocky De Nys. 2015. "From Waste Water Treatment to Land Management: Conversion of Aquatic Biomass to Biochar for Soil Amelioration and the Fortification of Crops with Essential Trace Elements." *Journal of Environmental Management* 157: 60–68. https://doi.org/10.1016/j.jenvman.2015.04.016.

Ruiz, Jesús, Giuseppe Olivieri, Jeroen De Vree, Rouke Bosma, Philippe Willems, J. Hans Reith, Michel H. M. Eppink, Dorinde M. M. Kleinegris, René H. Wijffels, and Maria J. Barbosa. 2016. "Towards Industrial Products from Microalgae." *Energy & Environmental Science* 9 (10): 3036–43. https://doi.org/10.1039/C6EE01493C.

Ruiz-Marin, Alejandro, Leopoldo G. Mendoza-Espinosa, and Tom Stephenson. 2010. "Growth and Nutrient Removal in Free and Immobilized Green Algae in Batch and Semi-Continuous Cultures Treating Real Wastewater." *Bioresource Technology* 101 (1): 58–64. https://doi.org/10.1016/j.biortech.2009.02.076.

Şen, S., and Demirer G. N. 2003. "Anaerobic Treatment of Real Textile Wastewater with a Fluidized Bed Reactor." *Water Research* 37 (8): 1868–78. https://doi.org/10.1016/S0043-1354(02)00577-8.

Sepúlveda, C., Acién F. G., C. Gómez, N. Jiménez-Ruíz, Riquelme C., and Molina-Grima E. 2015. "Utilization of Centrate for the Production of the Marine Microalgae Nannochloropsis Gaditana." *Algal Research* 9: 107–16. https://doi.org/10.1016/j.algal.2015.03.004.

Singh, Rajeev Pratap, Bhavisha Sharma, Abhijit Sarkar, Chandan Sengupta, Pooja Singh, and Mahamad Hakimi Ibrahim. 2014. "Biological Responses of Agricultural Soils to Fly-Ash Amendment." In *Reviews of Environmental Contamination and Toxicology Volume 232*, edited by David M. Whitacre, 232; 45–60. Reviews of Environmental Contamination and Toxicology. Cham: Springer International Publishing. https://doi.org/10.1007/978-3-319-06746-9_2.

Singh, Rajeev Pratap, Tyagi V. V., Tanu Allen, Hakimi Ibrahim M., and Richa Kothari. 2011. "An Overview for Exploring the Possibilities of Energy Generation from Municipal Solid Waste (MSW) in Indian Scenario." *Renewable and Sustainable Energy Reviews* 15 (9): 4797–4808. https://doi.org/10.1016/j.rser.2011.07.071.

Slade, A. H, Gapes D. J, Stuthridge T. R., Anderson S. M., Dare P. H., Pearson H. G. W., and Dennis M. 2004. "N-ViroTech—a Novel Process for the Treatment of Nutrient Limited Wastewaters." *Water Science and Technology: A Journal of the International Association on Water Pollution Research* 50 (3): 131–39.

Spolaore, Pauline, Claire Joannis-Cassan, Elie Duran, and Arsène Isambert. 2006. "Commercial Applications of Microalgae." *Journal of Bioscience and Bioengineering* 101 (2): 87–96. https://doi.org/10.1263/jbb.101.87.

Srivastava, Vaibhav, Sultan Ahmed Ismail, Pooja Singh, and Rajeev Pratap Singh. 2015. "Urban Solid Waste Management in the Developing World with Emphasis on India: Challenges and Opportunities." *Reviews in Environmental Science and Bio/Technology* 14 (2): 317–37. https://doi.org/10.1007/s11157-014-9352-4.

Su, Yanyan. 2021. "Revisiting Carbon, Nitrogen, and Phosphorus Metabolisms in Microalgae for Wastewater Treatment." *Science of The Total Environment* 762: 144590. https://doi.org/10.1016/j.scitotenv.2020.144590.

Suganya, T., Varman M., Masjuki H. H., and Renganathan S. 2016. "Macroalgae and Microalgae as a Potential Source for Commercial Applications along with Biofuels Production: A Biorefinery Approach." *Renewable and Sustainable Energy Reviews* 55: 909–41. https://doi.org/10.1016/j.rser.2015.11.026.

Sukačová, Kateřina, and Jan Červený. 2017. "Can Algal Biotechnology Bring Effective Solution for Closing the Phosphorus Cycle? Use of Algae for Nutrient Removal: Review of Past Trends and Future Perspectives in the Context of Nutrient Recovery." *European Journal Of Environmental Sciences* 7(1): 63–72. https://doi.org/10.14712/23361964.2017.6.

Tuantet, Kanjana, Marcel Janssen, Hardy Temmink, Grietje Zeeman, René H. Wijffels, and Cees J. N. Buisman. 2014. "Microalgae Growth on Concentrated Human Urine." *Journal of Applied Phycology* 26 (1): 287–97. https://doi.org/10.1007/s10811-013-0108-2.

USEPA. 2008. *Clean Watersheds Needs Survey 2008: Report to Congress* (EPA-832-R-10-002). Columbus: United States Environmental Protection Agency.

Vaish, Barkha, Pooja Singh, Prabhat Kumar Singh, and Rajeev Pratap Singh. 2017. "Biomethanation Potential of Algal Biomass." In *Algal Biofuels*, edited by Sanjay Kumar Gupta, Anushree Malik, and Faizal Bux, 331–46. Cham: Springer International Publishing. https://doi.org/10.1007/978-3-319-51010-1_16.

Vaish, Barkha, Pooja Singh, Richa Kothari, Vaibhav Srivastava, Prabhat Kumar Singh, and Rajeev Pratap Singh. 2016. "The Potential of Bioenergy Production from Marginalised Lands and Its Effect on Climate Change." *Climate Change and Environmental Sustainability* 4 (1): 7. https://doi.org/10.5958/2320-642X.2016.00002.8.

Vaish, Barkha, Vaibhav Srivastava, Prabhat Kumar Singh, Pooja Singh, and Rajeev Pratap Singh. 2019. "Energy and Nutrient Recovery from Agro-Wastes: Rethinking Their Potential Possibilities." *Environmental Engineering Research* 25 (5): 623–37. https://doi.org/10.4491/eer.2019.269.

Verstraete, Willy, Pieter Van De Caveye, and Vasileios Diamantis. 2009. "Maximum Use of Resources Present in Domestic 'Used Water.'" *Bioresource Technology* 100 (23): 5537–45. https://doi.org/10.1016/j.biortech.2009.05.047.

Wang, Junyao, Jun Zhao, Shuai Deng, Taiwei Sun, Yanping Du, Kaixiang Li, and Yaofeng Xu. 2019. "Integrated Assessment for Solar-Assisted Carbon Capture and Storage Power Plant by Adopting Resilience Thinking on Energy System." *Journal of Cleaner Production* 208: 1009–21. https://doi.org/10.1016/j.jclepro.2018.10.090.

Whitton, Rachel, Francesco Ometto, Marc Pidou, Peter Jarvis, Raffaella Villa, and Bruce Jefferson. 2015. "Microalgae for Municipal Wastewater Nutrient Remediation: Mechanisms, Reactors and Outlook for Tertiary Treatment." *Environmental Technology Reviews* 4 (1): 133–48. https://doi.org/10.1080/21622515.2015.1105308.

Wu, Yin-Hu, Yin Yu, Xin Li, Hong-Ying Hu, and Zhen-Feng Su. 2012. "Biomass Production of a Scenedesmus Sp. under Phosphorous-Starvation Cultivation Condition." *Bioresource Technology* 112: 193–98. https://doi.org/10.1016/j.biortech.2012.02.037.

Yetilmezsoy, Kaan, and Suleyman Sakar. 2008. "Development of Empirical Models for Performance Evaluation of UASB Reactors Treating Poultry Manure Wastewater under Different Operational Conditions." *Journal of Hazardous Materials* 153 (1–2): 532–43. https://doi.org/10.1016/j.jhazmat.2007.08.087.

Zou, Caineng, Qun Zhao, Guosheng Zhang, and Bo Xiong. 2016. "Energy Revolution: From a Fossil Energy Era to a New Energy Era." *Natural Gas Industry B* 3 (1): 1–11. https://doi.org/10.1016/j.ngib.2016.02.001.

5 Potential Applications of Algae Biomass for the Development of Natural Products

Getachew Tafere Abrha, Abdalah Makaranga, Bijaya Nag, Gourav Kumar, Neeru Gupta, Asha Arumugam Nesamma and Pannaga Pavan Jutur

5.1 INTRODUCTION

Society's need for energy from fossil fuels currently poses one of the significant threats to the long-term viability of the global environment and the stability of the economy (Kareya et al. 2022). Constant population growth will result in around 10 billion people by 2050, needing a continuous increase in food and energy supply. Hence, people and other animals inhabiting the globe will encounter many problems related to global warming and the energy crisis (Khan et al. 2018). Considering this, there is tremendous interest in bulk biomass produced by photosynthetic microalgae to manufacture biofuels and biorenewables (Cheng and He 2014). Bioproducts derived from microalgae biomass are already in great demand due to their nontoxic, biodegradable, and biorenewable resources. Their persistent use could decrease the release of dangerous pollutants (Gerotto et al. 2020). Microalgae can capture solar energy and transform it into bioenergy and other biological metabolites without harming the environment or interrupting the food supply. In addition, microalgae biomass is rich in biological components, such as proteins, carbohydrates, protein, carotenoids, vitamins, minerals, and other bioproducts for high-tech industries and processes like treating wastewater (Jacob-Lopes et al. 2009). Due to their efficiency, short life cycle, and relatively higher solar energy-to-biomass conversion rate than land plants, microalgae are appealing to industrial application targets. In addition, microalgae may grow in saltwater environments; metabolize CO_2, NO_x, and SO_x from various sources such as the atmosphere and industrial exhaust gases via photosynthesis; and produce an array of biorenewables (Singh and Dhar 2019; Makaranga and Jutur 2022).

Meanwhile, if the metabolism of photosynthetic microalgae is directed to the high-production bioenergy chemicals such as alcohols, alkanes, and fats, they could

DOI: 10.1201/9781003455950-5

serve as feedstock for bioenergy industries. Production from microalgae biomass is higher than first- or second-generation feedstocks, mainly produced from oleaginous crops, for example, rapeseed, soybean, sunflower, and palm (Gerotto et al. 2020). Land plants primarily store the energy as lignocellulose, a biopolymer not easily exploited as a biorenewable feedstock (Jacob-Lopes et al. 2009). Microalgae are unicellular, polyphyletic oxygenic phototrophs organisms that include, in a broad definition, both prokaryotic blue-green algae (cyanobacteria) and eukaryotic single-cell microalgae (green microalgae, red microalgae, and diatoms), and there are 72,500 species and 16 classes. Chlorophyceae, Chrysophyceae, and diatoms are the three major groups (Bacillariophyceae). Cyanobacteria, Cyanophyceae, and diatoms are most studied for biological implications. These microalgae can proliferate as a droopy stratum on the water's surface and cope with harsh environmental conditions. Several countries are involved in microalgae research and cultivation to synthesize diverse biochemicals. Even though microalgae have been grown for commercial purposes for more than 40 years, the market still has relatively little biomass from them, and microalgae biomass still needs to be higher in the market (Gerotto et al. 2020). An assembled yearly dry microalgae biomass ranges around 19,000 tonnes.y^{-1}, and a turnover of USD 5.7 billion has been reported from microalgae biomass-producing countries like Taiwan, China, Japan, and Germany (Jacob-Lopes et al. 2009).

Microalgae are considered promising biomass feedstock for bioenergy industries (Kareya et al. 2022). Despite having immense potential, microalgae-based biofuel production technologies need to be more focused on industrial use for a variety of reasons, including economic and energy viability (Gerotto et al. 2020). Therefore, researchers increasingly focus on the best utilization of microalgae biomass as a feedstock for high-value product synthesis or aquaculture feed to make biofuel production economically feasible for microalgae biorefinery processes (Singh and Dhar 2019). Bioprospecting of microalgae biorefineries must explore effective strategies to exploit all microalgal components to achieve the equilibrium of financial and environmental factors. This chapter focuses more on the microalgal production of essential chemicals and bioproducts. Initially, we examine microalgae biomass and the potential source to manufacture biofuels and high-value-added bioproducts (HVABs). We conclude by discussing recent advancements, challenges, and future potential.

5.2 ALGAE-BASED ENERGY PRODUCTION

According to the global carbon project, the continuous use of fossil fuels for energy production increased global CO_2 emissions to a record high of 36.6 billion tonnes as reported in 2022 (Friedlingstein et al. 2022). Microalgal biomass has been extensively researched as a potential source for producing biofuels and high-value-added bioproducts (HVABs) (Singh et al. 2020; Yap et al. 2021). Microalgal lipids are the most promising future renewable energy source (Rehmanji et al. 2022). Microalgal lipid biomass is being investigated as a possible energy source for producing environmentally friendly biofuels such as bioethanol, biomethane, biodiesel, biohydrogen, and biobutanol (Figure 5.1) (Cheng and He 2014). Under environmental stress conditions, microalgae can modify physiological changes such as nutrient content and biomass composition (Jiang et al. 2016).

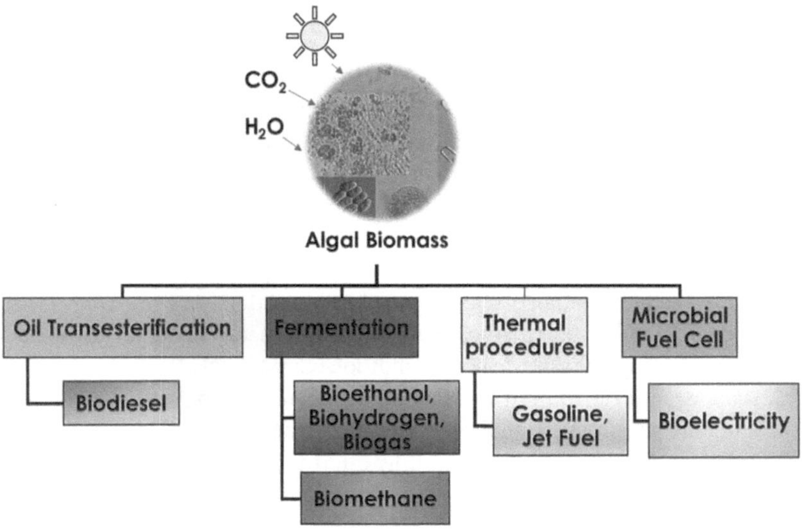

FIGURE 5.1 Schematic presentation of biofuel production from microalgae.

5.2.1 Biofuels

A continuous use of fossil fuels contributes to the formation of greenhouse gases, i.e., CO_2, in the environment, raising awareness of the need to find alternative and renewable sources (Paliwal et al. 2022). Biofuels are excellent substitutes for fossil fuels and can be produced from various biomass resources, have significant renewability advantages, and contribute significantly less to pollution and global warming (Khan et al. 2018). Microalgae can be a third-generation biofuel feedstock (Mofijur et al. 2019) with many environmental and economic advantages over other crops, such as rapid growth on non-arable land, the ability to grow in multifaceted trophic modes of nutrition, maximum oil yield, and high photosynthesis efficiency, which aids in CO_2 mitigation and wastewater treatment (Hussain et al. 2021; Razzak et al. 2017).

The lipid productivity of microalgae is 15–300 times that of common oil-producing crops (Zullaikah et al. 2019). The microalga cell stores lipid, which comes in two varieties: neutral lipid and polar lipid. Polar lipids are essential in cell membrane formation, whereas neutral lipids provide the most energy to microalgae cells. Due to their long chains of fatty acid field, polyunsaturated fatty acids (PUFAs) are essential for biofuel production (Alishah Aratboni et al. 2019). The proportion of saturated to unsaturated fatty acids is significant in determining microalgae's suitability as a biofuel feedstock. The main components of microalgae oil are unsaturated fatty acids, particularly palmitoleic (16:1), oleic (18:1), linoleic (18:2), linolenic acid (18:3), and palmitic (16:0) saturated fatty acids, with only a tiny proportion made up of stearic saturated fatty acids (18:0) (Sajjadi et al. 2018). Microalgae have different lipid productivity and lipid content depending on the culture medium and growth phase, that is, ranging from 20% to 50% of the total microalgal biomass (Yap et al. 2021). Microalga's biochemical composition consists of proteins, carbohydrates, nucleic acids, and lipids in different proportions depending on algae classes. The most energy-dense compound is a lipid (37.6 kJ g^{-1}), followed by proteins (16.7 kJ g^{-1}) and carbohydrates (15.7 kJ g^{-1} (Sajjadi et al. 2018). Microalgae biochemical composition, such as biofuel-related carbohydrates and lipids, varies with ecological conditions. Ecological stresses have been found to improve the efficiency of the microalgae biofuel production (Cheng and He 2014). Environmental changes can affect microalgal biomass and lipid content, and under nutrient deprivation conditions, the lipid concentration of microalgae can be greater than 50% (Zullaikah et al. 2019). Nitrogen (N) deficiency can enhance lipids in triacylglycerol (TAG) form, whereas phosphorus or sulfur deficiency can also affect neutral lipid production (Shaikh et al. 2019). Microalgal growth, other biochemical pathways, and metabolism shift toward lipid accumulation in the nitrogen deprivation condition (Conde et al. 2021). Environmental changes, such as culture medium and procedure, affect microalgae lipids, and more lipids accumulate in stressful environments due to nutrient deficiency (Sun et al. 2018). Microalgae biomass has been used to improve biofuel production, including species selection, nanotechnology, and pretreatment of biomass (Sydney et al. 2019; Sankaran et al. 2020). This review provides comprehensive information about the progress of various thermochemical pretreatment methods for converting microalgae biomass to biofuel (Mathimani et al. 2019).

Biological pretreatment employs natural microorganisms or their enzymes to convert microalgae biomass to biofuel, resulting in a low energy utilization, cost-effective method, and eco-friendly solution (Chen et al. 2013). Recently, much emphasis has been placed on how nanotechnology can improve biofuel production for fuel engines. Its advantages include robustness, recycling capability, high catalytic performance, cost-effectiveness, and an environmentally friendly technique (Hossain et al. 2019). Nanotechnology has a significant role in biofuel conversion, but its commercialization is limited. These include the nanomaterials chosen, the method of preparation, and the duration of the process. Furthermore, the availability of the desired nanoparticles for large-scale applications is a significant issue. To improve the application of nanotechnology in biofuel production, innovative designs and a sustainable approach are required (Yap et al. 2021).

5.2.2 Bioethanol

Microorganisms use the biological process of fermentation to turn sugars into bioethanol. As substrates for the procedure, fermentable sugars like glucose, fructose, maltose, and rhamnose are used (Kumar et al. 2018). Various microorganisms have been identified that can convert sugars into bioethanol, including bacteria, yeasts, and fungi (Nonklang et al. 2008). Bioethanol production can benefit significantly from using algae sugars, with some microalgae species being particularly advantageous because of their higher sugar content. Algal biomass contains polymeric carbohydrates, which must be converted to monomers before being fermented by microorganisms to produce bioethanol (Al Abdallah et al. 2016). Researchers found that different microalgae species and strains had varying amounts of fermentable sugars and ethanol (Khan et al. 2018). *Scenedesmus dimorphus* accumulated 53.7% w/w of carbohydrates, and when these carbohydrates were hydrolyzed with sulfuric acids, they produced 80% fermentable sugars, indicating that they could be used to make bioethanol (Chng et al. 2017). *Chlorella vulgaris* reported an ethanol yield of 11.7 g L^{-1} and an ethanol yield of 93% from the fermentation of sugars derived from *Scenedesmus* sp. (Sivaramakrishnan and Incharoensakdi 2018; Ho et al. 2013). Thus, resolving the problems pertaining to alcoholic fermentation makes bioethanol production profitable and marketable. Overall, the substrate maximizes yields since these fermenting bacteria are substrate-dependent. Since algae biomass contains various fermentable sugars, combining the fermenter microorganisms is a preferred and efficient (Khan et al. 2017).

5.2.3 Biohydrogen

Over the years, fossil fuels are getting depleted and causing environmental issues. In this context, hydrogen gas is a prospective energy source. Hydrogen gas is the only fuel devoid of carbon, having water as the by-product of combustion. Utilizing hydrogen will help reduce issues such as acid rain and emissions of greenhouse gases (Sampath et al. 2020; Wang and Yin 2017). "Biohydrogen" refers to hydrogen created biologically, most typically through metabolism by algae, bacteria, archaea, and waste organic sources (Wang and Wan 2009). Biohydrogen has shown great potential as a renewable energy source. The microalgal biomass can directly produce biohydrogen and indirectly by photolysis of water; fermentation in the dark along with the production of by-products such as volatile fatty acids and bioelectrochemical (electrolysis microbial fuel cells); and, finally, thermochemical transformation (pyrolysis and gasification) (Rajesh Banu et al. 2020; Arimbrathodi et al. 2023). Microalgae have emerged as suitable feedstock and superior method for H_2 production due to their high CO_2 fixation capacity, more extraordinary photosynthetic ability, faster growth rate, higher energy content, high level of lipid, increased nutrient uptake, and ability to prosper in a range of environments (Gholkar et al. 2021). Also, biohydrogen production from microalgal biomass depends upon various variables, such as nutrient content, light intensity, temperature, photoreactor arrangement, pH, the concentration of the substrate, and cell density. Microalgae like *Chlamydomonas* sp., *Monoraphidium* sp., *Tetraspora* sp., *Scenedesmus* sp., *Closterium* sp., and *Chlorella* sp. produce biohydrogen (Maswanna et al. 2020). Out of 70 species, approximately 30 genera have shown promising potential for utilization as a renewable energy source (Ahmed et al. 2022). In *C. reinhardtii*, the sulfur-deprived cells can produce H_2 continuously even if acetate or other organic substrate is absent in the medium under phototrophic conditions (Arimbrathodi et al. 2023). To stimulate optimum cell development and bioH_2 generation, compounds like thiamine, biotin, and cyanocobalamin must be supplemented with the culture and require a balance of carbohydrate-based substrate for efficient bioH_2 generation (Ahmed et al. 2022). The biohydrogen yield can be upregulated by limiting the O_2 generated by photosynthesis and reducing the lag period of the microalgae (Musa Ardo et al. 2022). The optimum cell density of the culture and active growth stage are critical for biohydrogen production. Biohydrogen production from microalgae is constrained due to insufficient techniques and infrastructure for proper distribution, capture, storage, and transformation.

Strategies for improvizing biohydrogen utilizing microalgae include a variety of techniques, like immobilization, pretreatment methods, and genetic engineering (Arimbrathodi et al. 2023). The method of entangling microalgal cells on or into solid support is known as microalgal immobilization. Its benefits include high cell density, easier culture manipulation, and simple microalgae cell harvesting (Maswanna et al. 2020). Genetic and metabolic engineering can change specific pathways to improve bioH_2 generation. It is possible to suppress the inhibitors and barriers to the photosynthetic growth (Li et al. 2022).

5.2.4 Biomethane

Microalgae cultured in sewage water are attractive candidates for anaerobic digestion-based biomethane generation. The hemicellulose content of the microalgal cell wall makes it compact, prevents the interior molecules from hydrolysis, and restricts the anaerobic digestion of these substances (Wang and Yin 2017). Hence, to extract the organic matter from microalgal cells and simultaneously accelerate the lysis process, it is necessary to pretreat the biomass from microalgae. Pretreatment processes exist like enzymatic, ultrasonic, thermochemical, and hydrothermal (Rodriguez et al. 2015). Other than pretreatment methods, a co-digestion method is also applicable to enhance methane production. By minimizing potential inhibitions and constraints, co-digestion adjusts the medium's carbon/nitrogen ratio to promote catabolic processes in the microorganism (González-Fernández et al. 2012).

Pretreating the biomass thermally in a range of 75–95°C proved advantageous over non-pretreated biomass as it efficiently enhanced microalgae anaerobic biodegradability and increased the value of methane generation by 70% (Passos and Ferrer 2014). Thermochemical processing significantly influenced the dissolution of biomass and increased methane generation in *Chlorella* and *Nannochloropsis* by 30% and 40%, respectively. Furthermore, the thermochemical processing field might break the stiff cell-wall composition of the microalgae (Veerabadhran et al. 2021). The other two methods of physical pretreatments use ultrasound and microwave for lysing the cell wall to make intracellular compounds easily accessible for anaerobic digestion. The energy given is crucial in both techniques.

A higher energy supply will increase biomass solubility and biogas field production (Rodriguez et al. 2015). The ultrasound pretreatment method uses sound waves to rupture the cell from the inside and break down the cell wall. It requires high electrical energy and can cause a negative energy balance. The basis for microwave pretreatment is the breakage of the hydrogen bond, which modifies the structures of proteins and lipids. The generated heat in this method also acts as a pretreatment method. The biggest drawback of this process is that the higher the biomass concentration, the greater the energy consumption. Compared to this method, biological pretreatment using bacteria, fungi, or enzyme consumes lesser energy and produces no inhibitory products; however, the drawbacks of this process are that it is a slower process and the cost of some enzymes are relatively high compared to the methane yield. Another way of enhancing the degradation process of microalgal biomass is the chemical pretreatment method by adding either acid or alkali. Pretreatment parameters employ a multi-objective optimization technique to achieve higher biogas yield with a positive energy balance (Rodriguez et al. 2015).

5.2.5 Biobutanol

Biobutanol is a practical alternative to traditional fuels. It is more effective as a biofuel than bio-methanol or bioethanol due to its similarity in energy density and chemical structure to the gasoline (Yeong et al. 2018). In addition to serving as a biofuel, it can be a solvent for the pharma and food industry (Huang et al. 2013). Reports on biobutanol production by fermenting microalgal biomass are limited. Most work is at a non-commercial scale or low-value proposition. The generation of biobutanol starts with a hydroformylation of alkene-like propene that is reduced with hydrogen to produce a mixture of butanol and iso-butanol (Jayaprakash 2016). The final product is mainly detrimental to the environment and causes economic hardships as it constitutes a residual gas flux that contains hydrogen, undissolved carbon monoxide, propylene, and propane (Yeong et al. 2018).

The general approach for biobutanol production occurs *via* two steps: (i) biomass production from microalgae; and (ii) biobutanol production from anaerobic fermentation. The most desired microalgae would be those with a high starch content and natural production like *Tetraselmis subcordiformis, C. vulgaris, Chlorella* sp. AE10, *and C. reinhardtii* (Yeong et al. 2018). The best phases to obtain the highest carbohydrate content from microalgae are during the nutrient deprivation condition, the initial stage of the stationary phase, and the end of the log phase (Ho et al. 2013). Microalgal biomass is used as a feedstock for the *Clostridia acetobutylicum* in the acetone–butanol–ethanol (ABE) fermentation ratio of 3:6:1 to produce biobutanol (Ranjan and Moholkar 2012). *C. acetobutylicum* is saccharolytic; hence, the pretreatment for amylase or saccharification reaction is unnecessary. The physiological structure of microalgae is more straightforward, requiring less pretreatment, which has an economic advantage over other feedstock types. ABE broth is mainly dilute, making the biobutanol separation process expensive. Purification methods must be stringent enough to avoid butanol toxicity on the fermentative bacterium by balancing the solvent concentration with bacterium butanol tolerance. ABE fermentation yielded approximately 0.15–0.25 g g^{-1} of butanol/sugar with productivity levels of 0.5 g h L^{-1} (Qureshi and Ezeji 2008). Butanol fermentation from microalgal biomass resulted in a butanol yield of around 0.02–0.29 g g^{-1} of butanol/sugar and a production titer between 0.8 and 13.2 g L^{-1} (Wang and Yin 2017). Immobilization of *Clostridial* cells and *in situ* product removal are two approaches that have been extensively investigated to improve the biobutanol production (Jiang et al. 2009). Out of the two, the immobilization of cell strategy is beneficial over the conventional method of isolated-cell fermentation in terms of the following factors: enhanced cell stability, high availability of cell densities, increased cell stability, less amount of microbial cell loss, a reduction in the negative impacts of metabolic side products produced during ABE fermentation, as well as a capacity to adapt to generated solvent (Jiang et al. 2009). To create highly productive carbohydrate-accumulating strains, a better understanding of algal glucose metabolism at the genome level is required, which can be conducted either by improving the microalgal strains or modifying the culturing parameters. Thus, the combination of algal cultivation and the biobutanol production process shows promising aspects, specifically in revitalizing the biofuel sector (Yeong et al. 2018).

5.3 BIOPOTENTIAL OF ALGAE-BASED PRODUCTS

The health concern is growing as the global population grows, owing to lifestyle changes resulting in various disease conditions. Because our body system is unable to synthesize essential compounds required to meet our nutritional needs, we must develop an alternative approach based on organic supplements that are easy to grow because, in a post-pandemic era, there is an increase in the consumption of organic supplements as opposed to chemically derived ones. A new generation of health supplements will only counteract this, and they should be more affordable so everyone can use them. We are rushing toward the subsequent generation development, resulting in the shrinkage of agricultural land. As a result, microalgae are becoming a more popular alternative tool for meeting the needs of food, nutraceuticals, pharmaceuticals, and cosmetics industries through a circular bio-economy approach, as microalgae no longer require agricultural land and are easily cultivated in the aquatic system (Russell et al. 2022; Kaur et al. 2023; Baldia et al. 2023). In this segment, we will majorly concentrate on the potential of microalgae toward the generation of pharmaceutical, nutraceutical, cosmetics, and food ingredients like HVABs products which can overcome the world's need for microalgae-based organic products and avoid chemical-based drugs and medicine in the generation of chemical-free health and nutritional supplements for the present and upcoming humanity and organisms in a green economy way.

5.3.1 Polyunsaturated Fatty Acids (PUFAs)

Docosahexaenoic acid (DHA) and eicosapentaenoic acid (EPA) belong to omega-based PUFAs, which are typical examples of dietary supplements which play a vital role in the anti-inflammation in several diseases like Alzheimer's and arthritis in which EPA and DHA act as a precursor for prostaglandins and eicosanoids to overcome the aid and also help in the brain development. Thus, most pediatricians refer to this supplement for pregnant women and babies in the early onset age (Gupta and Gupta 2020). As we know, fishes are the better source of such supplements, but it is not affordable for everyone to eat fish because of its scarcity; another reason is marine pollution, as there is accumulation of heavy metals in marine fishes (Barkia et al. 2019; Katiyar and Arora 2020). In the post-COVID era, people are very much health conscious, thus moving toward vegan-based food and nutraceuticals (Adarme-Vega et al. 2012). So, we must look for alternate approaches to overcome this scarcity of PUFA-based sources. In microalgae, we found certain species such as *Crypthecodinium, Isochrysis galbana,* and *Chlorella* sp. as sources of PUFAs, etc. Nevertheless, further studies need to explore it more to better understand and explore it on a broader scale (Matos et al. 2017; El Abed et al. 2008; Winwood 2013; Rosales-Mendoza 2016; Zárate et al. 2017).

5.3.2 Sterols

Fucosterol, sitosterol, campesterol, and chalinasterol are phytosterols with anti-inflammatory and anti-cancer properties and are used to treat neurological diseases such as Parkinson's (Kim et al. 2008; Devaraj et al. 2004). Microalgae like *Nannochloropsis* sp., *Tetraselmis* sp., and *Pavlola lutheri* on screening produce sterols with a net yield of 0.4–2.6% of dw, while *Chaetoceros* genus reported producing 27.7 ug sterols dw.g^{-1} (Ahmed et al. 2015). Sterol-like ergosterol in THe *C. pyranoidosa, Schizochytrium aggregatum*, and *Dunaliella tertiolecta* leads to an anti-inflammatory response by suppressing the LPS feedback (Francavilla et al. 2015).

5.3.3 Carotenoids

Microalgae are sources of carotenoids, the accessory pigments comprising an eight-isoprene unit structure containing five carbons that prevent photo-oxidative damage from reactive oxidants, free radicals, and other oxidative reactions. Carotenoids are essential in broad applications of pharmaceuticals, nutraceuticals, and the food industry, such as astaxanthin, fucoxanthin, β-carotene, lutein, zeaxanthin, lycopene, canthaxanthin, and neoxanthin (Henríquez et al. 2016; Chen et al. 2017; Tolpeznikaite et al. 2021; Orejuela-Escobar et al. 2021). Carotenoids (fucoxanthin, β-carotene, and lutein) are some applications of anti-obesity, anti-cancer, anti-hypertensive, and anti-diabetic effects, and they are way better than synthetic products. Thus, carotenoids appear in different food sources, such as cheese, cakes, and baked butter (Kim et al. 2012; Li et al. 2020; Cezare-Gomes et al. 2019).

These carotenoids are classified into carotenes and xanthophyll (Sathasivam and Ki 2018). Carotenes comprise β-carotenes and lycopene, which lack oxygen in their structure. While lutein and zeaxanthin have hydroxyl groups, canthaxanthin and echinenone have =O groups in their structures (Novoveská et al. 2019). Furthermore, astaxanthin comprises hydroxyl and =O groups (Shah et al. 2016). Due to the polar nature of astaxanthin, their therapeutic advantages prevent anti-inflammation, central nervous system disorders, reducing stomach-based inflammation, Parkinson's disease, Alzheimer's disease, depression, and neuropathic pain (Barkia et al. 2019; Villaró et al. 2021). Microalgae can produce more lutein than any other sources, and it has been shown to fight age-related macular degeneration (AMD), protect the human skin from free radicals, treat Alzheimer's disease (AD), protect the eyesight through retinal accumulation, and filter blue light (Singh et al. 2019; Ren et al. 2021).

5.3.4 Polysaccharides

Polysaccharides are considered the most diverse biopolymers on the earth and have a crucial role in safeguarding cellular structure, hereditary knowledge, and the energy storage (Ghosh et al. 2021). Commercially polysaccharides from microalgae have beneficial roles in improving human and animal health (Tang et al. 2020; Colusse et al. 2022). Polysaccharides consist of sugar molecules composed of monomers like glucose, xylose, and galactose—the biologically active polysaccharides reported in microalgae *Porphyridium* sp., *Chlorella* sp., and *Arthrospira* sp. (Jacob-Lopes et al.

2009; Dias et al. 2020), rich in sulphated polysaccharides, have demonstrated therapeutic potential as having anti-tumor, immune-modulating, anti-viral, and other applications in the pharmaceuticals (Xiao and Zheng 2016). Microalgae produce exopolysaccharides (EPSs) which prevent erosion from their natural environment and cluster together to form biofilms (Angelaalincy et al. 2017).

5.3.5 Vitamins

Vitamins typically fall into one of the two categories: fat-soluble or water-soluble. Vitamins are essential and act as proton carriers or electron acceptors in breaking down macronutrients. Vitamins are vital for preventing diseases like methyl-malonic acidemia, scurvy, and rickets. Vitamins from microalgae, including vitamins B12, C, E, A, and D, can protect the skin, fight cancer, and prevent cardiovascular diseases (Sathasivam et al. 2019; Del Mondo et al. 2020). Tocopherols (commonly known as vitamin E), is the only lipid-resolvable antioxidant biomolecule produced by these photosynthetic organisms (Maeda et al. 2005). Tocopherol is frequently used as a dietary supplement to prevent oxidative damage brought on by stress or pollution. It also benefits cardiovascular health, hypocholesterolemia, and antimutagenic properties (Kottuparambil et al. 2019).

5.3.6 Microalgal Proteins

Many species in microalgae consist of high protein content in them, which is around 40–60% of their dry weight, while some species have lower protein content, and they are useful in oil and biodiesel production. Some microalgae from *Chlorella* sp. and *Arthrospira* sp. are used in industries because of their high protein content, around 51–58% of dw (Becker 2007; Ismail et al. 2020). As our body system cannot synthesize some essential amino acids, we depend upon other sources. Microalgae are rich sources of proteins, and some species of microalgae produce more proteins than eggs, milk, and meat, etc. Microalgae productivity ranges from 2.5 to 7.5 tonnes he^{-1} $year^{-1}$ of proteins. Microalgae proteins have diverse pharmaceutical applications, such as activating cholecystokinin, which reduce cholesterol levels. Superoxide dismutase produced by microalgae *Porphyridium* sp. shields from oxidative damage (Gouveia et al. 2006; Smee et al. 2008; Bleakley and Hayes 2017).

5.3.7 Phycobiliproteins

Phycobiliproteins are the auxiliary photosynthetic pigments present in many algae forms, and their amount varies in algae depending upon the environment. Some microalgae, such as *Porphyridium* sp., are commercially exploited for producing phycobiliproteins (Borowitzka 1988). Phycobiliproteins, such as C-phycocyanin and R-phycoerythrin, are used in food and cosmetics and natural dyes (Sekar and Chandramohan 2008), while in some microalgae species, phycoerythrin has physical properties (highly resolvable in water and negatively charged at substantiAL pH, providing specificity) that make them suitable for labelling biological molecules in flow cytometry, fluorescence immunoassays, and prognostics (Glazer 1994; Spolaore et al. 2006) (Table 5.1).

TABLE 5.1
Bioactive Compounds with Potential Applications from Different Algae.

Algae	Bioactive compound	Potential application	Reference(s)
D. salina	β-Carotene	Anticancer, sunscreen, suppression of cholesterol synthesis, vitamin A activity	(Jurić et al. 20 22; Jacob-Lopes et al. 20 19; Beydoun et al. 2018)
H. pluvialis	Astaxanthin	Antioxidant, anti-inflammatory	(Ciccone et al. 2013; Markou and Nerantzis 2013)
P. cruentum	Polysaccharides	Pharmaceuticals	(Camacho et al. 2019)
C. vulgaris	Ascorbic acid	Health supplements	
Schizochytrium sp.	PUFA (DHA)	Nutraceuticals, food supplements	(Li et al. 20 08; Cavelius et al. 2023)
D. salina, H. pluvialis	Phycobiliproteins	Pigments, cosmetics, vitamins	
C. protothecoides	Lutein	Immune booster, anti-inflammatory, antioxidant	(Markou and Nerantzis 2013)
D. tertiolecta,	Vitamins and folic acid	Antioxidant, acts as precursor of some crucial enzyme cofactors	(Galasso et al. 2019)
Porphyridium sp.	Phycobiliproteins	Cytotoxic in nature, cause apoptosis, prevent Alzheimer's, antioxidant	(Jurić et al. 20 22; Jacob-Lopes et al. 20 19; Sigurdson et al. 20 17; Cuellar-Bermudez et al. 2015)

5.3.8 Livestock and Agriculture

Using microalgae or their extracts for livestock is an intelligent method for fostering public health due to their rich source of antioxidants and β-glucan, a natural compound possessing anti-inflammatory effects that stimulate immunity, thus being widely used as a dietary supplement; in agriculture in improving soil fertility as well as soil stabilizers. Microalgae give practically feasible solutions to treat the wastewater and can do a selective heavy metal removal from contaminated sites, resulting in a biomass as a biofuel feedstock. According to reports, adding microalgae extracts to animal feed can enhance intestinal health, boost animals' immune responses, and reduce enteric methane released from ruminant fermentation. Numerous microalgae strains represent prebiotic effects which improve host health in the animal feed industry. In *Tetraselmis* sp., cell walls also sustain the gut microbiota. *D. salina* is an interesting microalga known for having immunostimulant, anti-viral, and antitumor properties (Harvey and Ben-Amotz 2020). Hence, it is clear that microalgae species help animals indirectly by supporting the intestinal microbiota, which improves their health and immediately improves health and performance through nutrient supply.

For ruminant feed, *A. platensis* (Holman et al. 2012) and *Schizochytrium* sp. (Jeon et al. 2022) are the most popular algae. Supplementing broiler chickens with *A. platensis* (El-Hady et al. 2022), piglets with *A. platensis*, *C. vulgaris*, and *Chlorella* sp. (Furbeyre et al. 2017), results in some bioactive effects that are positive for productivity, inhibiting the growth of *E. coli*, and having some immunomodulating effects; these actions are similar to a step in the right direction towards reducing the overuse of antibiotics. In addition, seaweed food and feed are produced mainly from the kelps *Macrocystis* sp., *Fucus* sp., *Laminaria* sp., and *Ascophyllum nodosum*, used chiefly as additives for vitamins and minerals (Craigie 2011). In addition, due to the accessibility of various plant growth-based and stimulating products, seaweeds and related products made from seaweeds have been extensively utilized in agriculture to enhance the quality and production of cereal crops (Nabti et al. 2017). Depleted fucoidan, alginate, and other compounds can be added to the soil to enhance structure and aeration while stimulating bacteria in the root system to encourage plant development (Zodape et al. 2008). Green algae *U. lactuca*, *Codium* sp., brown algae *Ecklonia maxima*, *A. nodosum*, red algae *Gelidium* sp., and *C. crispus* can produce industrial extracts (Craigie 2011). Also, the microalgal-based biomass may be used to control some pathogens in certain crops due to their anti-fungal, anti-bacterial, and anti-viral properties (Prasanna et al. 2015). Microalgal biomass also appears to have nutrient characteristics comparable to agro-based chemicals with environmentally friendly and slow-releasing nutrient properties while also exhibiting potential in bio-based crop production for improved growth, germination, and floral traits (Garcia-Gonzalez and Sommerfeld 2016). Furthermore, the residual products left over will serve as the most valuable products, i.e., biofertilizers (Koller et al. 2014), which, although not a huge profit driver, will make the overall process more environmentally friendly. For example, Heliae (United States) company is selling and developing products from algae, such as a soil amendment (PhycoTerra) and plant stimulators, based on a more direct application of algal extracts than the use of the residual biomass in composted form.

5.4 ALGAE-BASED COMPANIES

In today's world, there is an ever-increasing request for environmentally friendly, healthy, and sustainable bio-inspired green products replacing synthetic and petroleum-derived products. In such a scenario, microalgae, using the power of photosynthesis, unequivocally the commercial market, primarily in the cosmetic, biofuel, and nutraceutical industries (Wang et al. 2015; Sathasivam and Ki 2018; Rehmanji et al. 2021), hold an upper hand and offer a plethora of biological products with unprecedented benefits for human health and society (Kareya et al. 2022). Numerous microalgal-based products are available; for example, depollutine (Givaudan active beauty), derived from *P. tricornutum* peptidic extract, is used for fighting wrinkles. Another biomolecule, to decrease wrinkles and make skin suppleness and hydration, Solasta Astaxanthin (Solix ingredients), employs a natural astaxanthin extract from *H. pluvialis* (Wang et al. 2015). Also, these microalgal-based extracts are part of health supplements and personal care products (Amaro et al. 2011). Algenol (Florida, United States) has developed one of the systems that largely omits the harvesting of biomass and instead focuses on recovering product (ethanol) through condensation on the bioreactor cover and processing of the water/ethanol product in their direct-to-ethanol process (Sarma et al. 2021). Furthermore, Sapphire Energy (Rajesh Banu et al. 2020), a company that produces algae-based biodiesel, has partnered with the Institute of Systems Biology to use models of biofuel production to establish engineering strategies as well. Algenol is one of several algal biofuel producers that have begun using modelling techniques to guide strain optimization with specialists to support their various initiatives (Rajesh Banu et al. 2020; Moriarty and Honnery 2021). Examining the Sapphire Energy green crude farm information, they use a DAF system and focus on biomass output, gasoline, and diesel production (Kumara Behera and Varma 2016). In Taiwan, one of the world's earliest companies, Taiwan Chlorella Manufacturing Company Co. Ltd, prominently Taiwan Chlorella, is one of the biggest manufacturers of products extracted from *Chlorella* sp., that is, providing fluids and other powder forms with the finest and purest grade of *Chlorella* products.

In conclusion, due to their unique bioactive molecules and nutritional content, the demand for biofuels, nutraceutical products, and dietary supplements from microalgae has increased in the past decades. There are several marketed microalgae-based biofuel and pharmaceutical products (Table 5.2) designed for biofuels, nutraceuticals, and dietary purposes, like gasoline, jet, and diesel from Algenol Biotech LLC, Algenol Biotech LLC, Solix Algredients, Inc., and Sapphire Energy, which includes astaxanthin from *H. pluvialis*, and TerraVia Holdings, Inc. produces several microalgae-based products rich in omega-9, i.e., AlgaPrime DHA (Dickinson et al. 2017; Rehmanji et al. 2021).

5.5 CONCLUSIONS, CHALLENGES, AND FUTURE PERSPECTIVES

Microalgae offer the unequalled potential to supply various sustainable products to humankind. In addition, they contribute to economic growth, mitigate elevated CO_2 levels, and reduce greenhouse gas emissions. They have demonstrated the potential to produce biofuels, proteins for food/feed and aquaculture, bioethanol, antioxidants, pigments, ω-3 FAs, and many medicinal biomolecules, as discussed before. However, an array of critical challenges remains to be addressed before the full potential of algae can be unleashed on a commercial scale. For instance, several initiatives to create economically viable biofuels and medications from microalgae to replace fossil fuels have contributed significantly to recent advancements in microalgae cultivation, preservation, dewatering, and harvesting. Despite the enormous efforts and scientific advances achieved in the past, several obstacles still need to be overcome before algae-based biofuels are commercially feasible. Although the challenges and bottlenecks will vary according to the product desired, the significant problems that obstruct the transition of microalgae from pilot to commercial scale are:

1. Lack of improved microalgae strains for the desired products to sustainably produce copious amounts of biomass on a large commercial scale
2. Maintaining the purity of the microalgae culture by preventing predator and other microalgal species contamination

TABLE 5.2
Various Companies Produce Algae-Based Biofuels and Value-Added Products.

Company	Country	Product	Microlgae	Functions	Manufacturing	Reference
Algenol Biotech LLC	United States	Ethanol, gasoline, jet fuel, and diesel	*Cyanobacteria*	Personal care, antioxidants, nutrients, amino acids, vitamins, and energy drinks	Photobioreactor (PBR): sunlight, high-pressure carbon dioxide, salt water	(Dickinson et al. 2017; Moriarty and Honnery 2021; Rajesh Banu et al. 2020)
Algatech	Israel	AstaPure®Astaxanthin, AstaPure®, BioGlenaa™, FucoVital™	*H. pluvialis, P. tricornutum, P. cruentum, Nannochloropsis*	Skin health, fertility, immunity, and brain health support	Photobioreactor (PBR)	(Li et al. 2020)
Seambiotic	Israel	Biofuel, animal and fish feed, food additives	*Nannochloropsis* sp., *P. tricornutum, Amphora* sp., *Navicula* sp., *Tetraselmis* sp.	Biodiesel, omega-3, facilitate productive processes	Open ponds over closed bioreactors	(Rajesh Banu et al. 2020)
Solix Algredients Inc.	United States	Astaxanthin, Solmega DHA omega-3, and other natural algal ingredients	*H. pluvialis*	Dietary supplements, personal care to reduce wrinkles and improve skin elasticity and moisture	Closed-growth PBR, supercritical carbon dioxide, supercritical fluid extraction	(Rehmanji et al. 2021)
Sapphire Energy	United States	High-value oils, aquaculture, and animal feed	*Spirulina strains*	Green crude, renewable crude oil, omega-3 for human nutrition	Open pond, photosynthetic growth system	(Rajesh Banu et al. 20 20; Gu et al. 2020)
Polaris Inc.	France	Omegavie algae oils, and Omegavie marine oils	*Marine microalgae*	Infant nutrition, nutraceuticals, brain vision maternity	Enzymatic synthesis of active ingredients	(Pandey and Pattnaik 2016)

Cellana Inc.	United States	Human nutritional oils, biofuel, animal nutritional feed	*Marine microalgae*	Increasing longevity, boosting brain function and concentration	PBR with open pond	(Dickinson et al. 2017; Moriarty and Honnery 2021; Rajesh Banu et al. 2020)
Cyanotech Corporation (Hawaii)	United States	BioAstin, Hawaiian Astaxanthi	*H. pluvalis*	Eye and brain health; joint and tendon health skin and cardiovascular health	Automated cell harvesting following flash evaporation	(Hoppe 2019)
TerraVia Holdings, Inc.	United States	Edible food oil, biofuel AlgaPrime DHA, food, AlgaVia, PURASORB	*Auxenochlorella protothecoides*	Personal care ingredients, healthy oils, fats, and proteins, fibers	Dark fermenters, heterotrophic algae	(AlProl and Elkatory 2022; Dickinson et al. 2017; Poonam and Nivedita 2017)
Earthrise Nutritional LLC.	United States	Spirulina, Natural® Spirulina Gold Plus® (increased phycocyanin content and vitamin C)	*A. platensis*	Support healthy immune system, eye health (high β-carotene content)	Automated cell harvesting following flash evaporation.	(Spolaore et al. 2006; Miao and Wu 2006)
Taiwan Chlorella Manufacturing Co.	China	Tablets, Chlorella extract powders, and noodles	*Chlorella* sp.	Detoxification, immunity support, anticarcinogenic,	PBR with open pond	

3. The development of energy-efficient harvesting, dewatering/drying, and economically sound culturing methods for treating microalgal biomass. For instance, large photobioreactors (PBRs) cultivate algae in a clean manner to produce proteins or bioactive biomolecules.
4. Lack of scalable extraction and purification methods, particularly for the purification of target nutraceutical compounds from the production strains
5. Producing valuable co-products; and
6. The reduction of environmental impacts by recycling wastewater, energy, and nutrients.

Some issues regarding strain suitability and robustness are already addressed by more intensive and intelligent bioprospecting, considering the product and cultivation conditions. For instance, in-depth bioprospecting and screening have established the potential for algae to produce several bioactive compounds and peptides. Similarly, screening of high-yielding, robust, or seasonal algal strains and the optimization of growth parameters to enhance algal pigment production should help improve the commercial feasibility of pigment production from algae. More in-depth omics, systems, and synthetic biology studies are required to shed light on pathways and processes regulating protein synthesis, cell division, signaling, and synthesis of nutraceutical and other bioactive metabolites. Moreover, industries and academic researchers must develop the biorefinery paradigm, which recovers a wide range of products from the algal feedstock to render the procedure commercially competitive.

ACKNOWLEDGMENTS

The work was supported by a grant from the Department of Biotechnology, Government of India, to Pannaga Pavan Jutur (Sanction No. BT/PB/Center/03/2011-Phase II). Fellowships awarded by the ICGEB-BIOTECHNET to Getachew Tafere Abrha; the ICGEB-Arturo Falaschi to Abdalah Makaranga; the Department of Biotechnology, India for Gourav Kumar and Asha Arumugam Nesamma; and the University Grants Commission, India to Neeru Gupta are duly acknowledged.

REFERENCES

Adarme-Vega, T. Catalina, David K. Y. Lim, Matthew Timmins, Felicitas Vernen, Yan Li, and Peer M. Schenk. 2012. "Microalgal biofactories: A promising approach towards sustainable omega-3 fatty acid production." *Microbial Cell Factories* 11 (1):96. doi: 10.1186/1475-2859-11-96.

Ahmed, Faruq, Wenxu Zhou, and Peer M. Schenk. 2015. "*Pavlova lutheri* is a high-level producer of phytosterols." *Algal Research* 10:210–217. doi: 10.1016/j.algal.2015.05.013.

Ahmed, Shams Forruque, M. Mofijur, Muntasha Nahrin, Sidratun Nur Chowdhury, Samiha Nuzhat, May Alherek, Nazifa Rafa, Hwai Chyuan Ong, L. D. Nghiem, and T. M. I. Mahlia. 2022. "Biohydrogen production from wastewater-based microalgae: Progresses and challenges." *International Journal of Hydrogen Energy* 47 (88):37321–37342. doi: 10.1016/j.ijhydene.2021.09.178.

Al Abdallah, Qusai, B. Tracy Nixon, and Jarrod R. Fortwendel. 2016. "The enzymatic conversion of major algal and cyanobacterial carbohydrates to bioethanol." *Frontiers in Energy Research* 4. doi: 10.3389/fenrg.2016.00036.

Alishah Aratboni, Hossein, Nahid Rafiei, Raul Garcia-Granados, Abbas Alemzadeh, and José Rubén Morones-Ramírez. 2019. "Biomass and lipid induction strategies in microalgae for biofuel production and other applications." *Microbial Cell Factories* 18 (1):178. doi: 10.1186/s12934-019-1228-4.

Alprol, Ahmed E., and Marwa R. Elkatory. 2022. "The current status of various algal industries." In *Handbook of Algal Biofuels*, edited by Mostafa El-Sheekh and Abd El-Fatah Abomohra, pp. 123–147. Elsevier.

Amaro, Helena M. A., Catarina Guedes, and F. Xavier Malcata. 2011. "Advances and perspectives in using microalgae to produce biodiesel." *Applied Energy* 88 (10):3402–3410. doi: 10.1016/j.apenergy.2010.12.014.

Angelaalincy, Mariajoseph, Nangan Senthilkumar, Rathinasamy Karpagam, Georgepeter Gnana Kumar, Balasubramaniem Ashokkumar, and Perumal Varalakshmi. 2017. "Enhanced extracellular polysaccharide production and self-sustainable electricity generation for PAMFCs by Scenedesmus sp. SB1." *ACS Omega* 2 (7):3754–3765. doi: 10.1021/acsomega.7b00326.

Arimbrathodi, Shirin P., Muhammad Asad Javed, Mohamed A. Hamouda, Ashraf Aly Hassan, and Mahmoud E. Ahmed. 2023. "BioH_2 production using microalgae: Highlights on recent advancements from a bibliometric analysis." *Water* 15 (1):185.

Baldia, Anshu, Deepanshi Rajput, Akshay Kumar, Ashok Pandey, and Kashyap Kumar Dubey. 2023. "Engineering microalgae as the next-generation food." *Systems Microbiology and Biomanufacturing* 3 (1):166–178. doi: 10.1007/s43393-022-00144-1.

Barkia, Ines, Nazamid Saari, and Schonna R. Manning. 2019. "Microalgae for high-value products towards human health and nutrition." *Marine Drugs* 17 (5):304.

Becker, E. W. 2007. "Micro-algae as a source of protein." *Biotechnology Advances* 25 (2): 207–210. doi: 10.1016/j.biotechadv.2006.11.002.

Beydoun, May A., Xiaoli Chen, Kanishk Jha, Hind A. Beydoun, Alan B. Zonderman, and Jose A. Canas. 2018. "Carotenoids, vitamin A, and their association with the metabolic syndrome: A systematic review and meta-analysis." *Nutrition Reviews* 77 (1):32–45. doi: 10.1093/nutrit/nuy044.

Bleakley, Stephen, and Maria Hayes. 2017. "Algal proteins: Extraction, application, and challenges concerning production." *Foods* 6 (5):33.

Borowitzka, Michael A. (eds.). 1988. "Vitamins and fine chemicals from micro-algae." In *Micro-algal Biotechnology*, pp. 173–196. Cambridge University Press.

Camacho, Franciele, Angela Macedo, and Francisco Malcata. 2019. "Potential industrial applications and commercialization of microalgae in the functional food and feed industries: A short review." *Marine Drugs* 17 (6):312.

Cavelius, Philipp, Selina Engelhart-Straub, Norbert Mehlmer, Johannes Lercher, Dania Awad, and Thomas Brück. 2023. "The potential of biofuels from first to fourth generation." *PLOS Biology* 21 (3):e3002063. doi: 10.1371/journal.pbio.3002063.

Cezare-Gomes, Eleane A., Lauris del Carmen Mejia-da-Silva, Lina S. Pérez-Mora, Marcelo C. Matsudo, Lívia S. Ferreira-Camargo, Anil Kumar Singh, and João Carlos Monteiro de Carvalho. 2019. "Potential of microalgae carotenoids for industrial application." *Applied Biochemistry and Biotechnology* 188 (3):602–634. doi: 10.1007/s12010-018-02945-4.

Chen, Bailing, Chun Wan, Muhammad Aamer Mehmood, Jo-Shu Chang, Fengwu Bai, and Xinqing Zhao. 2017. "Manipulating environmental stresses and stress tolerance of microalgae for enhanced production of lipids and value-added products–A review." *Bioresource Technology* 244:1198–1206. doi: 10.1016/j.biortech. 2017.05.170.

Chen, Chun-Yen, Ming-Der Bai, and Jo-Shu Chang. 2013. "Improving microalgal oil collecting efficiency by pretreating the microalgal cell wall with destructive bacteria." *Biochemical Engineering Journal* 81:170–176. doi: 10.1016/j.bej.2013. 10.014.

Cheng, Dan, and Qingfang He. 2014. "Assessment of environmental stresses for enhanced microalgal biofuel production – An overview." *Frontiers in Energy Research* 2. doi: 10.3389/fenrg.2014.00026.

Chng, L. M., K. T. Lee, and D. C. J. Chan. 2017. "Evaluation on Microalgae Biomass for Bioethanol Production." *IOP Conference Series: Materials Science and Engineering* 206 (1):012018. doi: 10.1088/1757-899X/206/1/012018.

Ciccone, Marco Matteo, Francesca Cortese, Michele Gesualdo, Santa Carbonara, Annapaola Zito, Gabriella Ricci, Francesca De Pascalis, Pietro Scicchitano, and Graziano Riccioni. 2013. "Dietary intake of carotenoids and their antioxidant and anti-inflammatory effects in cardiovascular care." *Mediators of Inflammation* 2013:782137. doi: 10.1155/2013/782137.

Colusse, Guilherme Augusto, Jaqueline Carneiro, Maria Eugênia Rabello Duarte, Julio Cesar de Carvalho, and Miguel Daniel Noseda. 2022. "Advances in microalgal cell wall polysaccharides: a review focused on structure, production, and biological application." *Critical Reviews in Biotechnology* 42 (4):562–577. doi: 10.1080/07388551.2021.1941750.

Conde, Tiago A., Bruna F. Neves, Daniela Couto, Tânia Melo, Bruno Neves, Margarida Costa, Joana Silva, Pedro Domingues, and M. Rosário Domingues. 2021. "Microalgae as sustainable bio-factories of healthy lipids: Evaluating fatty acid content and antioxidant activity." *Marine Drugs* 19 (7):357.

Craigie, James S. 2011. "Seaweed extract stimuli in plant science and agriculture." *Journal of Applied Phycology* 23 (3):371–393. doi: 10.1007/s10811-010-9560-4.

Cuellar-Bermudez, Sara P., Iris Aguilar-Hernandez, Diana L. Cardenas-Chavez, Nancy Ornelas-Soto, Miguel A. Romero-Ogawa, and Roberto Parra-Saldivar. 2015. "Extraction and purification of high-value metabolites from microalgae: essential lipids, astaxanthin and phycobiliproteins." *Microbial Biotechnology* 8 (2):190–209. doi: 10.1111/1751-7915.12167.

Del Mondo, Angelo, Arianna Smerilli, Elisabet Sané, Clementina Sansone, and Christophe Brunet. 2020. "Challenging microalgal vitamins for human health." *Microbial Cell Factories* 19 (1):201. doi: 10.1186/s12934-020-01459-1.

Devaraj, Sridevi, Ishwarlal Jialal, and Sonia Vega-López. 2004. "Plant sterol-fortified orange juice effectively lowers cholesterol levels in mildly hypercholesterolemic healthy individuals." *Arteriosclerosis, Thrombosis, and Vascular Biology* 24 (3):e25–e28. doi: 10.1161/01.ATV.0000120784.08823.99.

Dias, Rosangela R., Ihana A. Severo, Mariany C. Deprá, Mariana M. Maroneze, Leila Q. Zepka, and Eduardo Jacob-Lopes. 2020. "The next-generation of microalgae-based products." In *Microbial Enzymes and Biotechniques: Interdisciplinary Perspectives*, edited by Pratyoosh Shukla, pp. 15–42. Springer.

Dickinson, Selena, Miranda Mientus, Daniel Frey, Arsalon Amini-Hajibashi, Serdar Ozturk, Faisal Shaikh, Debalina Sengupta, and Mahmoud M. El-Halwagi. 2017. "A review of biodiesel production from microalgae." *Clean Technologies and Environmental Policy* 19 (3):637–668. doi: 10.1007/s10098-016-1309-6.

El Abed, M. M., B. Marzouk, M. N. Medhioub, A. N. Helal, and A. Medhioub. 2008. "Microalgae: A potential source of polyunsaturated fatty acids." *Nutrition and Health* 19 (3): 221–226. doi: 10.1177/026010600801900309.

El-Hady, Ahmed M. Abd, Osama A. Elghalid, Asmaa Sh Elnaggar, and Enass Abd El-khalek. 2022. "Growth performance and physiological status evaluation of Spirulina platensis algae supplementation in broiler chicken diet." *Livestock Science* 263:105009. doi: 10.1016/j.livsci.2022.105009.

Francavilla, M., P. Kamaterou, S. Intini, M. Monteleone, and A. Zabaniotou. 2015. "Cascading microalgae biorefinery: Fast pyrolysis of Dunaliella tertiolecta lipid extracted-residue." *Algal Research* 11:184–193. doi: 10.1016/j.algal. 2015.06.017.

Friedlingstein, P., M. O'Sullivan, M. W. Jones, R. M. Andrew, L. Gregor, J. Hauck, C. Le Quéré, I. T. Luijkx, A. Olsen, G. P. Peters, W. Peters, J. Pongratz, C. Schwingshackl, S. Sitch, J. G. Canadell, P. Ciais, R. B. Jackson, S. R. Alin, R. Alkama, A. Arneth, V. K. Arora, N. R. Bates, M. Becker, N. Bellouin, H. C. Bittig, L. Bopp, F. Chevallier, L. P.

Chini, M. Cronin, W. Evans, S. Falk, R. A. Feely, T. Gasser, M. Gehlen, T. Gkritzalis, L. Gloege, G. Grassi, N. Gruber, Ö Gürses, I. Harris, M. Hefner, R. A. Houghton, G. C. Hurtt, Y. Iida, T. Ilyina, A. K. Jain, A. Jersild, K. Kadono, E. Kato, D. Kennedy, K. Klein Goldewijk, J. Knauer, J. I. Korsbakken, P. Landschützer, N. Lefèvre, K. Lindsay, J. Liu, Z. Liu, G. Marland, N. Mayot, M. J. McGrath, N. Metzl, N. M. Monacci, D. R. Munro, S. I. Nakaoka, Y. Niwa, K. O'Brien, T. Ono, P. I. Palmer, N. Pan, D. Pierrot, K. Pocock, B. Poulter, L. Resplandy, E. Robertson, C. Rödenbeck, C. Rodriguez, T. M. Rosan, J. Schwinger, R. Séférian, J. D. Shutler, I. Skjelvan, T. Steinhoff, Q. Sun, A. J. Sutton, C. Sweeney, S. Takao, T. Tanhua, P. P. Tans, X. Tian, H. Tian, B. Tilbrook, H. Tsujino, F. Tubiello, G. R. van der Werf, A. P. Walker, R. Wanninkhof, C. Whitehead, A. Willstrand Wranne, R. Wright, W. Yuan, C. Yue, X. Yue, S. Zaehle, J. Zeng, and B. Zheng. 2022. "Global carbon budget 2022." *Earth System Science Data* 14 (11):4811–4900. doi: 10.5194/essd-14-4811-2022.

Furbeyre, H., J. van Milgen, T. Mener, M. Gloaguen, and E. Labussière. 2017. "Effects of dietary supplementation with freshwater microalgae on growth performance, nutrient digestibility and gut health in weaned piglets." *Animal* 11 (2):183–192. doi: 10.1017/S1751731116001543.

Galasso, Christian, Antonio Gentile, Ida Orefice, Adrianna Ianora, Antonino Bruno, Douglas M. Noonan, Clementina Sansone, Adriana Albini, and Christophe Brunet. 2019. "Microalgal derivatives as potential nutraceutical and food supplements for human health: A focus on cancer prevention and interception." *Nutrients* 11 (6):1226.

Garcia-Gonzalez, Jesus, and Milton Sommerfeld. 2016. "Biofertilizer and biostimulant properties of the microalga *Acutodesmus dimorphus*." *Journal of Applied Phycology* 28 (2):1051–1061. doi: 10.1007/s10811-015-0625-2.

Gerotto, Caterina, Alessandra Norici, and Mario Giordano. 2020. "Toward enhanced fixation of CO_2 in aquatic biomass: Focus on microalgae." *Frontiers in Energy Research* 8. doi: 10.3389/fenrg.2020.00213.

Gholkar, Pratik, Yogendra Shastri, and Akshat Tanksale. 2021. "Renewable hydrogen and methane production from microalgae: A techno-economic and life cycle assessment study." *Journal of Cleaner Production* 279:123726. https://doi.org/10.1016/j.jclepro.2020.123726.

Ghosh, Tonmoy, Rabinder Singh, Asha Arumugam Nesamma, and Pannaga Pavan Jutur. 2021. "Marine polysaccharides: Properties and applications." *Polysaccharides*, 37–60. doi: 10.1002/9781119711414.ch3.

Glazer, Alexander N. 1994. "Phycobiliproteins—A family of valuable, widely used fluorophores." *Journal of Applied Phycology* 6 (2):105–112. doi: 10.1007/BF02186064.

González-Fernández, Cristina, Bruno Sialve, Nicolas Bernet, and Jean-Philippe Steyer. 2012. "Impact of microalgae characteristics on their conversion to biofuel. Part II: Focus on biomethane production." *Biofuels, Bioproducts and Biorefining* 6 (2):205–218. doi: 10.1002/bbb.337.

Gouveia, Luísa, Anabela Raymundo, Ana Paula Batista, Isabel Sousa, and José Empis. 2006. "*Chlorella vulgaris* and *Haematococcus pluvialis* biomass as colouring and antioxidant in food emulsions." *European Food Research and Technology* 222 (3):362–367. doi: 10.1007/s00217-005-0105-z.

Gu, Xiangyu, Liang Yu, Na Pang, Jose Salomon Martinez-Fernandez, Xiao Fu, and Shulin Chen. 2020. "Comparative techno-economic analysis of algal biofuel production via hydrothermal liquefaction: One stage versus two stages." *Applied Energy* 259:114115. doi: 10.1016/j.apenergy.2019.114115.

Gupta, Jitendra, and Reena Gupta. 2020. "Nutraceutical status and scientific strategies for enhancing production of omega-3 fatty acids from microalgae and their role in healthcare." *Current Pharmaceutical Biotechnology* 21 (15):1616–1631. doi: 10.2174/1389201021666200703201014.

Harvey, Patricia J., and Ami Ben-Amotz. 2020. "Towards a sustainable *Dunaliella salina* microalgal biorefinery for 9-cis β-carotene production." *Algal Research* 50:102002. doi: 10.1016/j.algal.2020.102002.

Henríquez, Vitalia, Carolina Escobar, Janeth Galarza, and Javier Gimpel. 2016. "Carotenoids in microalgae." In *Carotenoids in Nature: Biosynthesis, Regulation and Function*, edited by Claudia Stange, pp. 219–237. Springer International Publishing.

Ho, Shih-Hsin, Shu-Wen Huang, Chun-Yen Chen, Tomohisa Hasunuma, Akihiko Kondo, and Jo-Shu Chang. 2013. "Bioethanol production using carbohydrate-rich microalgae biomass as feedstock." *Bioresource Technology* 135:191–198. doi: 10.1016/j.biortech.2012.10.015.

Holman, B. W. B., A. Kashani, and A. E. O. Malau-Aduli. 2012. "Growth and body conformation responses of genetically divergent Australian sheep to Spirulina (*Arthrospira platensis*) supplementation." *Journal of Experimental Agriculture International* 2 (2):160–173. doi: 10.9734/AJEA/2012/992.

Hoppe, Trent. 2019. *Cyanotech: A Strategic Audit*. Honors Theses, University of Nebraska-Lincoln, p. 133.

Hossain, Nazia, T. M. I. Mahlia, and R. Saidur. 2019. "Latest development in microalgae-biofuel production with nano-additives." *Biotechnology for Biofuels* 12 (1):125. doi: 10.1186/s13068-019-1465-0.

Huang, H. J., B. V. Ramarao, and S. Ramaswamy. 2013. *Separation and Purification Technologies in Biorefineries*. Wiley. doi: 10.1002/9781118493441.

Hussain, Fida, Syed Z. Shah, Habib Ahmad, Samar A. Abubshait, Haya A. Abubshait, A. Laref, A. Manikandan, Heri S. Kusuma, and Munawar Iqbal. 2021. "Microalgae an ecofriendly and sustainable wastewater treatment option: Biomass application in biofuel and biofertilizer production. A review." *Renewable and Sustainable Energy Reviews* 137:110603. https://doi.org/10.1016/j.rser.2020.110603.

Ismail, Ishamri, Young-Hwa Hwang, and Seon-Tea Joo. 2020. "Meat analog as future food: A review." *Journal of Animal Science and Technology* 62 (2):111–120. doi: 10.5187/jast.2020.62.2.111.

Jacob-Lopes, Eduardo, Carlos Henrique Gimenes Scoparo, Lucy Mara Cacia Ferreira Lacerda, and Telma Teixeira Franco. 2009. "Effect of light cycles (night/day) on CO_2 fixation and biomass production by microalgae in photobioreactors." *Chemical Engineering and Processing: Process Intensification* 48 (1):306–310. doi: 10.1016/j.cep.2008.04.007.

Jacob-Lopes, Eduardo, Mariana M. Maroneze, Mariany C. Deprá, Rafaela B. Sartori, Rosangela R. Dias, and Leila Q. Zepka. 2019. "Bioactive food compounds from microalgae: An innovative framework on industrial biorefineries." *Current Opinion in Food Science* 25:1–7. doi: 10.1016/j.cofs.2018.12.003.

Jayaprakash, Divya. 2016. *Drying Butanol Using Biosorbents in a Pressure Swing Adsorption Process*. Doctoral dissertation University of Saskatchewan SK, Canada.

Jeon, Jin-Joo, Hee-Jin Kim, Hwan-Ku Kang, Chan-Ho Kim, Hyun-Soo Kim, Eui-Chul Hong, Aera Jang, and Sang-Ho Kim. 2022. "Effects of dietary thraustochytrid *Schizochytrium* sp. and other omega-3 sources on growth performance, carcass characteristics, and meat quality of broilers." *Animals* 12 (9):1166.

Jiang, Ling, Jufang Wang, Shizhong Liang, Xiaoning Wang, Peilin Cen, and Zhinan Xu. 2009. "Butyric acid fermentation in a fibrous bed bioreactor with immobilized Clostridium tyrobutyricum from cane molasses." *Bioresource Technology* 100 (13):3403–3409. doi: 10.1016/j.biortech.2009.02.032.

Jiang, Liqun, Haiyan Pei, Wenrong Hu, Qingjie Hou, Fei Han, and Changliang Nie. 2016. "Biomass production and nutrient assimilation by a novel microalga, *Monoraphidium* spp. SDEC-17, cultivated in a high-ammonia wastewater." *Energy Conversion and Management* 123:423–430. doi: 10.1016/j.enconman.2016.06.060.

Jurić, Slaven, Marina Jurić, Żaneta Król-Kilińska, Kristina Vlahoviček-Kahlina, Marko Vinceković, Verica Dragović-Uzelac, and Francesco Donsì. 2022. "Sources, stability, encapsulation and application of natural pigments in foods." *Food Reviews International* 38 (8):1735–1790. doi: 10.1080/87559129.2020.1837862.

Kareya, Mukul Suresh, Iqra Mariam, Girish Halemirle Rajacharya, Asha Arumugam Nesamma, and Pannaga Pavan Jutur. 2022. "Valorization of carbon dioxide (CO_2) to enhance production of biomass, biofuels, and biorenewables (B^3) in *Chlorella saccharophila* UTEX247: A circular bioeconomy perspective." *Biofuels, Bioproducts and Biorefining* 16 (3):682–697. doi: 10.1002/bbb.2295.

Katiyar, Richa, and Amit Arora. 2020. "Health promoting functional lipids from microalgae pool: A review." *Algal Research* 46:101800. doi: 10.1016/j.algal.2020.101800.

Kaur, Manpreet, Surekha Bhatia, Urmila Gupta, Eric Decker, Yamini Tak, Manoj Bali, Vijai Kumar Gupta, Rouf Ahmad Dar, and Saroj Bala. 2023. "Microalgal bioactive metabolites as promising implements in nutraceuticals and pharmaceuticals: inspiring therapy for health benefits." *Phytochemistry Reviews*. doi: 10.1007/s11101-022-09848-7.

Khan, Muhammad Imran, Jin Hyuk Shin, and Jong Deog Kim. 2018. "The promising future of microalgae: current status, challenges, and optimization of a sustainable and renewable industry for biofuels, feed, and other products." *Microbial Cell Factories* 17 (1):36. doi: 10.1186/s12934-018-0879-x.

Khan, Muhammad Imran, Moon Geon Lee, Jin Hyuk Shin, and Jong Deog Kim. 2017. "Pretreatment optimization of the biomass of *Microcystis aeruginosa* for efficient bioethanol production." *AMB Express* 7 (1):19. doi: 10.1186/s13568-016-0320-y.

Kim, Hyun-Jin, Xiaotang Fan, Chiara Gabbi, Konstantin Yakimchuk, Paolo Parini, Margaret Warner, and Jan-Åke Gustafsson. 2008. "Liver X receptor β (LXRβ): A link between β-sitosterol and amyotrophic lateral sclerosis–Parkinson's dementia." *Proceedings of the National Academy of Sciences* 105 (6):2094–2099. doi: doi:10.1073/pnas.0711599105.

Kim, Sang Min, Suk-Woo Kang, O. Nam Kwon, Donghwa Chung, and Cheol-Ho Pan. 2012. "Fucoxanthin as a major carotenoid in Isochrysis aff. galbana: Characterization of extraction for commercial application." *Journal of the Korean Society for Applied Biological Chemistry* 55 (4):477–483. doi: 10.1007/s13765-012-2108-3.

Koller, Martin, Alexander Muhr, and Gerhart Braunegg. 2014. "Microalgae as versatile cellular factories for valued products." *Algal Research* 6:52–63. doi: 10.1016/j.algal.2014.09.002.

Kottuparambil, Sreejith, Roshni Lilly Thankamony, and Susana Agusti. 2019. "Euglena as a potential natural source of value-added metabolites. A review." *Algal Research* 37: 154–159. doi: 10.1016/j.algal.2018.11.024.

Kumar, Vinod, Manisha Nanda, H. C. Joshi, Ajay Singh, Sonal Sharma, and Monu Verma. 2018. "Production of biodiesel and bioethanol using algal biomass harvested from fresh water river." *Renewable Energy* 116:606–612. doi: 10.1016/j.renene.2017.10.016.

Kumara Behera, Basanta, and Ajit Varma. 2016. "From algae to liquid fuels." In *Microbial Resources for Sustainable Energy*, pp. 123–180. Springer International Publishing.

Li, Shengnan, Fanghua Li, Xun Zhu, Qiang Liao, Jo-Shu Chang, and Shih-Hsin Ho. 2022. "Biohydrogen production from microalgae for environmental sustainability." *Chemosphere* 291:132717. doi: 10.1016/j.chemosphere.2021.132717.

Li, Xin, Xiaoqian Wang, Chuanlan Duan, Shasha Yi, Zhengquan Gao, Chaowen Xiao, Spiros N. Agathos, Guangce Wang, and Jian Li. 2020. "Biotechnological production of astaxanthin from the microalga *Haematococcus pluvialis*." *Biotechnology Advances* 43:107602. doi: 10.1016/j.biotechadv.2020.107602.

Li, Yanqun, Mark Horsman, Nan Wu, Christopher Q. Lan, and Nathalie Dubois-Calero. 2008. "Biofuels from microalgae." *Biotechnology Progress* 24 (4):815–820. doi: 10.1021/bp070371k.

Maeda, Hiroshi, Yumiko Sakuragi, Donald A. Bryant, and Dean DellaPenna. 2005. "Tocopherols protect *Synechocystis* sp. strain PCC 6803 from lipid peroxidation." *Plant Physiology* 138 (3):1422–1435. doi: 10.1104/pp.105.061135.

Makaranga, Abdalah, and Pannaga P. Jutur. 2022. "Algae-bacteria interactomics unveils their role in growth and production of high-value biorenewables." In *Micro-algae: Next-generation Feedstock for Biorefineries: Contemporary Technologies and Future Outlook*, edited by Pradeep Verma, pp. 165–176. Springer Nature.

Markou, Giorgos, and Elias Nerantzis. 2013. "Microalgae for high-value compounds and biofuels production: A review with focus on cultivation under stress conditions." *Biotechnology Advances* 31 (8):1532–1542. doi: 10.1016/j.biotechadv. 2013.07.011.

Maswanna, Thanaporn, Peter Lindblad, and Cherdsak Maneeruttanarungroj. 2020. "Improved biohydrogen production by immobilized cells of the green alga *Tetraspora* Sp. CU2551 incubated under aerobic condition." *Journal of Applied Phycology* 32 (5):2937–2945. doi: 10.1007/s10811-020-02184-3.

Mathimani, Thangavel, Arianna Baldinelli, Karthik Rajendran, Desika Prabakar, Manickam Matheswaran, Richard Pieter van Leeuwen, and Arivalagan Pugazhendhi. 2019. "Review on cultivation and thermochemical conversion of microalgae to fuels and chemicals: Process evaluation and knowledge gaps." *Journal of Cleaner Production* 208:1053–1064. doi: 10.1016/j.jclepro.2018.10.096.

Matos, J., C. Cardoso, N. M. Bandarra, and C. Afonso. 2017. "Microalgae as healthy ingredients for functional food: a review." *Food & Function* 8 (8):2672–2685. doi: 10.1039/C7FO00409E.

Miao, Xiaoling, and Qingyu Wu. 2006. "Biodiesel production from heterotrophic microalgal oil." *Bioresource Technology* 97 (6):841–846. doi: 10.1016/j.biortech. 2005.04.008.

Mofijur, M., M. G. Rasul, N. M. S. Hassan, and M. N. Nabi. 2019. "Recent development in the production of third generation biodiesel from microalgae." *Energy Procedia* 156:53–58. doi: 10.1016/j.egypro.2018.11.088.

Moriarty, Patrick, and Damon Honnery. 2021. "New energy technologies: Microalgae, photolysis and airborne wind turbines." *Science* 3 (1):5.

Musa, Ardo, Fatima, Jun Wei Lim, Anita Ramli, Man Kee Lam, Worapon Kiatkittipong, Eman Alaaeldin Abdelfattah, Muhammad Kashif Shahid, Anwar Usman, Suwimol Wongsakulphasatch, and Nurul Tasnim Sahrin. 2022. "A review in redressing challenges to produce sustainable hydrogen from microalgae for aviation industry." *Fuel* 330:125646. doi: 10.1016/j.fuel.2022.125646.

Nabti, E., B. Jha, and A. Hartmann. 2017. "Impact of seaweeds on agricultural crop production as biofertilizer." *International Journal of Environmental Science and Technology* 14 (5):1119–1134. doi: 10.1007/s13762-016-1202-1.

Nonklang, Sanom, Babiker M. A. Abdel-Banat, Kamonchai Cha-aim, Nareerat Moonjai, Hisashi Hoshida, Savitree Limtong, Mamoru Yamada, and Rinji Akada. 2008. "High-temperature ethanol fermentation and transformation with linear DNA in the thermotolerant yeast *Kluyveromyces marxianus* DMKU3–1042." *Applied and Environmental Microbiology* 74 (24):7514–7521. doi: 10.1128/AEM.01854-08.

Novoveská, Lucie, Michael E. Ross, Michele S. Stanley, Rémi Pradelles, Virginie Wasiolek, and Jean-François Sassi. 2019. "Microalgal carotenoids: A review of production, current markets, regulations, and future direction." *Marine Drugs* 17 (11):640.

Orejuela-Escobar, Lourdes, Arleth Gualle, Valeria Ochoa-Herrera, and George P. Philippidis. 2021. "Prospects of microalgae for biomaterial production and environmental applications at biorefineries." *Sustainability* 13 (6):3063.

Paliwal, Chetan, Mohammed Rehmanji, Kashif Mohd Shaikh, Saeed Uz Zafar, and Pannaga Pavan Jutur. 2022. "Green extraction processing of lutein from Chlorella saccharophila in water-based ionic liquids as a sustainable innovation in algal biorefineries." *Algal Research* 66:102809. doi: 10.1016/j.algal.2022. 102809.

Pandey, Satyendra C., and Pinaki Nandan Pattnaik. 2016. "Polaris India Pvt. Ltd: A case of experiential marketing strategy." *South Asian Journal of Business and Management Cases* 5 (2):244–252. doi: 10.1177/2277977916665995.

Passos, Fabiana, and Ivet Ferrer. 2014. "Microalgae conversion to biogas: Thermal pretreatment contribution on net energy production." *Environmental Science & Technology* 48 (12):7171–7178. doi: 10.1021/es500982v.

Poonam, Sharma, and Sharma Nivedita. 2017. "Industrial and biotechnological applications of algae: A review." *Journal of Advances in Plant Biology* 1 (1):01–25. doi: 10.14302/issn.2638-4469.japb-17-1534.

Prasanna, Radha, Santosh Babu, Ngangom Bidyarani, Arun Kumar, Sodimalla Triveni, Dilip Monga, Arup Kumar Mukherjee, Sandhya Kranthi, Nandini Gokte-Narkhedkar, Anurup Adak, Kuldeep Yadav, Lata Nain, and Anil Kumar Saxena. 2015. "Prospecting cyanobacteria-fortified composts as plant growth promoting and biocontrol agents in cotton." *Experimental Agriculture* 51 (1):42–65. doi: 10.1017/S0014479714000143.

Qureshi, Nasib, and Thaddeus C. Ezeji. 2008. "Butanol, 'a superior biofuel' production from agricultural residues (renewable biomass): Recent progress in technology." *Biofuels, Bioproducts and Biorefining* 2 (4):319–330. doi: 10.1002/bbb.85.

Rajesh Banu, J., Preethi, S. Kavitha, M. Gunasekaran, and Gopalakrishnan Kumar. 2020. "Microalgae based biorefinery promoting circular bioeconomy-techno economic and life-cycle analysis." *Bioresource Technology* 302:122822. doi: 10.1016/j.biortech.2020.122822.

Ranjan, Amrita, and Vijayanand S. Moholkar. 2012. "Biobutanol: Science, engineering, and economics." *International Journal of Energy Research* 36 (3):277–323. doi: 10.1002/er.1948.

Razzak, Shaikh Abdur, Saad Aldin M. Ali, Mohammad Mozahar Hossain, and Hugo deLasa. 2017. "Biological CO_2 fixation with production of microalgae in wastewater – A review." *Renewable and Sustainable Energy Reviews* 76:379–390. doi: 10.1016/j.rser.2017.02.038.

Rehmanji, Mohammed, Asha Arumugam Nesamma, Nida Jamil Khan, Tasneem Fatma, and Pannaga Pavan Jutur. 2022. "Media engineering in marine diatom Phaeodactylum tricornutum employing cost-effective substrates for sustainable production of high-value renewables." *Biotechnology Journal* 17 (10):2100684. doi: 10.1002/biot.202100684.

Rehmanji, Mohammed, Sukannya Suresh, Asha A. Nesamma, and Pannaga P. Jutur. 2021. "Chapter 26 — Microalgal cell factories, a platform for high-value-added biorenewables to improve the economics of the biorefinery." In *Microbial and Natural Macromolecules*, edited by Surajit Das and Hirak Ranjan Dash, pp. 689–731. Academic Press.

Ren, Yuanyuan, Han Sun, Jinquan Deng, Junchao Huang, and Feng Chen. 2021. "Carotenoid Production from microalgae: Biosynthesis, salinity responses and novel biotechnologies." *Marine Drugs* 19 (12):713.

Rodriguez, C., A. Alaswad, J. Mooney, T. Prescott, and A. G. Olabi. 2015. "Pre-treatment techniques used for anaerobic digestion of algae." *Fuel Processing Technology* 138:765–779. doi: 10.1016/j.fuproc.2015.06.027.

Rosales-Mendoza, Sergio. 2016. "The Biopharmaceuticals Field and Algae as Expression Hosts." In *Algae-Based Biopharmaceuticals*, pp. 1–14. Springer International Publishing.

Russell, Callum, Cristina Rodriguez, and Mohammed Yaseen. 2022. "High-value biochemical products & applications of freshwater eukaryotic microalgae." *Science of The Total Environment* 809:151111. doi: 10.1016/j.scitotenv.2021. 151111.

Sajjadi, Baharak, Wei-Yin Chen, Abdul Aziz Abdul Raman, and Shaliza Ibrahim. 2018. "Microalgae lipid and biomass for biofuel production: A comprehensive review on lipid enhancement strategies and their effects on fatty acid composition." *Renewable and Sustainable Energy Reviews* 97:200–232. doi: 10.1016/j.rser.2018. 07.050.

Sampath, P., Brijesh, Kakarla Raghava Reddy, C. Venkata Reddy, Nagaraj P. Shetti, Raghavendra V. Kulkarni, and Anjanapura V. Raghu. 2020. "Biohydrogen production from organic waste – A review." *Chemical Engineering & Technology* 43 (7):1240–1248. doi: 10.1002/ceat.201900400.

Sankaran, Revathy, Ricardo Andres Parra Cruz, Harshini Pakalapati, Pau Loke Show, Tau Chuan Ling, Wei-Hsin Chen, and Yang Tao. 2020. "Recent advances in the pretreatment of microalgal and lignocellulosic biomass: A comprehensive review." *Bioresource Technology* 298:122476. doi: 10.1016/j.biortech.2019. 122476.

Sarma, Shyamali, Shaishav Sharma, Darshan Rudakiya, Jinal Upadhyay, Vinod Rathod, Aesha Patel, and Madhuri Narra. 2021. "Valorization of microalgae biomass into bioproducts promoting circular bioeconomy: A holistic approach of bioremediation and biorefinery." *3 Biotech* 11 (8):378. doi: 10.1007/s13205-021-02911-8.

Sathasivam, Ramaraj, and Jang-Seu Ki. 2018. "A review of the biological activities of microalgal carotenoids and their potential use in healthcare and cosmetic industries." *Marine Drugs* 16 (1):26.

Sathasivam, Ramaraj, Ramalingam Radhakrishnan, Abeer Hashem, and Elsayed F. Abd_Allah. 2019. "Microalgae metabolites: A rich source for food and medicine." *Saudi Journal of Biological Sciences* 26 (4):709–722. doi: 10.1016/j.sjbs.2017.11.003.

Sekar, Soundarapandian, and Muruganandham Chandramohan. 2008. "Phycobiliproteins as a commodity: trends in applied research, patents and commercialization." *Journal of Applied Phycology* 20 (2):113–136. doi: 10.1007/s10811-007-9188-1.

Shah, Md. Mahfuzur R., Yuanmei Liang, Jay J. Cheng, and Maurycy Daroch. 2016. "Astaxanthin-producing green microalga *Haematococcus pluvialis*: From single cell to high value commercial products." *Frontiers in Plant Science* 7. doi: 10.3389/fpls.2016.00531.

Shaikh, Kashif Mohd, Asha Arumugam Nesamma, Malik Zainul Abdin, and Pannaga Pavan Jutur. 2019. "Molecular profiling of an oleaginous trebouxiophycean alga *Parachlorella kessleri* subjected to nutrient deprivation for enhanced biofuel production." *Biotechnology for Biofuels* 12 (1):182. doi: 10.1186/s13068-019-1521-9.

Sigurdson, Gregory T., Peipei Tang, and M. Mónica Giusti. 2017. "Natural colorants: Food colorants from natural sources." *Annual Review of Food Science and Technology* 8 (1):261–280. doi: 10.1146/annurev-food-030216-025923.

Singh, Davinder Pal, Jasvirinder Singh Khattar, Alka Rajput, Rajni Chaudhary, and Ramsarup Singh. 2019. "High production of carotenoids by the green microalga *Asterarcys quadricellulare* PUMCC 5.1.1 under optimized culture conditions." *PLoS One* 14 (9):e0221930. doi: 10.1371/journal.pone.0221930.

Singh, Jyoti, and Dolly Wattal Dhar. 2019. "Overview of carbon capture technology: Microalgal biorefinery concept and state-of-the-art." *Frontiers in Marine Science* 6. doi: 10.3389/fmars.2019.00029.

Singh, Rabinder, Chetan Paliwal, Asha A. Nesamma, Alka Narula, and Pannaga P. Jutur. 2020. "Nutrient deprivation mobilizes the production of unique tocopherols as a stress-promoting response in a new indigenous isolate *Monoraphidium* Sp." *Frontiers in Marine Science* 7. doi: 10.3389/fmars.2020.575817.

Sivaramakrishnan, Ramachandran, and Aran Incharoensakdi. 2018. "Utilization of microalgae feedstock for concomitant production of bioethanol and biodiesel." *Fuel* 217:458–466. doi: 10.1016/j.fuel.2017.12.119.

Smee, Donald F., Kevin W. Bailey, Min-Hui Wong, Barry R. O'Keefe, Kirk R. Gustafson, Vasiliy P. Mishin, and Larisa V. Gubareva. 2008. "Treatment of influenza A (H1N1) virus infections in mice and ferrets with cyanovirin-N." *Antiviral Research* 80 (3): 266–271. doi: 10.1016/j.antiviral.2008.06.003.

Spolaore, Pauline, Claire Joannis-Cassan, Elie Duran, and Arsène Isambert. 2006. "Commercial applications of microalgae." *Journal of Bioscience and Bioengineering* 101 (2): 87–96. doi: 10.1263/jbb.101.87.

Sun, Xiao-Man, Lu-Jing Ren, Quan-Yu Zhao, Xiao-Jun Ji, and He Huang. 2018. "Microalgae for the production of lipid and carotenoids: A review with focus on stress regulation and adaptation." *Biotechnology for Biofuels* 11 (1):272. doi: 10.1186/s13068-018-1275-9.

Sydney, Eduardo Bittencourt, Alessandra Cristine Novak Sydney, Júlio Cesar de Carvalho, and Carlos Ricardo Soccol. 2019. "Microalgal strain selection for biofuel production." In *Biofuels from Algae (Second Edition)*, edited by Ashok Pandey, Jo-Shu Chang, Carlos Ricardo Soccol, Duu-Jong Lee and Yusuf Chisti, pp. 51–66. Elsevier.

Tang, Doris Ying, Kuan Shiong Khoo, Kit Wayne Chew, Yang Tao, Shih-Hsin Ho, and Pau Loke Show. 2020. "Potential utilization of bioproducts from microalgae for the quality enhancement of natural products." *Bioresource Technology* 304:122997. doi: 10.1016/j.biortech.2020.122997.

Tolpeznikaite, Ernesta, Vadims Bartkevics, Modestas Ruzauskas, Renata Pilkaityte, Pranas Viskelis, Dalia Urbonaviciene, Paulina Zavistanaviciute, Egle Zokaityte, Romas Ruibys, and Elena Bartkiene. 2021. "Characterization of macro- and microalgae extracts bioactive compounds and micro- and macroelements transition from algae to extract." *Foods* 10 (9):2226.

Veerabadhran, M., D. Gnanasekaran, J. Wei, and F. Yang. 2021. "Anaerobic digestion of microalgal biomass for bioenergy production, removal of nutrients and microcystin: current status." *Journal of Applied Microbiology* 131 (4):1639–1651. doi: 10.1111/jam.15000.

Villaró, Silvia, Martina Ciardi, Ainoa Morillas-España, Ana Sánchez-Zurano, Gabriel Acién-Fernández, and Tomas Lafarga. 2021. "Microalgae derived Astaxanthin: Research and consumer trends and industrial use as food." *Foods* 10 (10):2303.

Wang, Hui-Min David, Ching-Chun Chen, Pauline Huynh, and Jo-Shu Chang. 2015. "Exploring the potential of using algae in cosmetics." *Bioresource Technology* 184:355–362. doi: 10.1016/j.biortech.2014.12.001.

Wang, Jianlong, and Wei Wan. 2009. "Factors influencing fermentative hydrogen production: A review." *International Journal of Hydrogen Energy* 34 (2):799–811. doi: 10.1016/j.ijhydene.2008.11.015.

Wang, Jianlong, and Yanan Yin. 2017. *Biohydrogen Production from Organic Wastes*. Springer.

Winwood, Robert J. 2013. "Recent developments in the commercial production of DHA and EPA rich oils from micro-algae." *OCL* 20 (6):D604.

Xiao, Rui, and Yi Zheng. 2016. "Overview of microalgal extracellular polymeric substances (EPS) and their applications." *Biotechnology Advances* 34 (7):1225–1244. doi: 10.1016/j.biotechadv.2016.08.004.

Yap, Jiunn Kwok, Revathy Sankaran, Kit Wayne Chew, Heli Siti Halimatul Munawaroh, Shih-Hsin Ho, J. Rajesh Banu, and Pau Loke Show. 2021. "Advancement of green technologies: A comprehensive review on the potential application of microalgae biomass." *Chemosphere* 281:130886. doi: 10.1016/j.chemosphere. 2021.130886.

Yeong, Tong Kai, Kailin Jiao, Xianhai Zeng, Lu Lin, Sharadwata Pan, and Michael K. Danquah. 2018. "Microalgae for biobutanol production – Technology evaluation and value proposition." *Algal Research* 31:367–376. doi: 10.1016/j.algal.2018.02.029.

Zárate, Rafael, Nabil el Jaber-Vazdekis, Noemi Tejera, José A. Pérez, and Covadonga Rodríguez. 2017. "Significance of long chain polyunsaturated fatty acids in human health." *Clinical and Translational Medicine* 6 (1):e25. doi: 10.1186/s40169-017-0153-6.

Zodape, Sudhakar Tukaram, V. J. Kawarkhe, Jinalal Shambhubhai Patolia, and A. Warade. 2008. "Effect of liquid seaweed fertilizer on yield and quality of okra (*Abelmoschus esculentus* L.)." *Journal of Scientific & Industrial Research* 67:1115–1117.

Zullaikah, Siti, Adi Tjipto Utomo, Medina Yasmin, Lu Ki Ong, and Yi Hsu Ju. 2019. "9 — Ecofuel conversion technology of inedible lipid feedstocks to renewable fuel." In *Advances in Eco-Fuels for a Sustainable Environment*, edited by Kalam Azad, pp. 237–276. Woodhead Publishing.

6 Algal Metal Remediation for Contaminated Source

Alka Rani and Khem Chand Saini

6.1 INTRODUCTION

Despite improving the economy and fulfilling human needs, industrial development and improper disposal of wastewater have raised significant concerns about the availability of clean water (Kurniawan et al. 2020). Heavy metals (HMs) are the most threatening constituents in industrial wastewater because of their high toxicity and low degradability in nature, causing severe health concerns to humans, animals, and other living organisms. Nearly 80% of the untreated wastewater is discharged into water bodies, increasing the scarcity of freshwater and ultimately leading to global water stress (Jones et al. 2021). By 2025, approximately 60% of the world's population will suffer from a water crisis (Ungureanu et al. 2020). Industries such as dyes, mining, electroplating, automobiles, textiles, batteries, and many others also release HMs into the environment, from where HMs are dumped into the soil, air, and water bodies.

HMs such as copper (Cu), nickel (Ni), zinc (Zn), mercury (HG), iron (Fe), arsenic (As), chromium (Cr), lead (Pb), cadmium (Cd), and manganese (Mn) enter the food chain and, gathered in the human body, cause side effects on human health and influence the fauna and flora. Therefore, environmental deprivation forced the invention of remediation approaches. So, it is required to degrade metal ions and decontaminate polluted water before releasing into the environment.

Plentiful techniques, such as ion exchange, photocatalysis, electrochemical treatment, flotation, coagulation/flocculation, solvent extraction, adsorption, and biological sludge, have been used for metal ion abstraction. Most techniques possess several challenges to removing metals persistent in the environment. Therefore, modern techniques such as biological methods are used, which are easy to apply and are economical with maximum removal efficiency for the HMs (Anastopoulos and Kyzas 2015). Using biomass for the bioremediation of HMs offers countless benefits, including wide availability in nature, reusability, maximum metal uptake, cost-effectiveness, and the probability of their usage in batch and continuous modes. The term phycoremediation refers to using algae to remediate water pollutants. Algae constitute the base of energy for the food web of

DOI: 10.1201/9781003455950-6

all aquatic organisms. Apart from this, algae improve water quality, are used in sewage treatment and biomass generation, and produce clean fuels. Algae can sequester enormous volumes of CO_2, which are used as the best feedstocks (Rani et al. 2021b). The bioremediation of HMs using algae offers several benefits over other remediation methods, such as cost-effective and environmentally friendly tactics. This study aims to outline HMs as a latent pollutant in the environment, the potential of algae in phytoremediation, the core mechanism of HMs accretion, and also discuss the approaches to improve the HM removal efficiency of algae.

6.2 SOURCES OF HEAVY METALS (HMS)

Globally, HMs constitute one of the most hazardous pollutants causing environmental damage. Generally, about 67 chemical elements out of 118 with atomic numbers over 20 and a density of more than 5 g cm^{-3} are regarded as HMs. Automobiles, smelting of ores, batteries, mining, and agricultural waste such as fertilizers and pesticides release HMs into the environment (Abdullahi et al. 2021). Anthropogenic and natural sources, including the geological weathering of rocks and soils, also contribute to the release of HMs. The primary sources of HMs pollution are industrial wastewater, landfill leachates, municipal wastewater, and urban runoff, particularly in metal furnishing and electroplating industries. HMs are nonbiodegradable and negatively impact the ecosystem (Znad et al. 2022). Excessive uptake of lead (Pb), cadmium (Cd), chromium (Cr), and manganese (Mn) can elevate the transpiration rates (Chandra and Kang 2016). HMs increase the reactive oxygen species (ROS) that can cause detriment to fishes and other aquatic life (Berni et al. 2019; Yu et al. 2020). Nowadays, HMs above the acceptable amount in the soil, air, and water can adversely affect the ecological diversity of the Earth (Jaiswal et al. 2018). Aquatic ecosystem contaminated with HMs has intensified since the foundation of the industrial revolution, leading to biomagnification (Moiseenko and Gashkina 2020; Shah 2021).

Although traces of some metals like Fe, Cu, and Zn are essential for biological functions; yet, excessive quantity above the stated thresholds can cause serious health issues. Therefore, WHO has given directions regarding the permitted values of HMs content in clean water and wastewater, summarized in Table 6.1, along with the sources and harmful effects of the HMs.

6.3 IMPACT OF HMS

Although HMs do not contribute to metabolism, they accumulate via numerous mechanisms, including bioconcentration, bioaccumulation, and biomagnification. Organisms initiate a protective mechanism against the toxicity of HMs when exposed. Succeeding a threshold level, HMs cause toxicity and adverse impact on soil, air, water, aquatic life, and human beings. The impact of HMs on humans is depicted in Table 6.1.

TABLE 6.1
Sources of HMs, Their Acceptable Level, and Their Impact on Human Health.

Heavy metal	Source	Health effects	Permissible level (mg L^{-1})	Reference
Pb	Battery manufacturing industries, along with the ceramic and glass industries	Phytotoxic, metabolic poison, damage to the nervous, brain and circulatory systems	0.05	(Bhat et al. 2019)
Cd	Mining, smelting, production of fertilizer, metal plating and fuel combustion	Carcinogenic, bone lesions, anemia, bronchitis, emphysema, acute toxic effects in children	0.003	(Suhani et al. 2021)
Cu	Ore, mining, smelting	Liver damage and lung cancer	1.3	(Kumar et al. 2021)
Cr	Textile, steel, and ceramics industries	Vomiting, severe diarrhea, lung cancer and hemorrhage,	0.05	(Pratush et al. 2018)
Hg	Ore, paper, battery, pharmaceutical	Neurological diseases, paralysis, blindness	0.01	(Beckers and Rinklebe 2017)
As	Coal combustion and manufacturing processes	Lung dysfunction and cancer	0.01	(Rathi and Kumar 2021)
Ni	Mining, refining, battery, paint, and glass manufacturers	Chronic asthma, melanosis, lung cancer, keratosis, and dermatitis	0.015	(Luo et al. 2022)
Zn	Mining, galvanization, and coal industries	Neurological damage, gastrointestinal disorder, lethargy, nausea, and loss of appetite	3.0	(Aykol et al. 2003; Chakraborty et al. 2022)

6.3.1 Effects on Soil

Emission of HMs from industrial activities, mine tailings, paints, and use of pesticides, fertilizers, sewage sludge, wastewater irrigation, and spilling of petrochemicals can lead to soil contamination. HMs are neither degraded by microbial nor chemical mechanisms; their persistence in the soil is a serious problem due to their accumulation in the food chain. HMs alter soil's porosity, pH, color, texture, and natural chemistry, ultimately impacting soil quality. Humans are exposed to HMs primarily by accumulating these metals at high concentrations by ingesting foods grown in polluted soil. HMs are bio-accumulated in plants via absorption through the roots leading to photosynthesis inhibition, enzyme inhibition, decreased algal growth and loss of flagella, and increased permeability of the plasmalemma, which result in the breakdown of the cell membrane and loss of cell solutes, etc. (Alahabadi et al. 2017; Srivastava et al. 2017; Abdullahi et al. 2021).

6.3.2 Effects on Water

Industrialization and urbanization are major culprits for the increased water pollution, leading to poor water quality that imposes serious health issues on humans and other ecosystems. For example, humans feeding at the highest level of the food chain are more prone to health issues as the concentration of HMs increases in the food chain. HMs elevate reactive oxygen species (ROS) that might injure the fishes and aquatic life.

6.3.3 Effects on Air

HMs including group 1 metals, that is, Cd, Cu, Pb; group 2 metals, that is, Mn, Cr, Ni, V, and Zn; and group 3 metals, that is, Al, Ca, Fe, Na, K, Ti, and Mg, originating from industries, vehicles, and natural resources produce particulate matters, which cause eye irritation, cardiovascular diseases, respiratory infections, haze, premature mortality, eutrophication, and formation of acid rain.

6.3.4 Effects on Aquatic Ecosystem

The industrial revolution escalated the contamination of aquatic ecosystems with HMs. Nonbiodegradable HMs may present in the aquatic ecosystems, accumulate in the aquatic life, and raise the trophic levels, interfering with growth, development, and metabolism, which leads to numerous chronic impacts.

6.4 PHYCOREMEDIATION: AN ALGAL MECHANISM TO ERADICATE POLLUTION

Conventional techniques employed for HM removal include membrane filtration, chemical precipitation, electrochemical, floatation, ion exchange, and coagulation–flocculation methods. Moreover, these procedures have some drawbacks, such as costly maintenance and operational expenses and the generation of secondary toxic waste due to the formation of poisonous slurry.

Algae are photoautotrophic microorganisms, accounting for over 40% of the global key producers of the food chain (Akram and Suleria 2022). Algae are ubiquitous and well-adapted to live in many habitats, ranging from sea and freshwater to marshy and constructed wetlands and domestic to industrial effluents (Amaro et al. 2011). Algae can grow on non-arable lands or in wastewater by using sunlight and atmospheric CO_2 as the sources of energy and carbon, respectively. Algae have a comparatively higher growth rate, and cell division justifies their wide occurrence. The extraordinary withholding potential of HMs makes them an ideal podium to develop advanced technologies for the bioremediation of HMs (Ranjbar and Malcata 2022).

Up to 10% of biomass algae can retain HMs, due to their vast surface-to-volume ratio, along with their high metal uptake capacity, metabolization, binding, packing approaches, ready availability, environmentally friendliness, reusability/recyclability, perpetual occurrence, efficiency, non-toxic waste generation, and being beneficial in batch and continuous systems (Tripathi et al. 2019). However, the toxicity level of HMs varies among the species of the identical cluster; for example, a few diatoms can withstand 1.5–10 μM of dissolved Cu, whereas many Chlorophyta species can endure 15 μM Cu (Biswas and Bandyopadhyay 2017; Tripathi and Poluri 2021b).

The tendency of algae to accumulate metals within the tissues has led to an extensive usage of them as bioindicators for the availability of metals. Chlorophyta and Cyanophyta act as hyper-accumulators of B and As by absorbing them from the polluted sites because their cell walls have sufficient levels of sulphated polysaccharides and alginates; due to this, metals have higher affinity. Likewise, metals absorbed by *Chlamydomonas reinhardtii, Chlorella salina, C. sorokiniana, C. vulgaris, C. miniata, Spirulina platensis, Stichococcus bacillaris, Scenedesmus quadricauda, S. abundans, Spirogyra spp., S. subspicatus, Stigeoclonium tenue*, and *Porphyridium purpureum* include Al, Sr, Cu, V, As, Pb, Cd, Mg, Ni, Fe, K, Co, Ca, Mn, Mo, Se, and Zn (Biswas and Bandyopadhyay 2017; Tripathi and Poluri 2021b).

The application of algae for wastewater and industrial waste bioremediation accomplished the chance to produce an extensive range of bioactive compounds, such as proteins, pigments, and vitamins, which can be eventually used as food, feed additives, cosmetic ingredients, for biofuel production, as a nutritional supplement, and can also be used as a value-added by-product, such as biochar and biofertilizers. Algae follow numerous mechanisms, such as bioaccumulation, biosorption, and compartmentalization, for the remediation of toxic metals.

6.4.1 Extracellular Uptake (Biosorption)

Recently, biosorption appeared as an alternate and encouraging approach for removing HMs. In the mid-20th century, the initial research was published on biosorption. Meanwhile, notable contributions have been made to define and generate effective and low-cost biosorption biomaterials (Ordóñez et al. 2023). It is a clean, simple, alternate method for recovering metals in diluted solutions requiring minor investment. Biosorption employs inexpensive and renewable sorbents obtained from the cultures of bacteria, algae, fungi, and aquaculture wastes and agro-industrial (Beni and Esmaeili 2020).

Algae follow specific molecular mechanisms leading to the HM biosorption by the cell wall, extracellular polymeric substances (EPSs) such as biomineralization, ion exchange, hydroxide condensation, complexation, chelation, covalent binding, redox interaction, utilizing electrostatic, precipitation of unfathomable metal complexes, and hydrophobic interactions among the negatively charged functional groups such as polysaccharides, hydroxyl, carboxyl, proteins, amino, imidazole, thioether, sulfate, phosphate, and phenol groups existing on the algal cell wall and positively charged HM cations (Figure 6.1) (Sayadi and Shekari 2017; Danouche et al. 2021).

Therefore, algae might be considered cost-effective biosorbent ingredients, as no specific treatment is required (only base or acid modification is needed to alter or improve the negative charge) (Anastopoulos and Kyzas 2015). The algal cell wall's extracellular uptake/bioabsorption ability is a metabolism-independent method; therefore, it is pretentious by various environmental aspects such as contact time, temperature, pH, and HM concentration, as described in Table 6.2 (Salam 2019).

TABLE 6.2
Algal Species Used for the Removal of the Heavy Metals.

Algae	Growth condition	pH	Temperature	Contact time	Metals used for bioremediation	Metal removal efficiency	Reference
Stichococcus bacillaris CCAC 1896 B and CCAC 1898 B	Porous Substrate Bioreactor (PSBR) at 23°C and light intensity of 20–40 μmol photons m^{-2} s^{-1}	7.0 ± 0.1	25 ± 2 °C	12 h	Zn 2.0–3.0 L^{-1} from synthetic wastewater	Zn 15–19 mg g^{-1}	(Li et al. 2015)
Nannochloropsis sp.	Lab-scale growth, modified with silica and then coated with magnetite particle, and contact time 90 min	7	–	90 min	Cu (II) cations (0.1–1.0 mM)	Cu 56 mg g^{-1} (87.5%)	(Buhani et al. 2021)
Nannochloropsis oculata	Laboratory experiments and growth	8.5	25°C	–	Cu	89.29 ± 1.92% Cu removed by metabolism and 10.70 ± 1.92% adsorption	(Martínez-Macias et al. 2019)
Chlorella vulgaris	Laboratory-scale growth	4.0 and 6.0	25°C	–	SO_4 = 4,520 mg L^{-1}, Fe = 788 mg L^{-1}, Mn = 19.4 mg L^{-1}, Al = 310 mg L^{-1}	99.9% removal efficacy	(Brar et al. 2022)

(Continued)

TABLE 6.2
(Continued)

Algae	Growth condition	pH	Temperature	Contact time	Metals used for bioremediation	Metal removal efficiency	Reference
Spirulina platensis	Laboratory growth and dried at 100°C in oven	5.5, 6.0, and, 7.0	27° ± 0.5C	5–100 min	Al, Ni, and Cu	95.0 ± 0.3% Al, 87.0 ± 0.2% Ni, 63.0 ± 0.3% Cu	(Almomani and Bhosale 2021)
Chlorella vulgar		6				87.0% ± 0.2 Al, 79.1 ± 0.4% Ni, 80.0 ± 0.2% Cu	
Spirogyra verrucosa	Laboratory-scale growth	5	30°C	120 min	Mn (II) ions	Mn 40.7 mg g^{-1} (80.2%)	(Bansod and Nandkar 2016)
Scenedesmus sp.	Laboratory conditions	–	27±2°C	–	Cr, Cu, Pb, Zn	Pb: 75–98%, Cu:73.2–98%, Zn: 65–98%, Cr:81.2–96%	(Ajayan et al. 2015)
Scenedesmus obliquus FACHB-12	Laboratory conditions	7.1 ± 0.2	25°C	25 min	Cd (II)	Cd removal by 91.27%	(Ma et al. 2021)

Meanwhile, the biosorption of HMs toward the EPS is controlled through the cell by changing the properties of surface biopolymers depending on the type of metabolic stress, such as metal toxicity. Both EPS and cell wall act as a protective layer preventing the harmful impact of HMs, which enter the intracellular compartment without disturbing cellular integrity. Furthermore, the algal biomass' chemical modifications (such as oxidation with potassium permanganate or crosslinking with epichlorohydrin) can improve biosorption selectivity (Ubando et al. 2021).

Other mechanisms for extracellular bioremediation of HMs include producing specific organic acids, metal-chelating proteins, and consequent endocytosis of the formed organometallic complexes (Danouche et al. 2021). Transportation of HMs into the cells requires a particular carrier, known as a metal transporter. However, HM toxicity can be overcome inside the cytoplasm by maintaining or removing its adequate level. Metal efflux pumps were utilized to actively transport HMs into and out of the cell (Mantzorou et al. 2018).

Recently, the production of unique algae-based adsorbents using molecular, chemical, nanoparticles, and extraction modes has attained unexpected improvements amidst elucidating adsorption mechanisms and metal-binding capacity (Cheng et al. 2019). High uptake capacity and cheap and renewable algae-based biosorbents have gained attention for HM removal (Allahkarami and Rezai 2019; Emparan et al. 2020). After the completion of the biosorption process, the algal biomass can be transformed to bio-char or can be landfilled with a lesser environmental effect (Nair et al. 2019).

Cardoso et al. (2017) elucidated the biosorption capacity of alginate extracted from *Sargassum filipendula* with a maximum adsorption peak of Ni and Zn. Alginate extracted from *Callithamnion corymbosum* sp. exhibited maximum biosorption capacity upsurges as for mentioned order: Cu(II) (64.52 mg g^{-1}) > Zn(II) (37.04 mg g^{-1}) > Co(II) (18.79 mg g^{-1}) at operative pH of 4.4, 2.0 g L^{-1} biosorbent dose and a 22°C temperature (Lucaci et al. 2020).

Ali et al. (2020) evaluated *Cystoseira barbata* and *Hydroclathrus clathratus* (brown seaweed) for Cr and Pb biosorption and established an inverse effect of metal concentration on metal uptake. They proved that 10 g L^{-1} of algae achieved extreme biosorption efficiency at a contact time of 120 Min and pH 5 with the enhanced uptake of Cr and Pb biosorption, that is, 4.61, 7.30 mg g^{-1} on *C. barbata* and 4.97, 7.19 mg g^{-1} on *H. clathratus*, respectively. Likewise, Romera et al. (2007) evaluated the biosorption of Ca, Cu, Ni, Pb, and Zn from aqueous solutions using six algal species and reported that the unsurpassed outcomes were attained with *Fucus spiralis*. Moghazy (2019) studied *Chlamydomonas* sp. to eliminate methylene blue dye present in aqueous solutions with 80.8% elimination efficacy at a 1.5 g L^{-1} biosorbent dose. Sayadi and Shekari (2017) stated 85.6% removal of Cd by 2 mg L^{-1} of *Spirogyra* at a contact time of 30 Min and pH of 6. Fawzy (2020) explained that *Codium vermilara* is an excellent biosorbent of Cu^{2+} ions with more than 85% removal efficiency at 0.75 g L^{-1} algae dose, 5.3 pH, and contact time of 70.5 min. Indhumathi et al. (2018) reported Cu biosorbing by *Chlorella vulgaris* with a 90.3% rate of recovery at pH 7 and a contact time of 105 min with 20 mg L^{-1} of primary Cu concentration. Similarly, Jaafari and Yaghmaeian (2019) studied that *C. coloniales* could be an outstanding, cost-effective biosorbent and can confiscate approximately 95% Cd, As, Fe, Co, and

Cr present in wastewater. Cheng et al. (2019) investigated the deviations in the surface properties and the enhancement in the percentage of HM removal after using the chemical cure to the algal strains.

6.4.2 Intracellular Uptake (Bioaccumulation and Compartmentalization)

Bioaccumulation is a process in which HM ions are conveyed transversely the cell membranes via active and passive transportation systems and collected inside the cells. Both intracellular and extracellular metal-binding tactics, including physical adsorption, chelation, complexation, and ion exchanges, have been utilized by algae to reduce HM toxicity, thereby transforming toxic metals into least toxic or nontoxic forms (Mantzorou et al. 2018; Mustapha and Halimoon 2015).

Various approaches, like the binding of HMs to a specific intracellular organelle, its transportation to certain cellular constituents like vacuoles/polyphosphate bodies, flushing out HMs using an efflux pump into the solution, and synthesizing class III metallothioneins or phytochelatins (PCs), are used for HMs detoxification by the algae (Tripathi and Poluri 2021a). PCs are small-size peptides with molecular weights ranged about 2 to 10 kDa and have the ability to bind HMs and are produced by the constitutive trans-peptidase enzyme known as phytochelatin synthase (Torres et al. 2008) from glutathione, hydroxymethyl-glutathione, and γ-glutamylcysteine phytochelatin synthase required post-translational activation by HMs as described in Figure 6.1 (Henriques et al. 2017).

Apart from plants, algae can also synthesize phytochelatins quickly without depending on de novo protein synthesis (Cobbett and Goldsbrough 2002). Metals like Ag, As, Cu, Au, Pb, Cd, Hg, Sn, and Zn act as potential activators and activate phytochelatin synthase *in vivo* and *in vitro*. Gekeler et al. (1988) isolated Cd-binding phytochelating peptide complex $(\gamma\text{-Glu-Cys})_n$-Gly, $n = 2–5$, from *Chlorella fusca*. Whereas Gómez-Jacinto et al. (2015) studied microalgae *Chlorella sorokiniana* which was exposed to diverse concentrations of Hg and two Hg-binding phytochelatins, that is, Cu- and Hg-MTs and Zn- and Hg-GSH were extracted. Hirata et al. (2001) exposed marine alga, *Dunaliella tertiolecta*, to Zn^{2+}, producing an enhanced level of PCs, that Is, 200 μM, which was higher than that obtained after exposure to the 400 μM Cd^{2+}. Likewise, heavy metals Cd^{2+}, Hg^{2+}, and Ag^{+} induced peptide synthesis in *Chlamydomonas reinhardtii*, which sequestered 70% of the total Cd (Howe and Merchant 1992; Hu et al. 2001).

6.5 STRATEGIES TO IMPROVE THE BIOREMEDIATION ABILITY OF ALGAE

The major molecular tools were developed using *Chlamydomonas reinhardtii*, as a model microalga. Following recent advancements in genome sequencing and gene reconstruction, many strategies such as gene targeting, transformation, and bioinformatic tools and technologies have easier manipulating metabolic pathways in microalgae (Kumar et al. 2020). Gene editing approaches, including clustered regulatory interspaced short palindromic repeat (CRISPR)/Cas systems, transcription activator-like endonucleases (TALENs), zinc-finger nucleases (ZFNs), RNA interference (RNAi), Cre/*loxP* recombination systems, and integrated cloning, along with a number of vectors, promoters, regulatory elements and reporter genes, were used for the bioengineering of microalgae (Fajardo et al. 2020). Following the path of metagenomics, wastewater microbiota of a swine lagoon was analyzed using high-throughput sequencing methods, which unlocked the secret of N and P removal by *Chlorella* spp., along with the information regarding the interaction of various microbial communities with the ecosystem, leading to the advancement in the remediation efficacy of the microalgae (Ye et al. 2016; Sharma et al. 2022). Various genetic-engineering approaches to develop algal HMs bioremediation ability are described in Table 6.3.

TABLE 6.3
List of Transporter Families and Genetic Engineering Targets Foreseen to Improve the Capacity of Microalgae to Bioremediate HMs.

Approach	Targeted genes to improve HM bioremediation ability
Metal transporters	NRAMP, ZIP, MATE, CTR, PTA, AQP, ZRT, IRT, CDF, FPN, P1B-type ATPases, and Ccc1/VIT1
Transporters concerned with the sequestration of heavy metals in the subcellular compartments	
	ABC and CE
Transporters concerned with the efflux of heavy metals	
	ABC, ACR3, NRAMP, CE, MATE, and MTP
Metal chelation	PCs, MTs, VTC, GSH, PPK, PPX, Pro (P5CS), Cys (HAL2), His (HISN3), Ser (SDC1), glycine-betaine, and organic acids
Metal biotransformation	arsC, ChrR, CYPs CrACR2s, SMT, and MerA/B/P/C/T/F
Metal stress regulation	AP2, MYB, MTF-1, c2H2, bHLH, HB, yaBBY, bZIP, SBP, WRKY13, ABI5, CRR1, GATA, CKs (IPT and CKX1), GA, ABA, JA, BRs, ET, SA, miRNAs (miR319, miR390, miR393, miR171, miR395, miR397, miR408, miR398, and miR857)
Cell-surface bioengineering	PCs, MTs, GST, 6x-His, CXXEE, ModE MerR, and NikRm
Oxidative stress regulation	SOD, POD, GST CAT, HSPs, GPX, Trx, and carotenoids

Abbreviations: zinc-regulated transporters (ZRT), iron-regulated transporters (IRT), and Fe-transporter gene (FTR), ATP-binding cassette transporters (ABC transporters), MATE (multidrug and toxic compound extrusion); Zrt-Irt-like proteins (ZIP); metal tolerance protein (MTP); natural resistance-associated macrophage proteins (NRAMP); Cu transporter (CTR); and cation diffusion facilitators (CDF), arsenite efflux transporter (ACR3); FerroPortiN (FPN), the phosphate transporters (PTA), P1B-type ATPases, Ca^{2+}-sensitive cross-complementer1/Vacuolar iron transporter1 (Ccc1/VIT1), heat shock protein (HSPs), aquaglycoporin (AQP).
Source: Adapted from Ranjbar and Malcata (2022.

6.5.1 Algal Metal Transportation

Metal transporters were categorized into two groups in the *C. reinhardtii*: Group A contains transporters, such as zinc-regulated transporters (ZRT), natural resistance-associated macrophage proteins (NRAMP), Cu-transporter (CTR) families, Zrt-Irt-like proteins (ZIP), iron-regulated transporters (IRT), and Fe-transporter gene (FTR) confirming the metal transportation from the extracellular matrix toward the cytosol (i.e., HM uptake) and from cytosol toward vacuoles (i.e., HM storage) as described in Figure 6.1 (Blaby-Haas and Merchant 2012). These transporters utilized the trans-membrane proton gradient to mediate divalent cation transport to the cytoplasm (Nevo and Nelson 2006). Under high concentrations of Cd, NRAMP1 overexpression was reported in *Auxenochlorella protothecoides* (Lu et al. 2019). Likewise, upregulating the genes encoding ZIP, CTR, and NRAMP1 transporters in *Dunaliella acidophila* enhances Cd uptake (Puente-Sánchez et al. 2018). ZIP transporter genes play a role in the uptake and confiscating of Cd in *C. reinhardtii*; upregulated phosphate transporters (PTA) and aquaglycoporin (AQP) increased As uptake in *Microcystis aeruginosa* and *Chlamydomonas eustigma* ((Beauvais-Flück et al. 2017; Hirooka et al. 2017; Wang et al. 2019). Despite their extraordinary

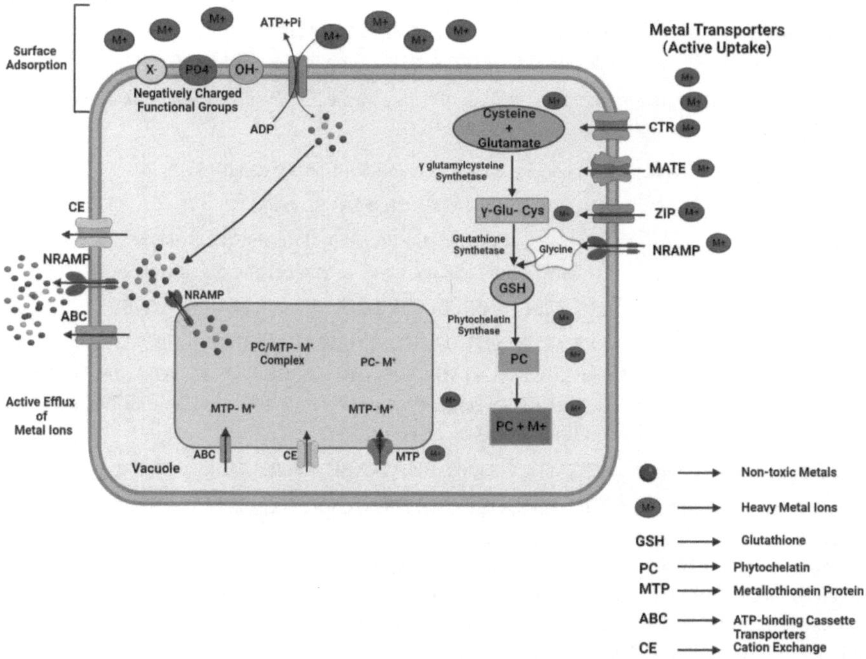

FIGURE 6.1 Bioaccumulation and detoxification of heavy metals in algae using transporters. *Abbreviations*: MTP, metal tolerance protein; ZIP, Zrt-Irt-like proteins; NRAMP, natural resistance-associated macrophage proteins; MATE, multidrug and toxic compound extrusion; ABC transporters, ATP-binding cassette transporters; CTR, Cu transporter.

potential, only a few researches were conducted to increase the phycoremediation via metal transporter engineering. For instance, Ibuot et al. (2017) revealed the metal-tolerant protein (MTP) CrMTP4 from the Mn-CDF clade of the cation dispersal facilitator group of metal transporters in the *C. reinhardtii*. MTP remarkably raised bioaccumulation efficiency and Cd toxicity resistance due to amplified transfer and Cd storage in the acidic vacuoles. MTP1 present in *C. reinhardtii* genome, restricted in the vacuolar membranes, exhibits a perilous role in Cd detoxification and Zn homeostasis (Blaby-Haas and Merchant 2012).

Group B transporters include P1B-type ATPases members belonging to FerroPortiN (FPN) families, cation diffusion facilitators (CDF) families, and the Ca^{2+} sensitive cross-complementer1/Vacuolar iron transporter1 (Ccc1/VIT1). Group B transporters result in the reduction of metal concentration within cytosolic via organometallic complexes and active metal ions efflux into the extracellular environment (Tripathi et al. 2019). The P1B-type ATPases, known as heavy metal ATPases (HMAs), are important in metal trafficking through cell membranes (Rosenzweig and Argüello 2012). Both group A and group B transporters are responsible for the metals' internalization within the algal cell and also play a role in cytoplasmic uptake of the metals, thus maintaining metal ion concentrations at equilibrium within the cell. In contrast, Group B transporters assimilate the metals to intracellular compartments like vacuoles.

Ibuot et al. (2017) proved that the overexpression of AtHMA4 (Zn and Cd transporter in plant) into the *C. reinhardtii* retrieved from *Arabidopsis thaliana* can result in increased bioaccumulation and uptake of Zn and Cd in the transformed microalgae. Another study led by Ramírez-Rodríguez et al. (2019) also overexpressed the Acr3 (arsenic (As) hyperaccumulator) confined within vacuolar membrane and acted as an efflux propel leading to 1.5- to 3-fold rise in the capacity of As removal. Plasma membrane H^+-ATPase (PMA) expression in *C. reinhardtii* resulted in a 3.2-fold-raised photoautotrophic yield of CO_2 concentrations from the flue gas that is highly toxic (Choi et al. 2021) thus emphasizing the extraordinary role of efflux pumps in microalgal cell bioengineering.

6.5.2 Metal Chelation

Metallothioneins (MTs), a ubiquitous class of enzymatically synthesized (Class III) or genetically encoded (Class I and II) polypeptides, which aid in metal homeostasis and trafficking (Capdevila and Atrian 2011) and contain a higher amount of cysteine (15–35%), a few aromatic amino acids (<10%), and lesser extent of histidine residues, accounted for the higher metal-binding capacity of MTs (Ziller and Fraissinet-Tachet 2018). Class I MTs have two slighter Cys-rich dominions with a bulky spacer domain among them. Whereas low-molecular-weight Class II (6–7 kDa) MT proteins located in the cytosol have three Cys-rich dominions constituting 10–15 residues that control the metal's intracellular concentrations. Class III MTs are called phytochelatins (PCs) which are thiol-containing oligopeptides containing three amino acids, that is, Cys, γ-Glu, and Gly, and are synthesized enzymatically (Gaur and Rai 2001).

Numerous microalgae species exhibited an elevated activity of the PCs biosynthetic enzymes and noticeable upsurge in the PCs supply and GSH when exposed to HMs (Suárez et al. 2010; Li et al. 2021; Abboud and Wilkinson 2013; Gómez-Jacinto et al. 2015). Cd tolerance of *C. reinhardtii* was significantly increased due to the overexpression of a synthetic gene *gshA* which was encoded by the γ-glutamylcysteine synthetase (Piña-Olavide et al. 2020). Furthermore, under Cu toxicity, *CrGNAT* gene encoded an acetyltransferase enzyme, which is involved in the chromatin remodeling and histone methylation, was overexpressed in *C. reinhardtii* resulting in an inconsiderable upsurge in the chlorophyll accumulation, photosynthetic efficiency, and cell population. In contrast, *CrGNAT* knockdown antisense lines displayed sensitivity toward Cu stress. Furthermore, it was supported that pursuing phytochelatin synthase to precise organelles, like sensitivity induced by chloroplast, whereas cytosol targeting can induces forbearance against the As stress (Picault et al. 2006).

In microalgae, heterologous expression of MTs was first reported by Cai et al. (1999); they encouraged that chicken MT-II overexpression in *C. reinhardtii* cell wall lacking mutants amplified Cd tolerance and boosted Cd sequestering up to two-fold in the transgenic microalgae. Some studies evaluated that the metal recovery ability of transgenic microalgae recovered from polluted deposits, utilizing *in situ* sonication, was doubled compared to its wild counterpart (Domínguez-Solís et al. 2004). Such biotechnological methods seem to upsurge microalgal accumulation, intending to improve HM phycoremediation.

6.5.3 Metal Biotransformation

Toxic HMs in cells are digested resulting in less toxic by-products through biotransformation. Chromate reductase (ChrR) in *C. vulgaris* changes Cr (VI) to Cr (III) following an enzymatic response concerning the GSH oxidation, conferring higher Cr toxicity tolerance (Yen et al. 2017; Lee et al. 2017). Similarly, arsenate reductase (CrACR2s) from *C. reinhardtii* reduces arsenate to arsenite using additional electrons that are transported to the glutaredoxin (Yin et al. 2011). Likewise, mercuric reductase expression in *Chlorella fusca*, *Selenastrum minutum*, and *Galdiera sulphuraria* allows Hg^{2+} biotransformation to elemental Hg and metacinnabar (β-HgS) (Kelly et al. 2007). To improve the mercury-detoxifying potential of *Chlorella* spp., mercuric reductase (*merA*) gene isolated from the *Bacillus megaterium* was upregulated, and the hybrids obtained displayed a two-fold escalated Hg^{2+} detoxification than wild type (Huang et al. 2006).

Photoreduction pathway is another HM volatilization mechanism defined in *C. vulgaris* for the Cr bioconversion (Deng et al. 2006), extracellular and intracellular biosynthesis of metal nanoparticles, and reduced contacts with functional biomolecules groups present within and external to the cell (Chaudhary et al. 2020; Hamida et al. 2020; Priyadarshini et al. 2019). Although HM-volatilizing proteins and genes suggest improved efficiency to the progression of phycoremediation, basic pathways for HMs biotransformation remain unknown; therefore, extensive in-depth research is needed.

6.5.4 Oxidative Stress Response Regulation

Under toxic levels of Cu, the biochemical, physiological, and expression gene characteristics of *C. reinhardtii* were assessed, resulting in the inhibition of photosynthesis, diverse chlorophyll content, cell growth, and increased lipid peroxidation (Jiang et al. 2016). Correspondingly, transcriptional analysis of *C. reinhardtii* and *D. salina* exposed to Pb and Cd expressed the genes encoding the antioxidant enzymes that were increased in microalgae exposed to HM stress (Zhu et al. 2019). Along with antioxidant enzymes' upregulation, heat shock proteins (HSPs), thioredoxin (Trx), and carotenoid overexpression were recognized in *C. vulgaris*, *C. reinhardtii*, and *A. protothecoides* in retort to the toxic range of HMs (Jamers et al. 2013; Simon et al. 2008).

Algae have also developed anti-oxidative stress management strategies such as antioxidant enzymes, including superoxide dismutase (SOD), catalase (CAT), peroxidase (POD), glutathione-S-transferase (GST), and glutathione peroxidase (GPX) converting superoxide radicals (O^{2-}) to H_2O_2, and consequently to O_2 and H_2O. In comparison, nonenzymatic antioxidants like GSH and ascorbic acid quench ROS via complexation (Danouche et al. 2021; Rani et al. 2021a; Rani et al. 2022). Likewise, HSPs are molecular chaperones that guard and repair proteins underneath the HM stress (Ireland et al. 2004; Lewis et al. 2001).

6.5.5 Metal Stress Response Regulation

Phytohormones such as auxins, abscisic acid (ABA), cytokinins (CKs), ethylene, gibberellic acid (GA), brassinosteroids (BRs), jasmonic acid (JA), and salicylic acid (SA) are signaling molecules coordinating the cellular response to HMs toxicity. However, their precise mode of action is still unknown; yet, they stop the deprivation of photosynthetic pigments, proteins, and monosaccharides and also trigger antioxidant defence reactions needed for the algal growth during stress circumstances (Pal and Rai 2010; Leszczyszyn et al. 2013).

Gas work as HM stress-tolerants in microalgae due to their impact on ROS networks and photosynthetic pathways. GA-treated *C. vulgaris* exhibited a high capacity of HMs biosorption and raised lipid and fatty acid accumulation (Udayan et al. 2018; Yu et al. 2016). The intracellular ABA level was increased by 111% in *A. obliquus* on Pb exposure; exogenous use of auxins, CKs, BR, and SA upgraded tolerance to HM but reduced endogenous ABA, which enhanced 2.1-fold cell growth in *Scenedesmus quadricauda*, accumulating 12% of saturated fatty acids during nitrogen famishment (Sulochana and Arumugam 2016). Under the HM stress, the ABA treatments show similar results in the *Chlorella saccharophila, C. saccharophila*, and *E. gracilis* (Noble et al. 2014).

6.5.6 Bioengineering of Algal Cell Surface

As aforementioned, the ability of cell wall and extracellular algal components to adsorb metals plays a crucial role in detoxifying HMs. Many studies suggested that cell surface metallosorbing properties might be efficiently altered to increase the HM phycoremediation ability of algae and is termed cell surface engineering (Danouche et al. 2021; Wang et al. 2021). Such engineering is possible by the expression and fusion of metal-binding proteins, that is, PCs and MTs, with an affixing motif on the cell surface (Yang et al. 2015; Kuroda and Ueda 2011).

Transgenic *C. reinhardtii* exhibited an increased Hg removal capacity due to membrane-anchored MT polymer compared to the wild type (He et al. 2011). Cell surface engineering permits the retrieval of valuable metals from wastewater and is a less expensive approach that improves the discernment for the target metal as related to conservative methods.

6.6 CONCLUSION AND FUTURE PERSPECTIVE

HM contamination is a serious apprehension due to its high toxicity to human and all biotas. Numerous algal species have been acknowledged as encouraging aspirants for removing or detoxifying HMs and are economic alternatives for physicochemical remediation practices. Bioaccumulation and biosorption methods can be employed to remove HMs. Algae possess a different mechanism for HM sequestering, acting as potential biosorbents and hyperaccumulators due to their high affinity for heavy metals.

The major obstacle is the low concentration attained by the algae, which reduces the process efficiency. Therefore, the initial issue is to improve algal strains bearing advanced confrontation contrary to the metallic stresses and concurrently increase biomass harvest in wastewater. Numerous genetic engineering methods have been conferred to augment the formation of transgenic algae that express PCs and MTs forming multiplexes with HMs, translocating HMs into the vacuoles that help in phytoaccumulation and removal of a risky components from the contaminated environment. It is essential to carry out genetic studies and modify the algae's genetic makeup to enhance the phycoremediation process. Intermediate products produced after the transformation of HMs need to be observed along with the characterization of the strains. Using metabolic and genetic engineering, the algal gene can be altered to overexpress the protein, lipids, enzymes, etc., used for the remediation. Transgenic algal species can boost the efficiency of the by-products such as proteins, vitamins, and nutraceuticals.

Another major issue is extracting HMs from the bioremediators since they have less desorption efficacy. Therefore, to achieve these, advanced techniques are used, including the use of alginates, which are extracellular polymeric substances used to modify the algal biomass. Meanwhile, molecular-level studies are required in the arena of forbearance of HMs using algae and resistance approaches followed by them. Studies to design techniques that simultaneously detect as well as abate the HMs from the environment are needed.

REFERENCES

Abboud P, Wilkinson KJ (2013) Role of metal mixtures (Ca, Cu and Pb) on Cd bioaccumulation and phytochelatin production by *Chlamydomonas reinhardtii*. *Environmental Pollution* 179:33–38

Abdullahi UA, Khandaker MM, Shaari NEM, Alias N (2021) Biological processes of heavy metals-contaminated environmental remediation: A review. *Journal of Environmental Treatment Techniques* 9 (3):601–608

Ajayan KV, Selvaraju M, Unnikannan P, Sruthi P (2015) Phycoremediation of tannery wastewater using microalgae *Scenedesmus* species. *International Journal of Phytoremediation* 17 (10):907–916. doi:10.1080/15226514.2014.989313

Akram A, Suleria HAR (2022) Functional and nutraceutical significance of macro and micro algae. In *Bioactive Compounds from Multifarious Natural Foods for Human Health*. Apple Academic Press, pp. 119–159

Alahabadi A, EhraMpoush MH, Miri M, Aval HE, YouSefzadeh S, Ghaffari HR, Ahmadi E, Talebi P, Fathabadi ZA, Babai F (2017) A comparative study on capability of different tree species in accumulating heavy metals from soil and ambient air. *Chemosphere* 172:459–467

Ali HS, Kandil N, Ibraheem I (2020) Biosorption of Pb^{2+} and Cr^{3+} ions from aqueous solution by two brown marine macroalgae: an equilibrium and kinetic study. *Desalination and Water Treatment* 206:250–262

Allahkarami E, Rezai B (2019) Removal of cerium from different aqueous solutions using different adsorbents: A review. *Process Safety and Environmental Protection* 124:345–362

Almomani F, Bhosale RR (2021) Bio-sorption of toxic metals from industrial wastewater by algae strains *Spirulina platensis* and *Chlorella vulgaris*: Application of isotherm, kinetic models and process optimization. *Science of The Total Environment* 755:142654. doi:https://doi.org/10.1016/j.scitotenv.2020.142654

Amaro HM, Guedes AC, Malcata FX (2011) Antimicrobial activities of microalgae: An invited review. *Science against Microbial Pathogens: Communicating Current Research and Technological Advances* 3:1272–1284

Anastopoulos I, Kyzas GZ (2015) Progress in batch biosorption of heavy metals onto algae. *Journal of Molecular Liquids* 209:77–86. doi:https://doi.org/10.1016/j.molliq.2015.05.023

Aykol A, Budakoglu M, Kumral M, H. Gultekin A, Turhan M, Esenli V, Yavuz F, Orgun Y (2003) Heavy metal pollution and acid drainage from the abandoned Balya Pb-Zn sulfide Mine, NW Anatolia, Turkey. *Environmental Geology* 45:198–208

Bansod SR, Nandkar P (2016) Biosorption of Mn (II) by *Spirogyra verrucosa* collected from manganese mine water. *Plant Science Today* 3 (3):282–292

Beauvais-Flück R, Slaveykova VI, Cosio C (2017) Cellular toxicity pathways of inorganic and methyl mercury in the green microalga *Chlamydomonas reinhardtii*. *Scientific Reports* 7 (1):8034

Beckers F, Rinklebe J (2017) Cycling of mercury in the environment: Sources, fate, and human health implications: A review. *Critical Reviews in Environmental Science and Technology* 47 (9):693–794

Beni AA, Esmaeili A (2020) Biosorption, an efficient method for removing heavy metals from industrial effluents: A review. *Environmental Technology & Innovation* 17:100503

Berni R, LuyckX M, Xu X, Legay S, Sergeant K, Hausman J-F, LuttS S, Cai G, Guerriero G (2019) Reactive oxygen species and heavy metal stress in plants: Impact on the cell wall and secondary metabolism. *Environmental and Experimental Botany* 161:98–106

Bhat SA, Hassan T, Majid S (2019) Heavy metal toxicity and their harmful effects on living organisms–a review. *International Journal of Medical Science and Diagnosis Research* 3 (1):106–122

Biswas H, Bandyopadhyay D (2017) Physiological responses of coastal phytoplankton (Visakhapatnam, SW Bay of Bengal, India) to experimental copper addition. *Marine Environmental Research* 131:19–31

Blaby-Haas CE, Merchant SS (2012) The ins and outs of algal metal transport. *Biochimica et Biophysica Acta (BBA)-Molecular Cell Research* 1823 (9):1531–1552

Brar KK, Etteieb S, MagdouLi S, Calugaru L, Brar SK (2022) Novel approach for the management of acid mine drainage (AMD) for the recovery of heavy metals along with lipid production by *Chlorella vulgaris*. *Journal of Environmental Management* 308:114507. doi:https://doi.org/10.1016/j.jenvman.2022.114507

Buhani, Wijayanti TA, Suharso, Sumadi, Ansori M (2021) Application of modified green algae *Nannochloropsis* sp. as adsorbent in the simultaneous adsorption of Methylene Blue and Cu(II) cations in solution. *Sustainable Environment Research* 31 (1):17. doi:https://doi.org/10.1186/s42834-021-00090-y

Cai X-H, Brown C, Adhiya J, Traina SJ, Sayre RT (1999) Growth and heavy metal binding properties of transgenic *Chlamydomonas* expressing a foreign metallothionein gene. *International Journal of Phytoremediation* 1 (1):53–65

Capdevila M, Atrian S (2011) Metallothionein protein evolastewayminiassay. *JBIC Journal of Biological Inorganic Chemistry* 16:977–989

Cardoso SL, Costa CSD, Nishikawa E, da Silva MGC, Vieira MGA (2017) Biosorption of toxic metals using the alginate extraction residue from the brown algae *Sargassum filipendula* as a natural ion-exchanger. *Journal of Cleaner Production* 165:491–499. doi:https://doi.org/10.1016/j.jclepro.2017.07.114

Chakraborty R, Asthana A, Singh AK, Jain B, Susan ABH (2022) Adsorption of heavy metal ions by various low-cost adsorbents: A review. *International Journal of Environmental Analytical Chemistry* 102 (2):342–379

Chandra R, Kang H (2016) Mixed heavy metal stress on photosynthesis, transpiration rate, and chlorophyll content in poplar hybrids. *Forest Science and Technology* 12 (2):55–61

Chaudhary R, Nawaz K, Khan AK, Hano C, Abbasi BH, Anjum S (2020) An overview of the algae-mediated biosynthesis of nanoparticles and their biomedical applications. *Biomolecules* 10 (11):1498

Cheng SY, Show P-L, Lau BF, Chang J-S, Ling TC (2019) New prospects for modified algae in heavy metal adsorption. *Trends in Biotechnology* 37 (11):1255–1268

Choi HI, Hwang S-W, Kim J, Park B, Jin E, Choi I-G, Sim SJ (2021) Augmented CO_2 tolerance by expressing a single H+-pump enables microalgal valorization of industrial flue gas. *Nature Communications* 12 (1):6049

Cobbett C, Goldsbrough P (2002) Phytochelatins and metallothioneins: Roles in heavy metal detoxification and homeostasis. *Annual Review of Plant Biology* 53 (1):159–182. doi:https://doi.org/10.1146/annurev.arplant.53.100301.135154

Danouche M, El Ghachtouli N, El Arroussi H (2021) Phycoremediation mechanisms of heavy metals using living green microalgae: physicochemical and molecular approaches for enhancing selectivity and removal capacity. *Heliyon* 7 (7):e07609

Deng L, Wang H, Deng N (2006) Photoreduction of chromium (VI) in the presence of algae, *Chlorella vulgaris*. *Journal of Hazardous Materials* 138 (2):288–292

Domínguez-Solís J, López-Martín M, Ager F, Ynsa M, Romero L, Gotor C (2004) Increased cysteine availability is essential for cadmium tolerance and accumulation in *Arabidopsis thaliana. Plant Biotechnology Journal* 2(6): 469–476.

Emparan Q, Jye YS, Danquah MK, Harun R (2020) Cultivation of *Nannochloropsis* sp. microalgae in palm oil mill effluent (POME) media for phycoremediation and biomass production: Effect of microalgae cells with and without beads. *Journal of Water Process Engineering* 33:101043

Fajardo C, De Donato M, CaRrasco R, Martínez-RodríGuez G, Mancera JM, Fernández-Acero FJ (2020) Advances and challenges in genetic engineering of microalgae. *Reviews in Aquaculture* 12 (1):365–381

Fawzy MA (2020) Biosorption of copper ions from aqueous solution by *Codium vermilara*: Optimization, kinetic, isotherm and thermodynamic studies. *Advanced Powder Technology* 31 (9):3724–3735. doi:https://doi.org/10.1016/j.apt.2020.07.014

Gaur J, Rai L (2001) Heavy metal tolerance in algae. *Algal Adaptation to Environmental Stresses: Physiological, Biochemical and Molecular Mechanisms*:363–388

Gekeler W, Grill E, Winnacker E-L, Zenk MH (1988) Algae sequester heavy metals via synthesis of phytochelatin complexes. *Archives of Microbiology* 150 (2):197–202. doi:https://doi.org/10.1007/BF00425162

Gómez-Jacinto V, García-Barrera T, Gómez-Ariza JL, Garbayo-Nores I, Vílchez-Lobato C (2015) Elucidation of the defence mechanism in microalgae *Chlorella sorokiniana* under mercury exposure. Identification of Hg–phytochelatins. *Chemico-Biological Interactions* 238:82–90. doi:https://doi.org/10.1016/j.cbi.2015.06.013

Hamida RS, Ali MA, Redhwan A, Bin-Meferij MM (2020) Cyanobacteria–a promising platform in green nanotechnology: A review on nanoparticles fabrication and their prospective applications. *International Journal of Nanomedicine*:6033–6066

He Z, Siripornadulsil S, Sayre RT, Traina SJ, Weavers LK (2011) Removal of mercury from sediment by ultrasound combined with biomass (transgenic *Chlamydomonas reinhardtii*). *Chemosphere* 83 (9):1249–1254. doi:https://doi.org/10.1016/j.chemosphere.2011.03.004

Henriques B, LoPes CB, Figueira P, Rocha LS, Duarte AC, Vale C, Pardal MA, Pereira E (2017) Bioaccumulation of Hg, Cd and Pb by *Fucus vesiculosus* in single and multi-metal contamination scenarios and its effect on growth rate. *Chemosphere* 171:208–222. doi:https://doi.org/10.1016/j.chemosphere.2016.12.086

Hirata K, TsujimoTo Y, Namba T, OHta T, Hirayanagi N, Miyasaka H, Zenk MH, Miyamoto K (2001) Strong induction of phytochelatin synthesis by zinc in marine green alga, *Dunaliella tertiolecta*. *Journal of Bioscience and Bioengineering* 92 (1):24–29. doi:https://doi.org/10.1016/S1389-1723(01)80193-6

Hirooka S, Hirose Y, KaneSaki Y, Higuchi S, Fujiwara T, Onuma R, ErA A, Ohbayashi R, UzukA A, Nozaki H (2017) Acidophilic green algal genome provides insights into adaptation to an acidic environment. *Proceedings of the National Academy of Sciences* 114 (39):E8304–E8313

Howe G, Merchant S (1992) Heavy metal-activated synthesis of peptides in *Chlamydomonas reinhardtii* 1. *Plant Physiology* 98 (1):127–136. doi:https://doi.org/10.1104/pp.98.1.127

Hu S, Lau KWK, Wu M (2001) Cadmium sequestration in *Chlamydomonas reinhardtii*. *Plant Science* 161 (5):987–996. doi:https://doi.org/10.1016/S0168-9452(01)00501-5

Huang C-C, Chen M-W, Hsieh J-L, Lin W-H, Chen P-C, Chien L-F (2006) Expression of mercuric reductase from *Bacillus megaterium* MB1 in eukaryotic microalga *Chlorella* sp. DT: an approach for mercury phytoremediation. *Applied Microbiology and Biotechnology* 72:197–205

Ibuot A, Dean AP, McIntosh OA, Pittman JK (2017) Metal bioremediation by CrMTP4 over-expressing *Chlamydomonas reinhardtii* in comparison to natural wastewater-tolerant microalgae strains. *Algal Research* 24:89–96

Indhumathi P, Sathiyaraj S, Koelmel JP, Shoba SU, Jayabalakrishnan C, Saravanabhavan M (2018) The efficient removal of heavy metal ions from industry effluents using waste biomass as low-cost adsorbent: Thermodynamic and kinetic models. *Zeitschrift für Physikalische Chemie* 232 (4):527–543

Ireland HE, Harding SJ, Bonwick GA, Jones M, Smith CJ, Williams JHH (2004) Evaluation of heat shock protein 70 as a biomarker of environmental stress in *Fucus serratus* and *Lemna minor*. *Biomarkers* 9 (2):139–155. doi:https://doi.org/10.1080/13547500410001732610

Jaafari J, Yaghmaeian K (2019) Optimization of heavy metal biosorption onto freshwater algae (*Chlorella coloniales*) using response surface methodology (RSM). *Chemosphere* 217:447–455

Jaiswal A, Verma A, Jaiswal P (2018) Detrimental effects of heavy metals in soil, plants, and aquatic ecosystems and in humans. *Journal of Environmental Pathology, Toxicology and Oncology* 37 (3)

Jamers A, Blust R, De Coen W, Griffin JL, Jones OA (2013) An omics based assessment of cadmium toxicity in the green alga *Chlamydomonas reinhardtii*. *Aquatic Toxicology* 126:355–364

Jiang Y, Zhu Y, Hu Z, Lei A, Wang J (2016) Towards elucidation of the toxic mechanism of copper on the model green alga *Chlamydomonas reinhardtii*. *Ecotoxicology* 25 (7):1417–1425

Jones ER, Van Vliet MT, Qadir M, Bierkens MF (2021) Country-level and gridded estimates of wastewater production, collection, treatment and reuse. *Earth System Science Data* 13 (2):237–254

Kelly DJ, Budd K, Lefebvre DD (2007) Biotransformation of mercury in pH-stat cultures of eukaryotic freshwater algae. *Archives of Microbiology* 187:45–53

Kumar G, Shekh A, Jakhu S, Sharma Y, Kapoor R, Sharma TR (2020) Bioengineering of microalgae: recent advances, perspectives, and regulatory challenges for industrial application. *Frontiers in Bioengineering and Biotechnology* 8:914

Kumar V, Pandita S, Sidhu GPS, Sharma A, Khanna K, Kaur P, Bali AS, Setia R (2021) Copper bioavailability, uptake, toxicity and tolerance in plants: A comprehensive review. *Chemosphere* 262:127810

Kurniawan SB, Abdullah SRS, Imron MF, Said NSM, Ismail NI, Hasan HA, Othman AR, Purwanti IF (2020) Challenges and opportunities of biocoagulant/bioflocculant application for drinking water and wastewater treatment and its potential for sludge recovery. *International Journal of Environmental Research and Public Health* 17 (24):9312

Kuroda K, Ueda M (2011) Molecular design of the microbial cell surface toward the recovery of metal ions. *Current Opinion in Biotechnology* 22 (3):427–433. doi:https://doi.org/10.1016/j.copbio.2010.12.006

Lee L, Hsu C-Y, Yen H-W (2017) The effects of hydraulic retention time (HRT) on chromium (VI) reduction using autotrophic cultivation of *Chlorella vulgaris. Bioprocess and Biosystems Engineering* 40:1725–1731

Leszczyszyn OI, Imam HT, Blindauer CA (2013) Diversity and distribution of plant metallothioneins: a review of structure, properties and functions. *Metallomics* 5 (9):1146–1169. doi:https://doi.org/10.1039/c3mt00072a

Lewis S, Donkin ME, Depledge MH (2001) Hsp70 expression in *Enteromorpha intestinalis* (Chlorophyta) exposed to environmental stressors. *Aquatic Toxicology* 51 (3):277–291. doi:https://doi.org/10.1016/S0166-445X(00)00119-3

Li C, Zheng C, Fu H, Zhai S, Hu F, Naveed S, Zhang C, Ge Y (2021) Contrasting detoxification mechanisms of *Chlamydomonas reinhardtii* under Cd and Pb stress. *Chemosphere* 274:129771

Li T, Lin G, Podola B, Melkonian M (2015) Continuous removal of zinc from wastewater and mine dump leachate by a microalgal biofilm PSBR. *Journal of Hazardous Materials* 297:112–118. doi:https://doi.org/10.1016/j.jhazmat.2015.04.080

Lu J, Ma Y, XinG G, Li W, Kong X, Li J, Wang L, Yuan H, Yang J (2019) Revelation of microalgae's lipid production and resistance mechanism to ultra-high Cd stress by integrated transcriptome and physiochemical analyses. *Environmental Pollution* 250:186–195

Lucaci AR, BulgarIu D, Ahmad I, Bulgariu L (2020) Equilibrium and kinetics studies of metal ions biosorption on alginate extracted from marine red algae biomass (*Callithamnion corymbosum* sp.). *Polymers (Basel)* 12 (9):1888

Luo H, Wang Q, Guan Q, Ma Y, Ni F, Yang E, Zhang J (2022) Heavy metal pollution levels, source apportionment and risk assessment in dust storms in key cities in Northwest China. *Journal of Hazardous Materials* 422:126878

Ma X, Chen Y, Liu F, Zhang S, Wei Q (2021) Enhanced tolerance and resistance characteristics of *Scenedesmus obliquus* FACHB-12 with K3 carrier in cadmium polluted water. *Algal Research* 55:102267. doi:https://doi.org/10.1016/j.algal.2021.102267

Mantzorou A, NavaKoudis E, Paschalidis K, Ververidis F (2018) Microalgae: a potential tool for remediating aquatic environments from toxic metals. *International Journal of Environmental Science and Technology* 15:1815–1830

Martínez-Macias MdR, Correa-Murrieta MA, Villegas-Peralta Y, Dévora-Isiordia GE, Álvarez-Sánchez J, Saldivar-Cabrales J, Sánchez-Duarte RG (2019) Uptake of copper from acid mine drainage by the microalgae *Nannochloropsis oculata. Environmental Science and Pollution Research* 26 (7):6311–6318. doi:https://doi.org/10.1007/s11356-018-3963-1

Moghazy RM (2019) Activated biomass of the green microalga *Chlamydomonas variabilis* as an efficient biosorbent to remove methylene blue dye from aqueous solutions. *Water Sa* 45 (1):20–28

Moiseenko T, Gashkina N (2020) Distribution and bioaccumulation of heavy metals (Hg, Cd and Pb) in fish: Influence of the aquatic environment and climate. *Environmental Research Letters* 15 (11):115013

Mustapha MU, Halimoon N (2015) Microorganisms and biosorption of heavy metals in the environment: A review paper. *Journal of Microbial and Biochemical Technology* 7 (5):253–256

Nair AT, Senthilnathan J, Nagendra SS (2019) Application of the phycoremediation process for tertiary treatment of landfill leachate and carbon dioxide mitigation. *Journal of Water Process Engineering* 28:322–330

Nevo Y, Nelson N (2006) The NRAMP family of metal-ion transporters. *Biochimica et Biophysica Acta (BBA)-Molecular Cell Research* 1763 (7):609–620

Noble A, KisiAla A, Galer A, ClysDale D, Emery RJN (2014) *Euglena gracilis* (Euglenophyceae) produces abscisic acid and cytokinins and responds to their exogenous application singly and in combination with other growth regulators. *European Journal of Phycology* 49 (2):244–254. doi:https://doi.org/10.1080/09670262.2014.911353

Ordóñez JI, Cortés S, Maluenda P, Soto I (2023) Biosorption of heavy metals with algae: Critical review of its application in real effluents. *Sustainability* 15 (6):5521

Pal R, Rai JPN (2010) Phytochelatins: Peptides involved in heavy metal detoxification. *Applied Biochemistry and Biotechnology* 160 (3):945–963. doi:https://doi.org/10.1007/s12010-009-8565-4

Picault N, CAzalé A, Beyly A, Cuiné S, Carrier P, Luu D, Forestier C, Peltier G (2006) Chloroplast targeting of phytochelatin synthase in *Arabidopsis*: Effects on heavy metal tolerance and accumulation. *Biochimie* 88 (11):1743–1750

Piña-Olavide R, Paz-Maldonado LM, Alfaro-De La Torre MC, García-Soto MJ, Ramírez-Rodríguez AE, RoSales-Mendoza S, Bañuelos-Hernández B, García De la-Cruz RF (2020) Increased removal of cadmium by *Chlamydomonas reinhardtii* modified with a synthetic gene for γ-glutamylcysteine synthetase. *International Journal of Phytoremediation* 22 (12):1269–1277

Pratush A, Kumar A, Hu Z (2018) Adverse effect of heavy metals (As, Pb, Hg, and Cr) on health and their bioremediation strategies: A review. *International Microbiology* 21:97–106

Priyadarshini E, Priyadarshini SS, Pradhan N (2019) Heavy metal resistance in algae and its application for metal nanoparticle synthesis. *Applied Microbiology and Biotechnology* 103:3297–3316

Puente-Sánchez F, Díaz S, Penacho V, Aguilera A, Olsson S (2018) Basis of genetic adaptation to heavy metal stress in the acidophilic green alga *Chlamydomonas acidophila*. *Aquatic Toxicology* 200:62–72

Ramírez-Rodríguez AE, Bañuelos-Hernández B, García-Soto MJ, Govea-Alonso DG, RoSales-Mendoza S, Alfaro de la Torre MC, MonrEal-Escalante E, Paz-Maldonado LM (2019) Arsenic removal using *Chlamydomonas reinhardtii* modified with the gene acr3 and enhancement of its performance by decreasing phosphate in the growing media. *International Journal of Phytoremediation* 21 (7):617–623

Rani A, Saini KC, Bast F, Mehariya S, Bhatia SK, Lavecchia R, Zuorro A (2021a) Microorganisms: A Potential Source of Bioactive Molecules for Antioxidant Applications. *Molecules* 26 (4):1142

Rani A, Saini KC, Bast F, Varjani S, Mehariya S, Bhatia SK, Sharma N, Funk C (2021b) A review on microbial products and their perspective application as antimicrobial agents. *Biomolecules* 11 (12):1860

Rani A, YAdaV DS, Kumar A, Jaitak V, Bast F (2022) In vitro evaluation of antiproliferative and antioxidant activities of methanolic extracts of *Gracilaria corticata* and *Gracilaria foliifera* against breast cancer cells. *Applied Biological Research* 24 (3):298–308

Ranjbar S, Malcata FX (2022) Is genetic engineering a route to enhance microalgae-mediated bioremediation of heavy metal-containing effluents? *Molecules* 27 (5):1473

Rathi BS, Kumar PS (2021) A review on sources, identification and treatment strategies for the removal of toxic Arsenic from water system. *Journal of Hazardous Materials* 418:126299

Romera E, González F, BAllester A, Blázquez ML, Muñoz JA (2007) Comparative study of biosorption of heavy metals using different types of algae. *Bioresource Technology* 98 (17):3344–3353. doi:https://doi.org/10.1016/j.biortech.2006.09.026

Rosenzweig AC, Argüello JM (2012) Toward a molecular understanding of metal transport by P1B-Type ATPases. In *Current Topics in Membranes*, vol 69. Elsevier, pp. 113–136

Salam KA (2019) Towards sustainable development of microalgal biosorption for treating effluents containing heavy metals. *Biofuel Research Journa*l 6 (2):948

Sayadi MH, Shekari H (2017) Biosorption of cadmium and lead from aqueous solutions using *Spirogyra*. *Journal of Environmental Studies* 43 (3):379–390

Shah SB (2021) Heavy metals in the marine environment—An overview. *Heavy Metals in Scleractinian Corals*:1–26

Sharma R, Saini KC, Rajput S, Kumar M, Mehariya S, Karthikeyan OP, Bast F (2022) Environmental friendly technologies for remediation of toxic heavy metals: Pragmatic approaches for environmental management. In Aravind J, Kamaraj M, Karthikeyan S (eds) *Strategies and Tools for Pollutant Mitigation: Research Trends in Developing Nations*. Springer International Publishing, pp. 199–223. doi:https://doi.org/10.1007/978-3-030-98241-6_10

Simon DF, Descombes P, Zerges W, Wilkinson KJ (2008) Global expression profiling of *Chlamydomonas reinhardtii* exposed to trace levels of free cadmium. *Environmental Toxicology and Chemistry: An International Journal* 27 (8):1668–1675

Srivastava V, Sarkar A, Singh S, Singh P, De Araujo AS, Singh RP (2017) Agroecological responses of heavy metal pollution with special emphasis on soil health and plant performances. *Frontiers in Environmental Science* 5:64

Suárez C, Torres E, Pérez-RaMa M, Herrero C, Abalde J (2010) Cadmium toxicity on the freshwater microalga *Chlamydomonas moewusii* Gerloff: Biosynthesis of thiol compounds. *Environmental Toxicology and Chemistry* 29 (9):2009–2015

Suhani I, Sahab S, Srivastava V, Singh RP (2021) Impact of cadmium pollution on food safety and human health. *Current Opinion in Toxicology* 27:1–7

Sulochana SB, Arumugam M (2016) Influence of abscisic acid on growth, biomass and lipid yield of *Scenedesmus quadricauda* under nitrogen starved condition. *Bioresource Technology* 213:198–203. doi:https://doi.org/10.1016/j.biortech.2016.02.078

Torres MA, Barros MP, Campos SC, Pinto E, Rajamani S, Sayre RT, Colepicolo P (2008) Biochemical biomarkers in algae and marine pollution: A review. *Ecotoxicology and Environmental Safety* 71 (1):1–15

Tripathi S, Arora N, GuPta P, Pruthi PA, Poluri KM, Pruthi V (2019) Microalgae: An emerging source for mitigation of heavy metals and their potential implications for biodiesel production. In *Advanced Biofuels*. Elsevier, pp. 97–128

Tripathi S, Poluri KM (2021a) Adaptive and tolerance mechanism of microalgae in removal of cadmium from wastewater. *Algae: Multifarious Applications for a Sustainable World*:63–88

Tripathi S, Poluri KM (2021b) Heavy metal detoxification mechanisms by microalgae: Insights from transcriptomics analysis. *Environmental Pollution* 285:117443

Ubando AT, Africa ADM, Maniquiz-Redillas MC, Culabaa AB, Chen W-H, Chang J-S (2021) Microalgal biosorption of heavy metals: a comprehensive bibliometric review. *Journal of Hazardous Materials* 402:123431

Udayan A, Kathiresan S, Arumugam M (2018) Kinetin and Gibberellic acid (GA3) act synergistically to produce high value polyunsaturated fatty acids in *Nannochloropsis oceanica* CASA CC201. *Algal Research* 32:182–192. doi:https://doi.org/10.1016/j.algal.2018.03.007

Ungureanu N, Vlăduţ V, Voicu G (2020) Water scarcity and wastewater reuse in crop irrigation. *Sustainability* 12 (21):9055

Wang Y, Selvamani V, Yoo I-K, Kim TW, Hong SH (2021) A novel strategy for the microbial removal of heavy metals: Cell-surface display of peptides. *Biotechnology and Bioprocess Engineering* 26 (1):1–9. doi:https://doi.org/10.1007/s12257-020-0218-z

Wang Z, Gui H, Luo Z, Zhen Z, Yan C, Xing B (2019) Dissolved organic phosphorus enhances arsenate bioaccumulation and biotransformation in *Microcystis aeruginosa. Environmental Pollution* 252:1755–1763

Yang T, Chen M-L, Wang J-H (2015) Genetic and chemical modification of cells for selective separation and analysis of heavy metals of biological or environmental significance. *TrAC Trends in Analytical Chemistry* 66:90–102. doi:https://doi.org/10.1016/j.trac.2014.11.016

Ye J, Song Z, Wang L, Zhu J (2016) Metagenomic analysis of microbiota structure evolution in phytoremediation of a swine lagoon wastewater. *Bioresource Technology* 219:439–444

Yen H-W, Chen P-W, Hsu C-Y, Lee L (2017) The use of autotrophic *Chlorella vulgaris* in chromium (VI) reduction under different reduction conditions. *Journal of the Taiwan Institute of Chemical Engineers* 74:1–6

Yin X, Wang L, Duan G, Sun G (2011) Characterization of arsenate transformation and identification of arsenate reductase in a green alga *Chlamydomonas reinhardtii. Journal of Environmental Sciences* 23 (7):1186–1193

Yu B, Wang X, Dong KF, Xiao G, Ma D (2020) Heavy metal concentrations in aquatic organisms (fishes, shrimp and crabs) and health risk assessment in China. *Marine Pollution Bulletin* 159:111505

Yu X-J, Sun J, Sun Y-Q, Zheng J-Y, Wang Z (2016) Metabolomics analysis of phytohormone gibberellin improving lipid and DHA accumulation in *Aurantiochytrium* sp. *Biochemical Engineering Journal* 112:258–268. doi:https://doi.org/10.1016/j.bej.2016.05.002

Zhu Q-L, Guo S-N, Wen F, Zhang X-L, Wang C-C, Si L-F, Zheng J-L, Liu J (2019) Transcriptional and physiological responses of *Dunaliella salina* to cadmium reveals time-dependent turnover of ribosome, photosystem, and ROS-scavenging pathways. *Aquatic Toxicology* 207:153–162

Ziller A, Fraissinet-Tachet L (2018) Metallothionein diversity and distribution in the tree of life: A multifunctional protein. *Metallomics* 10 (11):1549–1559

Znad H, Awual MR, Martini S (2022) The utilization of algae and seaweed biomass for bioremediation of heavy metal-contaminated wastewater. *Molecules* 27 (4):1275

7 Algal–Bacterial Interactions in Environment
Emerging Applications

Imran Pancha

7.1 INTRODUCTION

Microalgae are considered as important component of aquatic ecosystems as primary producers as well as also they play an important role in various biogeochemical cycles. Apart from normal environments, it is also reported that microalgae also have the ability to survive under various extreme environmental conditions such as hot springs, deserts, and deep ocean. Recently, microalgae are being considered as potential bioresource for the production of biofuels and industrially relevant compounds (Pancha et al. 2020). Microalgae have high growth rate and higher photosynthetic ability compared to other photosynthetic eukaryotes. Additionally, microalgae also have simple growth requirements and have the ability to grow in seawater and various types of wastewaters (Pancha et al. 2019). However, industrial scale production as well as commercialization of microalgal based products are still in their initial phases. One of the major limitations is slow growth rate and high cost of downstream processing associated with microalgal-based products. Various studies indicate that the use of various physico-chemical stresses such as nitrogen starvation, salinity stress, high light intensity, and high temperature affect the production of desirable compounds in microalgae (Paliwal et al. 2017). Apart from physico-chemical optimization, various genetic engineering techniques are also used to enhance the production of biomass as well as for the improvements of bioproducts in the microalgal cells (Fayyaz et al. 2020). Similar to other organisms, microalgae are also associated with bacteria in natural environments. Associated bacteria have positive, negative, or no effect on the growth and accumulation of various metabolites in the microalgae. Exploitation of ecological interactions between microalgae and its associated bacteria is of prime importance by various research groups throughout the globe for various reasons such as understanding the role of such interaction for biogeochemical cycling, enhancement of microalgal biomass, use of such interaction for wastewater treatment, and understanding the evolutionary aspects of plant–bacteria interactions. Three major terms are used in the literature to define the microalgae–bacteria interactions.

DOI: 10.1201/9781003455950-7

a. Phycosphere: area surrounded by microalgae cell, which is generally rich in various biochemicals and where most of the microalgae–bacteria interaction occurs (Seymour et al. 2017)
b. Microalgae Growth Promoting Bacteria (MGPB): the bacteria which generally help in the growth of microalgae for their biotechnological applications (Palacios et al. 2022)
c. Polltantbiome: interaction of microalgae–bacteria under various wastewaters along with various pollutants (Ashraf et al. 2023)

In nature, microalgae–bacteria interaction results in mutualistic, commensalism, or parasitic interactions. In mutualistic interaction, both the partners, that is, microalgae and its associated bacteria, are benefited from the interactions; such a study was carried out by Kim et al. (2014) that indicates that the co-cultivation of plant-growth-promoting bacterium *Rhizobium* sp. with microalga *Chlorella vulgaris* results in mutualism in which bacterium provides nitrogen to the microalgae and in turn microalgae provide fixed carbon for the bacterial growth (Kim et al. 2014). However, Guo et al. reported that the co-cultivation of microalga *C. vulgaris* with bacterium *Pseudomonas* sp. was mutualistic during photoautotrophic conditions, while same partner becomes competitors during heterotrophic conditions. Such results indicate that cultural conditions and probably physico-chemical parameters are responsible for various types of interactions between microalgae and bacteria in the nature (Guo et al. 2014). More in-depth studies are required to uncouple such interactions. Commensalism is another type of ecological interaction in which only one partner is benefited during the relationship. However, many times it is very difficult to distinguish between commensalism and mutualism in nature. Very rare reports are available in microalgae–bacteria relationship which clearly describes commensalism. The interaction between nitrogen-fixing bacterium *Azotobacter vinelandii* and microalga *Neochloris oleoabundans* indicates that bacterium provides siderophore azotobactin which will help in microalgal growth; however, in return, no benefits such as providing fixed carbon by microalgae to the bacteria were reported (Villa et al. 2014). Similarly, in another study, it was reported that exopolysaccharide secreted by bacterium *Variovorax paradoxus* IS1 enhances the growth as well the uptake of nutrients by microalgae *Coelastrella* sp. IS3. The results also indicate that microalgae also accumulate high amount of carbohydrates and protein in the cell. However, it is not very clear how bacteria were benefited from such interactions (Perera et al. 2022). The study between microalgae–bacteria is always difficult and dynamic as the interaction between these two partners is dependent on many factors such as availability of nutrients, crosstalk using signalling molecules, and age of the microalgae. Many times, at first these relationship are mutualistic; however, during the long term, the same relationship will become parasitic or antagonistic in nature (Segev et al. 2016). Therefore, more studies are required in order to understand such complex and dynamic relationship. The final type of relationship between microalgae and associated bacteria is parasitism in which one partner is negatively affected by the interaction. Many studies are available in which anti-algal compounds which are secreted by the phycospheric bacteria have a growth inhibition effect on microalgae. These types of relationships are mainly explored in controlling the harmful algal or cyanobacterial blooms in the natural environment. Microalgae-associated bacterium *Aeromonas* sp. L23 secrete various enzymes which hydrolyse the microalgal cell wall, inhibiting the growth of microalgae such as *Microcystis* sp. and *Scendesmus* sp. (Nishu et al. 2019). Most of the time, bacteria associated with microalgae secrete various hydrolytic enzymes such as amylase and cellulase which hydrolyse the microalgal cell walls and are released into the intracellular compartments. Then this bacterium utilizes the microalgal biomass for the own growth. However, it is also reported that, sometimes, microalgae synthesize few compounds such as PUFAs and few chemicals which have antibacterial activity and inhibit the growth of microalgae (Little et al. 2021). Generally, in case of higher plants, it has been reported that bacteria inject few effectors which interrupt the normal cell signalling; however, such type of mechanism is not documented in case of microalgae. As microalgae are similar to plants, the understanding of and establishing such host–bacteria interaction will enhance our evolutionary understanding about host–microbe interactions.

In the present book chapter, we are planning to overview the mechanisms between microalgae and associated bacteria, and how such interaction will change with the change in the environment surrounding. Finally, we will be explaining the biotechnological application of such microalgal–bacterial interaction majorly in wastewater treatment, aquaculture, production of algal biomass, and accumulation of industrially important compounds. We will also provide few main unanswered research questions in the domain of microalgal–bacterial interactions, which will help the researcher to build the major gaps in the area.

7.2 MICROALGAL BACTERIA INTERACTIONS IN NATURAL ENVIRONMENTS

Nutrient exchange between the microalgae and its associated bacteria is one of the major driving forces to establish algal–bacterial interactions in natural environment. In natural environment, majorly, microalgae provide molecular oxygen and dissolve organic compounds to bacterial partner, while bacteria generally provide various type of phytohormone and growth factors such as IAA and vitamin B12, etc., help in dissolving the nutrients as well as help in making certain elements such as Fe for microalgal growth (Kim et al. 2022). Among various nutrients, nitrogen is considered as the most important element for growth of microalgae because nitrogen is important for synthesis of protein and also plays an important role in microalgal growth and metabolism. Generally, microalgae obtain nitrogen in the form of nitrate, nitrite, or ammonia from the environment. It has been reported that the co-cultivation of microalga *Chlorella vulgaris* and nitrogen-fixing and microalgal-growth-promoting bacterium *Azospirillum brasilense* improves the microalgal growth in various conditions and also helps in improving the nitrogen uptake in the cell (de-Bashan et al. 2005). In another study, *nifL* mutant of *Azotobacter vinelandii*, which accumulates very high amounts of ammonium in the cultivation medium compared to wild-type strain, enhances more algal biomass, and such synthetic symbiosis is further utilized to enhance microalgal growth and lipid accumulation for the production of microalgal-based biodiesel (Ortiz-Marquez et al. 2012). Iron is another element, which plays an important role in various physiological processes such as electron transport chain and chlorophyll metabolism (Hutchins et al. 1995). However, most of the time in aquatic environment, iron is not available in the usable form for the microalgae. On the other hand, bacteria have ability to synthesize the compounds known as siderophores, a high-affinity ferric ion chelator, which helps make the iron into a soluble form of it. Association of siderophore-producing bacteria with microalgae will help the microalgae to obtain bound iron form from the environment. For example Rajapitamahuni et al. reported that the co-cultivation of siderophore-producing bacterium *Idiomarina loihiensis* with microalga *Chlorella variabilis* enhances the iron uptake by the microalgae and improves the biomass and lipid production in microalgal cells (Rajapitamahuni et al. 2019). In another example, Marinobacter produces an unusual lower-affinity dicitrate siderophore, vibrioferrin (VF). Photolysis of VF–Fe complex releases Fe in the surroundings which is utilized by dinoflagellate partner, *Scrippsiella trochoidea* (Amin et al. 2009). Microalgae, during their growth, generally release dissolve organic matter (DOM) into the surroundings. These DOMs are generally rich in carbohydrates consisting of sucrose, ribose, xylitol, galacturonic acid, etc. (Watanabe et al. 2006). These sugars work as carbon source for the growth of microalgae-associated bacteria, and bacteria benefit from such interaction in the natural environment. Among various nutrients, CO_2 is considered to be one of the most important nutrients for the growth of microalgae as microalgae are photosynthetic organisms. Generally due to low solubility of CO_2 in water, in most of the aquatic systems, CO_2 is limiting factor. Heterotrophic bacteria, associated with microalgae, generally metabolize complex organic matter, and during their metabolism, CO_2 is released as by-product. Incorporation of such approach will be important for the production of higher biomass from microalgae. Bai et al reported that co-cultivation of microalga Chlorella with heterotrophic bacteria during domestic sewage treatment enhances the microalgal growth by almost 4.8-fold (Bai et al. 2015). Application of these approaches at a large scale in wastewater treatment plants reduces GHG emission as well as also provides a sustainable way to provide oxygenation in activation sludge process. Apart from this nutrient exchange, microalgae–bacteria interaction also plays an important role in sulphur cycle in the nature. In ocean, different types of microalgae are producers for the sulphur-containing compound dimethylsulphoniopropionate (DMSP). Many *Roseobacter* groups of bacteria which are mainly associated with the microalgae convert DMSP to dimethyl sulphide (Geng and Belas 2010). So, in ocean, algae provide DMSP to the bacteria, while certain bacteria in turn produce certain antibacterial compounds such as Roseobacticides which prevent the growth of algicidal bacteria, and, thereby, such mutualistic interaction in the ocean helps to run the sulphur biogeochemical cycle (Seyedsayamdost et al. 2011). There are many different studies that are available on role of nutrients' exchange basis for establishing the microalgal–bacterial interactions in the nature. The advancement of various omics approach such as transcriptomics and metabolomics along with the recent use of isotope labelling helps to understand the molecular basis of such interaction and exploitations for biomass and production of industrially important compounds from the microalgae.

In nature, various groups of bacteria produce different types of plant-growth-promoting hormone known as phytohormones. Plant–bacteria interaction through various types of phytohormones is a well-documented phenomenon in the plant–bacteria interactions. However, is the role of phytohormone in microalgal–bacterial interaction is not well documented. Indole-3-acetic acid (IAA) is most abundantly produced phytohormone by plant-associated bacteria and plays a crucial role in plant growth. Amin et al. reported that diatom-associated *Sulfitobacter* sp., which has ability to synthesise IAA, enhances the cell-division in diatom *Pseudonitzschia multiseries* PC9; on the other hand, tryptophan produced by diatom is used by bacteria for the synthesis of IAA (Amin et al. 2015). It has been documented extensively by the research group of de-Bashan that the co-cultivation of IAA-producing bacteria with microalgae enhances the growth of microalgae and also produces high lipids in the cells (de-Bashan et al. 2008; De-Bashan and Bashan 2008). They have also documented that mutant of bacterium A. Brasiliense, which was unable to synthesize IAA, had no growth promotion effect on *C. vulgaris* cells indicating that phytohormone IAA produced by bacteria is important for microalgal growth (De-Bashan and Bashan 2008). There are few reports also indicating that a phytohormone such as IAA will involve not only in symbiotic interaction between microalgae and associated bacteria but also during some phase of life cycle, where it works as a growth inhibitor. In the case of marine model coccolithophore microalga *Emiliania huxleyi* and *Phaeobacter inhibens* interaction, initially IAA secreted by the bacterial partner allows the attachment of bacteria to the microalgal cells leading to their growth enhancement. While the production and secretion of tryptophan by microalgae serve as a precursor for the synthesis of IAA by bacteria; however, it has been shown that during the late phase of microalgal growth, the higher production of IAA by bacteria activates the pathways unique to oxidative stress response and programmed cell death (Segev et al. 2016). Similar reports are also found for the higher plant–bacteria interaction where high production of phytohormone is negative for the plant growth. Various recent studies indicate that the exogenous addition of phytohormone to the microalgal culture will help in the enhancement of microalgal growth, improvement in nutrient acquisition from the growth medium, and mitigating the oxidative stress generated through nutrient starvation and other stresses (Zhao et al. 2019). Such chemical modulator-based approach is very easy to implement at a large scale and helpful for the generation of microalgal biomass and biofuel production. However, the cost associated with the exogenous addition of phytohormone is a major limitation in its large-scale application. Co-cultivation of phytohormone-producing bacteria with microalgae will be cheap and win-win approach for generation of higher microalgal biomass without addition of costly phytohormone. Apart from phytohormone, microalgae also require small amount of vitamin B12 and vitamin B1; however, many microalgae do not lead to a complete pathway for the synthesis of these vitamins and generally vitamin auxotrophs. It has been reported that out of 306 microalgae selected, almost 50% require vitamin B12, 22% require thiamine, and almost 5% of them require biotin for their growth (Croft et al. 2005). In the same study, it has been reported that bacterium *Halomonas* which was isolated from the culture of microalgae enhances the growth of *Amphidinium operculatum* and *Porphyrdium purpureum* by providing vitamin B12 required by microalgae, and the growth is almost similar to the exogenous addition of vitamin B12 to the culture medium (Croft et al. 2005). Further, when fucidin, a commercial algal extract, was added to the bacterial growth medium, it showed significant amount of vitamin B12 synthesis indicating that vitamin B12 plays a crucial role in microalgal–bacterial interactions (Croft et al. 2005). In another study, it was showed that *C. reinhardtii* generally does not require vitamins as it possesses vitamin-B12-independent methionine synthase (MetE); however, the co-cultivation of microalga *C. reinhardtii* with heterotrophic bacterium *Mesorhizobium* sp. reduced the expression of MetE. However, the co-cultivation of same bacterium with micro alga *Lobomonas rostrata* results in higher growth as this alga is unable to produce vitamin B12 (Kazamia et al. 2012).

Many bacteria have the ability to interact with other microorganisms through signalling molecules such as N-acyl-homoserine lactone (AHL). It has been also documented that microalgae-associated bacteria also have the ability to produce AHLs, and maybe they also participate in microalgal–bacterial interactions. It has been observed that the treatment of diatom *Seminavis robusta* with C14-AHL enhances the diatom growth; further, transcriptome analysis indicates that differential expression of few important group of genes such as glutathione might be responsible for the enhancement of the growth of diatoms (Stock et al. 2020). Additionally, it has been reported that the addition of DMSP to the bacterial culture *Ruegeria pomeroyi* upregulates oxo-C14-AHL biosynthesis, which indicates quorum-sensing-based crosstalk between microalgae and bacteria (Johnson et al. 2016).

7.3 BIOTECHNOLOGICAL APPLICATIONS OF MICROALGAL–BACTERIAL INTERACTIONS

In recent years, microalgae gained importance due to their vast applications in various industrial sectors such as food industry, pharma industry, production of renewabgy, and wastewater remediation (Chokshi et al. 2017; Pancha et al. 2019). In natural environment, microalgae generally exist with other microorganisms such as heterotrophic bacteria. Therefore, various types of interactions such as mutualism, commensalism, and antagonist occur (Cooper and Smith 2015). Understanding and the utilization of microalgae–bacteria interaction for the production of industrially relevant products and environmental remediation are main objectives of this section. Figure 7.1 indicates the mechanism behind the interaction of microalgae and bacteria along with its various biotechnological applications.

Microalgae are potential resources for the production of biofuel production. However, the commercialization of such an approach is still in its infant stage due to lower biomass production and high downstream processing cost. Apart from photosynthetic growth, microalgae can also directly assimilate organic compounds such as glucose into their central carbon metabolism through hetero and mixotrophic growth. One of the ways to improve the microalgal growth enhancement is through mixotrophic and heterotrophic growth of microalgae using organic compounds (Pancha et al. 2015). Various types of wastewaters such as municipality wastewater and dairy wastewater contain huge amount of organic compounds. The treatment of various types of wastewaters through microalgal cultivation is a promising alternative, and the use of such approach is win-win situation for the treatment of hazardous wastewater and

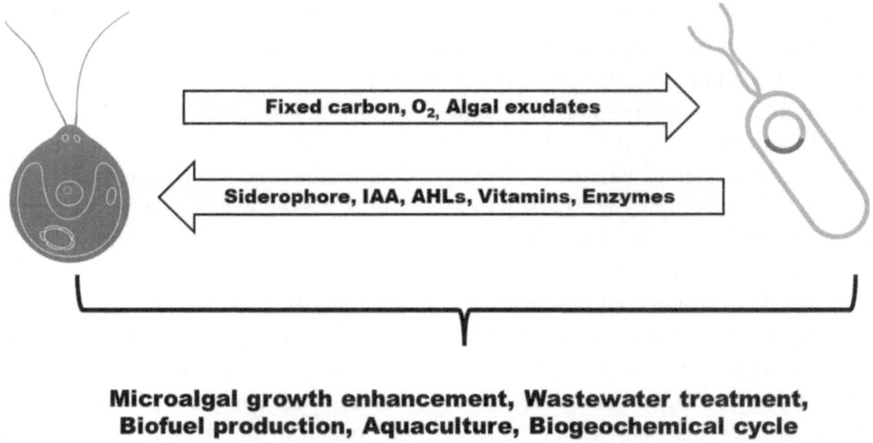

FIGURE 7.1 Microalgal–bacterial interactions and its potential applications.

generation of algal biomass (Pancha et al. 2019). On the other hand, the use of algae or bacteria alone is not a good alternative as various types of wastewaters are heterogenous mixtures which contain various toxic compounds which might inhibit the growth of single species; however, the interaction of both the partners will help to mitigate toxic effects and oxidative stress (Vo et al. 2019). Co-cultivation of microalgae–bacteria not only improves the wastewater efficiency but also helps in harvesting of biomass as well as improve the microalgal biomass production (Ashraf et al. 2022; Jiang et al. 2021). The utilization of microalgae–bacteria for wastewater treatment also reduced the economic burden of running wastewater treatment plant, such as production of oxygen by algae reduces almost 50% total energy inputs in activated sludge process (Rawat et al. 2011). In wastewater treatment plants, bacteria generally oxidize the chemicals and reduce the COD of the wastewater. During this oxidization, they release CO_2 which is utilized by microalgae and in return alga provide O_2 to the bacteria, and such an approach will reduce the aeration cost in the wastewater treatment plants. Under anaerobic wastewater treatment plants, microalgae and their associated methane-oxidizing bacteria reduce the methane generated during the process and hence also reduce the amount of released methane gas to the environment, thereby reducing the pollution (Nagarajan et al. 2019). Recently, a new term "PollutantBiome" is coined to explain the interaction between microbes and microalgae under wastewater conditions which will help to improve the health of aquatic life and enhance the environmental sustainability (Ashraf et al. 2022). The association of microalgae with bacteria has a significant impact on nutrient removal efficiency; for example almost 82% nitrate and 90% phosphate removal efficiency were observed when microalgae *C. vulgaris*, *Acutodesmus obliquus*, and *Oscillatoria* sp. were cultivated with bacteria (García et al. 2018). The growth of microalgae-–bacteria consortia in wastewater is greatly influenced by various factors such as Light intensity, pH of the wastewater, and temperature of the environment. Lee et al. reported a significant removal of nutrients from wastewater by two-phase photoperiod, first 12 hours of light/60 hours of darkness followed by 12 hours of light/12 hours of darkness (Lee et al. 2016). Heavy metal precent in the wastewater is one of the toxic pollutants which can affect normal aquatic microbial flora and fauna. Munoz et al. reported that microalgae–bacteria association has ability to remove toxic metals such as Cd(II), Ni(II), Cu(II), and Zn(II) through adsorption mechanism for the wastewater (Muñoz et al. 2006). Mostly, microalgae–bacteria remove toxic heavy metal through biosorption, apart from its sometime bioaccumulation and precipitation are also involved in removal of toxic metals from the various environment such as acid mine drainage. Almost 50% removal of various toxic metals such as Cu, Ni, Mn, Zn, and Co were reported to be done by native microalgae–bacteria consortia (Orandi et al. 2012).

As mentioned earlier, microalgae are considered as excellent bioresource for the production of renewable energy sources such as biodiesel, bioethanol, and biohydrogen. As microalgae have high growth rate and photosynthetic capacity, they also accumulate high amount of energy-reserved compounds such as lipids and carbohydrates into the cell (Chisti 2007). Co-cultivation of microalgae with its associated bacteria enhances the overall biomass production as well as the accumulation of energy-reserved compounds in the cells. Microalgae-associated bacteria enhance the microalgal growth through the generation of phytohormone and vitamins, helping in the uptake of nutrients such as nitrate and phosphate (Palacios et al. 2022). Biohydrogen is an important renewable energy resource, and hydrogenase enzyme is important for the biohydrogen production. However, hydrogenase is oxygen-sensitive enzyme. Co-cultivation of microalga *C. reinhardtii* with its associated bacteria *Brevundimonas* sp., *Rhodococcus* sp., and *Leifsonia* sp. enhances the biohydrogen production by eliminating oxygen through respiration (Lakatos et al. 2014). As per another study with microalga Chlorella vulgaris and its association of four different bacteria, there was an enhancement of biomass from 1.3 gm L^{-1} to 3.31 gm L^{-1}, and also lipid content was improved from 22.4% to 28% which indicates such interactions are beneficial for the enhancement of microalgal biomass as well as for its renewable energy production capabilities (Cho et al. 2014). Generally, microalgae accumulate high amount of energy-reserved compounds such as lipids and carbohydrates under various stress conditions such as nitrogen starvation, salinity stress, and high temperature (Paliwal et al. 2017). However, such stress conditions result in lower biomass production due to the generation of oxidative stress. Mitigation of oxidative stress would be a good strategy to enhance biomass production along with the generation of renewable energy from microalgae. Microalga *C. reinhardtii* routinely is cultivated at 25°C; however, at large scale, temperature control is difficult. Higher temperature cultivation of microalga *C. reinhardtii* results in growth inhibition and subsequently chlorosis and death due to the inhibition of cobalamin-independent methionine synthase. Co-cultivation of cobalamin-producing bacterium *Sinorhizobium meliloti* with microalga *C. reinhardtii* results in high-temperature tolerance, and such an interaction will be helpful to large-scale cultivation of microalgae (Xie et al. 2013). Apart from fresh *Tetraselmis striata* water microalgae, it has been reported that the co-cultivation of *Pelagibaca bermudensis* and *Stappia* sp., isolated from mass-cultivated *Tetraselmis striata* culture enhance the growth of microalgae (Park et al. 2017). Apart from co-cultivation for the enhancement of biomass and biofuel production potential, enzymes secreted by bacterium *Bacillus licheniformis* will be used for hydrolysis and saccharification of microalga *C. reinhardtii*, resulting in sugar fermented by brewer's yeast *Saccharomyces cerevisiae* to produce bioethanol (de Farias Silva and Bertucco 2016). Algae-associated bacteria may also play an important role in downstream processes of biofuel production from microalgae such as harvesting. Harvesting of algal cells from the culture is one of the very high-energy-intensive processes. Microalgae-associated bacteria are good producers of exopolysaccharide. Wang et al. have reported the bacterium which has ability to flocculate *Nannochloropsis oceanica* IMET1 biofuel-producing microalgal strain (Wang et al. 2012). In another study, it was reported that the co-cultivation of fungus *Aspergillus fumigatus* with various microalgae enhances the aggregation of microalgal cultures (Muradov et al. 2015). Microalgae are rich source of proteins, vitamins, essential trace elements, and other growth-promoting substances, and due to all these characteristics, microalgae are also considered as best resource for aqua feed (Matias-Peralta et al. 2012). Cyanobacteria such as Spirulina are considered good human nutraceuticals; however, more study is required to understand how prebiotic-bacteria and such microalgal will use in future for animal, human, and aquafeed.

7.4 CONCLUSION AND FUTURE PROSPECTS

In conclusion, microalgae are primary producers in many ecological processes and also recently are being considered to be best resource for biomass production as well as natural cleaners for environment. In nature, microalgae and bacteria interact with each other through various chemical signals such as phytohormone, vitamins, and chemicals. Understanding their interaction with bacteria will help to maximize their potential for bioremediation and the production of green and sustainable bioproducts. The exact diversity and the functional role of microalgae-associated bacteria are hardly known. In contrast, the microbial studies associated with seaweed, human gut, and plants have significantly advanced with their microbiome and its functional role. Therefore, it is important to investigate about the microalgae- associated bacteria and their functional role through high-throughput methods like metabolomics and transcriptomics. There are still many questions which are unanswered in the microalgal–bacterial interactions such as i) Does the same group of microalgae have similar microbial diversity? ii) How algal–bacterial interaction will change according to environmental conditions? iii) How about microbial attack and immunity in microalgae case? iv) What about microalgae–bacteria interactions at molecular level? v) What about techno-economic analysis of such interactions for bioproduct development and wastewater remediation? Answering these and many other questions will enhance our understanding about such phenomena and their biotechnological applications.

ACKNOWLEDGEMENTS

Financial support by the Science and Engineering Research Board (SERB), Department of Science and Technology (DST), Government of India, through a start-up Research Grant (SRG/2020/000165LS) and by Gujarat State Biotechnology Mission (GSBTM), Department of Science and Technology (DST), Government of Gujarat, through advanced biofuel network project, is sanctioned to Dr. Imran Pancha.

REFERENCES

Amin, Shady A., David H. Green, Mark C. Hart, Frithjof C. Küpper, William G. Sunda, and Carl J. Carrano. 2009. "Photolysis of iron–siderophore chelates promotes bacterial–Algal mutualism." *Proceedings of the National Academy of Sciences* 106: 17071–17076. https://doi.org/10.1073/pnas.090551210

Amin, Shady A., L. R. Hmelo, H. M. Van Tol, B. P. Durham, L. T. Carlson, K. R. Heal, R. L. Morales et al. 2015. "Interaction and signalling between a cosmopolitan phytoplankton and associated bacteria." *Nature* 522: 98–101. https://doi.org/10.1038/nature14488

Ashraf, Noreen, Fiaz Ahmad, and Yandu Lu. 2023. "Synergy between microalgae and microbiome in polluted waters." *Trends in Microbiology* 31: 9–21. https://doi.org/10.1016/j.tim.2022.06.004

Bai, Xue, Paul Lant, and Steven Pratt. 2015. "The contribution of bacteria to algal growth by carbon cycling." *Biotechnology and Bioengineering* 112: 688–695. https://doi.org/10.1002/bit.25475

Chisti, Yusuf. 2007. "Biodiesel from microalgae." *Biotechnology Advances* 25: 294–306. https://doi.org/10.1016/j.biotechadv.2007.02.001

Cho, Dae-Hyun, Rishiram Ramanan, Jina Heo, Jimin Lee, Byung-Hyuk Kim, Hee-Mock Oh, and Hee-Sik Kim. 2015. "Enhancing microalgal biomass productivity by engineering a microalgal–bacterial community." *Bioresource Technology* 175: 578–585. https://doi.org/10.1016/j.biortech.2014.10.159

Chokshi, Kaumeel, Imran Pancha, Arup Ghosh, and Sandhya Mishra. 2017. "Oxidative stress-induced bioprospecting of microalgae." *Systems Biology of Marine Ecosystems*: 251–276. https://doi.org/10.1007/978-3-319-62094-7_13

Cooper, Matthew B., and Alison G. Smith. 2015. "Exploring mutualistic interactions between microalgae and bacteria in the omics age." *Current Opinion in Plant Biology* 26: 147–153. https://doi.org/10.1016/j.pbi.2015.07.003

Croft, Martin T., Andrew D. Lawrence, Evelyne Raux-Deery, Martin J. Warren, and Alison G. Smith. 2005. "Algae acquire vitamin B12 through a symbiotic relationship with bacteria." *Nature* 438: 90–93. https://doi.org/10.1038/nature04056

De-Bashan, Luz E., Hani Antoun, and Yoav Bashan. 2005. "Cultivation factors and population size control the uptake of nitrogen by the microalgae *Chlorella vulgaris* when interacting with the microalgae growth-promoting bacterium *Azospirillum brasilense*." *FEMS Microbiology Ecology* 54: 197–203. https://doi.org/10.1016/j.femsec.2005.03.014

De-Bashan, Luz E., and Yoav Bashan. 2008. "Joint immobilization of plant growth-promoting bacteria and green microalgae in alginate beads as an experimental model for studying plant-bacterium interactions." *Applied and Environmental Microbiology* 74: 6797–6802. https://doi.org/10.1128/AEM.00518-08

de Farias Silva, Carlos Eduardo, and Alberto Bertucco. 2016. "Bioethanol from microalgae and cyanobacteria: A review and technological outlook." *Process Biochemistry* 51: 1833–1842. https://doi.org/10.1016/j.procbio.2016.02.016

Fayyaz, Mehmooda, Kit Wayne Chew, Pau Loke Show, Tau Chuan Ling, I-Son Ng, and Jo-Shu Chang. 2020. "Genetic engineering of microalgae for enhanced biorefinery capabilities." *Biotechnology Advances* 43: 107554. https://doi.org/10.1016/j.biotechadv.2020.107554

García, Dimas, Esther Posadas, Saúl Blanco, Gabriel Acién, Pedro García-Encina, Silvia Bolado, and Raúl Muñoz. 2018. "Evaluation of the dynamics of microalgae population structure and process performance during piggery wastewater treatment in algal-bacterial photobioreactors." *Bioresource Technology* 248: 120–126. https://doi.org/10.1016/j.biortech.2017.06.079

Geng, Haifeng, and Robert Belas. 2010. "Molecular mechanisms underlying Roseobacter–Phytoplankton symbioses." *Current Opinion in Biotechnology* 21: 332–338. https://doi.org/10.1016/j.copbio.2010.03.013

Guo, Zhi, and Yen Wah Tong. 2014. "The interactions between *Chlorella vulgaris* and algal symbiotic bacteria under photoautotrophic and photoheterotrophic conditions." *Journal of Applied Phycology* 26: 1483–1492. https://doi.org/10.1007/s10811-013-0186-1

Hutchins, David A., Wen-Xiong Wang, and Nicholas S. Fisher. 1995. "Copepod grazing and the biogeochemical fate of diatom iron." *Limnology and Oceanography* 40: 989–994. https://doi.org/10.4319/lo.1995.40.5.0989

Jiang, Liqun, Yizhen Li, and Haiyan Pei. 2021. "Algal–bacterial consortia for bioproduct generation and wastewater treatment." *Renewable and Sustainable Energy Reviews* 149: 111395. https://doi.org/10.1016/j.rser.2021.111395

Johnson, Winifred M., Melissa C. Kido Soule, and Elizabeth B. Kujawinski. 2016. "Evidence for quorum sensing and differential metabolite production by a marine bacterium in response to DMSP." *The ISME Journal* 10: 2304–2316. https://doi.org/10.1038/ismej.2016.6

Kazamia, Elena, Hjördis Czesnick, Thi Thanh Van Nguyen, Martin Tom Croft, Emma Sherwood, Severin Sasso, Sarah James Hodson, Martin James Warren, and Alison Gail Smith. 2012. "Mutualistic interactions between vitamin B12-dependent algae and heterotrophic bacteria exhibit regulation." *Environmental Microbiology* 14: 1466–1476. https://doi.org/10.1111/j.1462-2920.2012.02733.x

Kim, Byung-Hyuk, Rishiram Ramanan, Dae-Hyun Cho, Hee-Mock Oh, and Hee-Sik Kim. 2014. "Role of Rhizobium, a plant growth promoting bacterium, in enhancing algal biomass through mutualistic interaction." *Biomass and Bioenergy* 69: 95–105. https://doi.org/10.1016/j.biombioe.2014.07.015

Kim, Hyungseok, Jeffrey A. Kimbrel, Christopher A. Vaiana, Jessica R. Wollard, Xavier Mayali, and Cullen R. Buie. 2022. "Bacterial response to spatial gradients of algal-derived nutrients in a porous microplate." *The ISME Journal* 16: 1036–1045. https://doi.org/10.1038/s41396-021-01147-x

Lakatos, Gergely, Zsuzsanna Deák, Imre Vass, Tamás Rétfalvi, Szabolcs Rozgonyi, Gábor Rákhely, Vince Ördög, Éva Kondorosi, and Gergely Maróti. 2014. "Bacterial symbionts enhance photo-fermentative hydrogen evolution of *Chlamydomonasalgae*." *Green Chemist* ry 16: 4716–4727. https://doi.org/10.1039/C4GC00745J

Lee, Chang Soo, Hyung-Seok Oh, Hee-Mock Oh, Hee-Sik Kim, and Chi-Yong Ahn. 2016. "Two-phase photoperiodic cultivation of algal–bacterial consortia for high biomass production and efficient nutrient removal from municipal wastewater." *Bioresource Technology* 200: 867–875. https://doi.org/10.1016/j.biortech.2015.11.007

Little, Shannon M., Gerusa N. A. Senhorinho, Mazen Saleh, Nathan Basiliko, and John A. Scott. 2021. "Antibacterial compounds in green microalgae from extreme environments: A review." *Algae* 36: 61–72.

Matias-Peralta, Hazel Monica, Fatimah Md Yusoff, Mohamed Shariff, and Suhaila Mohamed. 2012. "Reproductive performance, growth and development time of a tropical harpacticoid copepod, *Nitocra affinis* californica Lang, 1965 fed with different microalgal diets." *Aquacult ure* 344: 168–173. https://doi.org/10.1016/j.aquaculture.2012.02.026

Muñoz, Raul, Maria Teresa Alvarez, Adriana Muñoz, Enrique Terrazas, Benoit Guieysse, and Bo Mattiasson. 2006. "Sequential removal of heavy metals ions and organic pollutants using an algal-bacterial consortium." *Chemosphere* 63: 903–911. https://doi.org/10.1016/j.chemosphere.2005.09.062

Muradov, Nazim, Mohamed Taha, Ana F. Miranda, Digby Wrede, Krishna Kadali, Amit Gujar, Trevor Stevenson, Andrew S. Ball, and Aidyn Mouradov. 2015. "Fungal-assisted algal flocculation: Application in wastewater treatment and biofuel production." *Biotechnology for Biofuels* 8: 1–23. https://doi.org/10.1186/s13068-015-0210-6

Nagarajan, Dillirani, Duu-Jong Lee, and Jo-Shu Chang. 2019. "Integration of anaerobic digestion and microalgal cultivation for digestate bioremediation and biogas upgrading." *Bioresource Technology* 290: 121804. https://doi.org/10.1016/j.biortech.2019.121804

Nishu, Susmita Das, Yunhee Kang, Il Han, Tae Young Jung, and Tae Kwon Lee. 2019. "Nutritional status regulates algicidal activity of *Aeromonas* sp. L23 against cyanobacteria and green algae.: *PlosOne*14:e0213370. https://doi.org/10.1371/journal.pone.0213370

Orandi, S., D. M. Lewis, and N. R. Moheimani. 2012. "Biofilm establishment and heavy metal removal capacity of an indigenous mining algal-microbial consortium in a photo-rotating biological contactor." *Journal of Industrial Microbiology and Biotechnolo* gy 39: 1321–1331. https://doi.org/10.1007/s10295-012-1142-9

Ortiz-Marquez, Juan Cesar Federico, Mauro Do Nascimento, Maria de los Angeles Dublan, and Leonardo Curatti. 2012. "Association with an ammonium-excreting bacterium allows diazotrophic culture of oil-rich eukaryotic microalgae." *Applied and Environmental Microbiology* 78: 2345–2352. https://doi.org/10.1128/AEM.06260-11

Palacios, Oskar A., Blanca R. López, and Luz E. de-Bashan. 2022. "Microalga Growth-Promoting Bacteria (MGPB): A formal term proposed for beneficial bacteria involved in microalgal–bacterial interactions." *Algal Research* 61: 102585. https://doi.org/10.1016/j.algal.2021.102585

Paliwal, Chetan, Madhusree Mitra, Khushbu Bhayani, S. V. Vamsi Bharadwaj, Tonmoy Ghosh, Sonam Dubey, and Sandhya Mishra. 2017. "Abiotic stresses as tools for metabolites in microalgae." *Bioresource Technology* 244: 1216–1226. https://doi.org/10.1016/j.biortech.2017.05.058

Pancha, Imran, Kaumeel Chokshi, Kan Tanaka, and Sousuke Imamura. 2020. "Microalgal target of rapamycin (TOR): A central regulatory hub for growth, stress response and biomass production." *Plant and Cell Physiology* 61: 675–684. https://doi.org/10.1093/pcp/pcaa023

Pancha, Imran, Kaumeel Chokshi, and Sandhya Mishra. 2015. "Enhanced biofuel production potential with nutritional stress amelioration through optimization of carbon source and light intensity in *Scenedesmus* sp. CCNM 1077." *Bioresource Technology* 179: 565–572. https://doi.org/10.1016/j.biortech.2014.12.079

Pancha, Imran, Kaumeel Chokshi, and Sandhya Mishra. 2019. "Industrial wastewater-based microalgal biorefinery: A dual strategy to remediate waste and produce microalgal bioproducts." *Application of Microalgae in Wastewater Treatment: Biorefinery Approaches of Wastewater Treatment* 2: 173–193. https://doi.org/10.1007/978-3-030-13909-4_8

Park, Jungsoo, Bum Soo Park, Pengbin Wang, Shailesh K. Patidar, Jin Ho Kim, Sae-Hee Kim, and Myung-Soo Han. 2017. "Phycospheric native bacteria *Pelagibaca bermudensis* and *Stappia* sp. ameliorate biomass productivity of *Tetraselmis striata* (KCTC1432BP) in co-cultivation system through mutualistic interaction." *Frontiers in Plant* Science 8: 289. https://doi.org/10.3389/fpls.2017.00289

Perera, Isiri Adhiwarie, Sudharsanam Abinandan, Suresh R. Subashchandrabose, Kadiyala Venkateswarlu, Nicole Cole, Ravi Naidu, and Mallavarapu Megharaj. 2022. "Extracellular polymeric substances drive symbiotic interactions in bacterial–Microalgal consortia." *Microbial Ecology* 83: 596–607. https://doi.org/10.1007/s00248-021-01772-1

Rajapitamahuni, Soundarya, Pooja Bachani, Raj Kumar Sardar, and Sandhya Mishra. 2019. "Co-cultivation of siderophore-producing bacteria *Idiomarina loihiensis* RS14 with *Chlorella variabilis* ATCC 12198, evaluation of micro-algal growth, lipid, and protein content under iron starvation." *Journal of Applied Phycology* 31: 29–39. https://doi.org/10.1007/s10811-018-1591-2

Rawat, I., R. Ranjith Kumar, T. Mutanda, and F. Bux. 2011. "Dual role of microalgae: phycoremediation of domestic wastewater and biomass production for sustainable biofuels production." *Applied Energy* 88: 3411–3424. https://doi.org/10.1016/j.apenergy.2010.11.025

Segev, Einat, Thomas P. Wyche, Ki Hyun Kim, Jörn Petersen, Claire Ellebrandt, Hera Vlamakis, Natasha Barteneva et al. 2016. "Dynamic metabolic exchange governs a marine algal-bacterial interaction." Elife 5: e17473. https://doi.org/10.7554/eLife.17473

Seyedsayamdost, Mohammad R., Gavin Carr, Roberto Kolter, and Jon Clardy. 2011. "Roseobacticides: small molecule modulators of an algal-bacterial symbiosis." *Journal of the American Chemical Society* 133: 18343–18349. https://doi.org/10.1021/ja207172s

Seymour, Justin R., Shady A. Amin, Jean-Baptiste Raina, and Roman Stocker. 2017. "Zooming in on the phycosphere: The ecological interface for phytoplankton–bacteria relationships." *Nature Microbiology* 2: 1–12. https://doi.org/10.1038/nmicrobiol.2017.65

Stock, Frederike, Gust Bilcke, Sam De Decker, Cristina Maria Osuna-Cruz, Koen Van den Berge, Emmelien Vancaester, Lieven De Veylder, Klaas Vandepoele, Sven Mangelinckx, and Wim Vyverman. 2020. "Distinctive growth and transcriptional changes of the diatom *Seminavis robusta* in response to quorum sensing related compounds." *Frontiers in Microbiology* 11: 1240. https://doi.org/10.3389/fmicb.2020.01240

Villa, Juan A., Erin E. Ray, and Brett M. Barney. 2014. "*Azotobacter vinelandii* siderophore can provide nitrogen to support the culture of the green algae *Neochloris oleoabundans* and *Scenedesmus* Sp. BA032." *FEMS Microbiology Letters* 351: 70–77. https://doi.org/10.1111/1574-6968.12347

Vo, Hoang Nhat Phong, Huu Hao Ngo, Wenshan Guo, Thi Minh Hong Nguyen, Yiwen Liu, Yi Liu, Dinh Duc Nguyen, and Soon Woong Chang. 2019. "A critical review on designs and applications of microalgae-based photobioreactors for pollutants treatment." *Science of the Total Environment* 651: 1549–1568. https://doi.org/10.1016/j.scitotenv.2018.09.282

Wang, Hui, Haywood D. Laughinghouse IV, Matthew A. Anderson, Feng Chen, Ernest Willliams, Allen R. Place, Odi Zmora, Yonathan Zohar, Tianling Zheng, and Russell T. Hill. 2012. "Novel bacterial isolate from Permian groundwater, capable of aggregating potential biofuel-producing microalga *Nannochloropsis oceanica* IMET1." *Applied and Environmental Microbiology* 78: 1445–1453. https://doi.org/10.1128/AEM.06474-11

Watanabe, K., M. Imase, K. Sasaki, N. Ohmura, H. Saiki, and H. Tanaka. 2006. "Composition of the sheath produced by the green alga *Chlorella sorokiniana*." *Letters in Applied Microbiology* 42: 538–543. https://doi.org/10.1111/j.1472-765X.2006.01886.x

Xie, Bo, Shawn Bishop, Dan Stessman, David Wright, Martin H. Spalding, and Larry J. Halverson. 2013. "*Chlamydomonas reinhardtii* thermal tolerance enhancement mediated by a mutualistic interaction with vitamin B12-producing bacteria." *The ISME Journal* 7: 1544–1555. https://doi.org/10.1038/ismej.2013.43

Zhao, Yongteng, Hui-Ping Wang, Benyong Han, and Xuya Yu. 2019. "Coupling of abiotic stresses and phytohormones for the production of lipids and high-value by-products by microalgae: A review." *Bioresource Technology* 274: 549–556. https://doi.org/10.1016/j.biortech.2018.12.030

8 Sustainable Bio-Applications of Diatom Silica as Nanoarchitectonic Material

Sahil Kapoor, Meenakshi Singh, Sanchita Paul, Surojit Kar, Trisha Bagchi, Murthy Chavali and K. Chandrasekhar

8.1 INTRODUCTION

Fossil evidence suggests that diatoms are part of the earth's evolutionary history since the Jurassic period. They have sustained millions of years in marine and freshwater environments by phototrophic and heterotrophic mechanisms (Belegratis et al. 2013). They act as oxygen synthesizers and form a major carbon sequestration algal biomass community. The research on diatom nanostructure gained a significant acceptance after a successful phylogenetic characterization that depicts great diversification in body organization (Leynaert et al. 2018). However, nowadays, diatom nanotechnology has multiple applications in drug delivery, biosensors, biophotonics, and cell labelling because silica-based frustules have supramolecular assemblies (Ariga and Fakhrullin 2021). According to Leynaert et al. (2018), silica derived from diatoms can easily be converted into silicon by magnesiothermic reduction to prepare mesoporous silica-based nanomaterials. These nanomaterials are environment-friendly sources of nanoscale silica structures that are currently used in optical biosensors, heavy metals adsorbers, drug delivery, and biomolecule purification (Terracciano et al. 2018).

The fabrication of fine structures is crucial for the production of highly efficient and highly specific functional materials (Tramontano et al. 2020). The nanoarchitectonics approach is a promising alternative for fabricating functional materials from nanoscale components that encompass self-organization, microfabrication, organic synthesis, field-assisted assembly, bio-related processes, and molecular alterations (Figure 8.1). Nanoarchitectonics methodology includes several permutations, combinations, and rational and methodological applications of diverse unit processes, namely energy-intensive non-equilibrium processes and energy-free equilibrium processes (Ariga and Horiz 2021). The diatom nanoporous cell walls can efficiently be complex with silicon

DOI: 10.1201/9781003455950-8

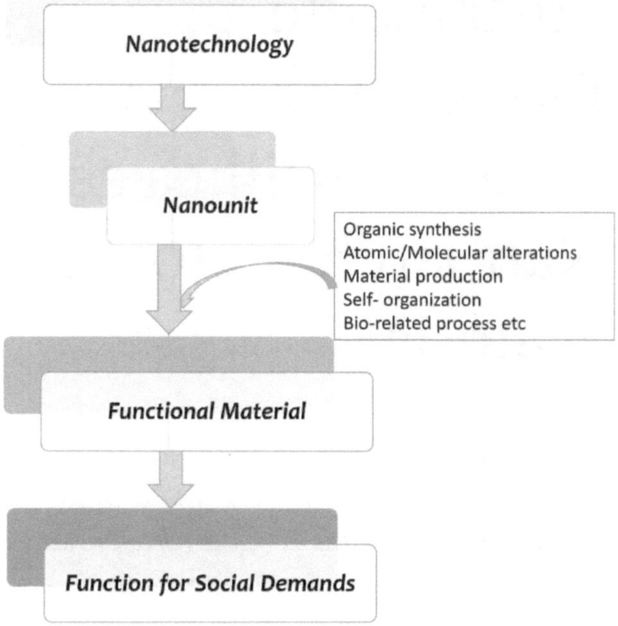

FIGURE 8.1 The nanoarchitectonics methodology for fabricating functional materials from nanoscale components.

substrate to form an integrated membrane platform. They are ideal candidates to be combined with existing devices featuring micron-sized structures such as microfluidics. Silicon provides an ideal substrate because micropores can be precisely manufactured using photolithography and deep reactive ion etching and integrated with microfluidics and microelectronics. The peculiar properties of silica make it a suitable material due to its chemical inertness and thermal stability. According to Ariga et al. (2020), the novel concept of nanoarchitectonic development functional material by combining nano unit components, comprising organic synthesis, biological processes, atomic or molecular alterations, and other supramolecular chemistry systems. Therefore, this review collates the existing literature on diatom nanotechnological applications for making biotemplates, bioprinting, biosensors, biofiltration, biocomposites, biomimetic analogues, and biomanufacturing technology. These biofunctional materials can prove useful to solve existing energy crises and environmental problems based on diatoms' biological, physical, chemical, and optical characteristics.

The scientometric analysis of biogenic silica uses advanced nanotechnological techniques to create various nanofabrication materials. Atomic force microscopy (AFM), transmission electron microscopy (TEM), X-ray photoelectron spectroscopy (XPS), surface-enhanced Raman scattering (SERS), Fourier-transform infrared spectroscopy (FT-IR), and X-ray powder diffraction (XRD) discussed here demonstrate the structural feasibility of biosilica to form a sustainable matrix. Moreover, the challenges faced in diatom-inspired nanotechnologies like photonic bioreactors are discussed in brief to prepare diatom biomaterial. Further, Table 8.1 enlists the bio-functional properties of diatoms and their multidimensional roles in various applications.

TABLE 8.1
Bio-Functional Properties of Diatom-Based Silica Applications.

Applications	Bio-functional properties	References
Biotemplates	• Highly precise, cheap, simply handled, high efficiency. • Nature-inspired material was used. • Lithography, X-ray, and electron beam techniques are compared to biotemplating.	Singh and Chakravarti 2016 Chao et al. 2014
Biosensors	• Biomedical diagnosis at rapid detection is easier and at low cost. • Able to convert bio-recognition into a measurable signal, that is transducer activity. • Detect a specific analyte, that is selectivity • Generate identical responses for a corresponding experimental set-up, that is reproducibility. • Usable in biosensing devices, electrochemical sensors, field-effect transistors, optical sensors, and acoustic-sensitive sensors.	Bhalla et al. 2016 Jadhav et al. 2021
Bioprinting	• 3D assemblage for 3D artificial scaffold. • It was able to form a bone tissue scaffold by combining it with silk fibroin. • It has regenerative capability and uses in bone tissue engineering.	Lowe et al. 2019
Biofiltration	• The 3D pore analogy creates an asymmetric diffusion in diameter that shows a potential for particle separation. • Pores are having the properties of mechanical defense, colloidal sorting, and light harvesting. • The inertness, angularity, and rigidity of diatomaceous biosilica enable it to manipulate microfluidic and nanofluidic particles with sizes ranging from tens of nanometers to micrometers. Examples include viruses, bacteria, and ammonium ions, among others. • Functions as optical filters; protects living cells from excessive-high intensity and harmful wavelengths. • Inhibits ROS-producing excessive blue lights. • Easy multiplication, small transportation, homogenous permeability, encapsulation capability and pore size make them fabricate liquid and microbes-filtrating and DNA-purifying nanodevices. • Microorganisms that are mostly used for volatile compounds are heterotrophs. • Used technology – biofilters (immobilized). • Biotrickling filters (immobilized). • Bioscrubbers (suspended).	Sivangi et al. 2008 Losaic et al. 2009 Mcheik et al. 2018 Terraciano et al. 2018

(Continued)

TABLE 8.1
(Continued)

Applications	Bio-functional properties	References
Biocomposites	• Diatomaceous earth is cost-effective. • Used in food, agriculture, beverages, natural medicines, construction stuffs, cosmetics, lubricating agents, and waterproof adhesives. • Also used as an electrode in lithium-ion batteries for high thermostable capabilities, tough FTIR scattering, periodic pore networking, and amorphous silica composition like opal-A. • Promising nanofabrication compound due to its 3D structural features. • Produces high fatty acids and oil-containing biofuels by the fossilized minerals. • DE fractionation different from base diatomite can be used as an epoxy resin filler.	Losic et al. 2009 Maher et al. 2018 Nowak et al. 2019 Dobrosielska et al. 2021
Biomimetic analogues	• Enzyme immobilization using its bio-silicification properties. • Rapid precipitation of nanoparticles insisted with silaffins and other polyamines in *in vitro* silicic acid solution. • Fabrication of sophisticated nanodevices through biomineralization and mimicking properties.	Ariga et al. 2013, 2021
Biomanufacturing technology	• Used in drug delivery devices, biosensor biocarrier, solar cells, integrated circuits, multiple channel fuel cells, logically operated light processors, battery electrodes and electroluminescent display devices due to exclusive shape, ornamentation, and optical properties. • Compatible to tissue engineering and bio-detection as it can fix itself well to a creature body part without showing any allergic reaction. • Diatoms have naturally carbon-sinking properties used in carbon sequestration (mainly CO_2, the greenhouse gases) by cultivating artificially in autotrophic and mixotrophic modes.	De Yuan et al. 2012 Sethi et al. 2020 Sharma et al. 2021

8.2 DIATOMACEOUS NANOSTRUCTURES – A LIVING SOURCE OF BIOGENIC SILICA

8.2.1 Biophysical Properties

Diatoms play an important role in agriculture due to their biophysical properties. Diatomite, the under-ocean deposition of dead diatom cells, is used as desiccator (silica aerogel), being a renewable dehydrating agent. A type of diatom species

(Mamaghan diatom) is considered dominant due to its biophysical features like 89.896% SiO_2 contents, 515.63 g L^{-1} density, and 2.42 mS/cm electrical conductivity. The increasing level of silica content increases its activity while a decrease in density and EC increases the dynamism of the diatom. This feature increases the kinetic energy itself. The high networking potency of crystalline silica, lacking affinity, and high surface tension co-efficiency make diatomite suitable for emulsion spray-making components used in pesticides (applied uniformly under pressure nozzled) that is cost-effective and economically beneficial.

8.2.2 Mechanical Properties

The diatom frustules have raphe in the middle region that parts along the length with blunt ends, which helps in alleviating the stress concentrations (Hamm et al. 2007). The distance between the hexagon ledge and hexagon centroid in the hexagonal pattern of the frustules provides an elevated moment of inertia in diatoms. A high level of inactivity gives an increase to the enduring capacity of structural deformation (Hamm et al. 2003). The composing material of the diatom frustules is an important feature for its suppleness and rigidity. To form the organic components of the frustules, the event of the co-precipitation of silaffins, polyamine proteins, and silicic acid is as important in the initial growth stage as it is in the marine environment (Kroger 2007, Schroder et al. 2008). The wall that connects the cribrum and the areola is evocative of an I-beam. This I-beam shape is a common technique in nanoscale engineering (Figure 8.2); as an example, the I-beam made up of steel is used in the building construction to offer a competent design because of high lethargy. It allows an increase in bending rigidity and larger shear confrontation. After revealing the mechanical and structural analyses, it is confirmed that diatoms have a mesmerizing nanoscale fabrication to integrate multiple designing concepts ensuring their mechanical constancy optimization (Garcia 2010).

8.2.3 Chemical Properties

To modify some properties of interest, germanium is inserted metabolically into the frustules (experimented on *Pinnularia* sp.) (Jeffryes et al. 2008a). First, the cell

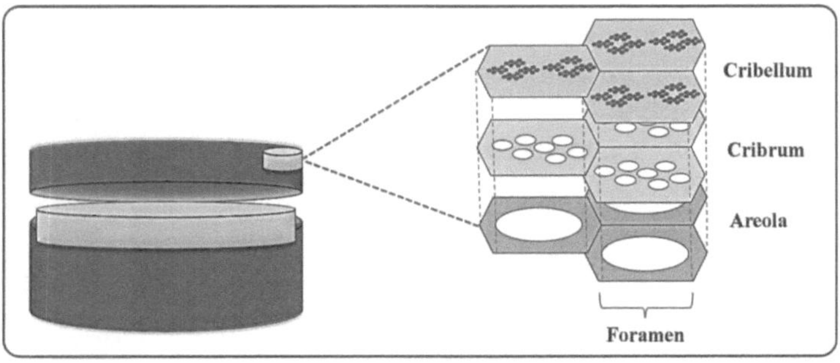

FIGURE 8.2 The internal structural arrangement of a diatom.

suspension is allowed to grow on silicon starvation, co-adding silicon and germanium to allow the dividing cells to take in germanium and incorporate it in the newly divided frustules. No inhibition in cell division is noticed until the molar ratio of Si:Ge is 14:1. Using ICP analysis, silica is alloyed with germanium in the matrix of the frustules to form Ge-O or Si-O-Ge, which cannot precipitate randomly onto the cell surface. The insertion of germanium metabolically can modify the shape of the pore in a controlled way (Rogato and De Tommasi 2020). The diatom biosilica shows a specific blue spectral range peak due to organic residues and silica hydroxide groups (De Tommasi et al. 2018). Besides that, the Ge-doped frustules show electroluminescence featured with narrow peaks constant with considered resonance modes modeled as PhC slab (De Tommasi 2016; Gössling 2017). The doping alloyed with Nickel results in photoluminescence quenching detected at the range of 544 nm after blue light excitation, which is due to the overlapping of frustules with nickel ions and biosilica emission spectra (Townley et al. 2007). The incorporation of Ni in the frustule makes it an efficient component of sensing aquatic pollution. When the alloy is altered with titanium, it lines the base of each pore converting it into TiO_2 (Jeffryes et al. 2008b). It increases the efficiency of dye-sensitized solar cells (DSSCs) (Xiao et al. 2020).

8.2.4 Optical Properties

The biomaterials that are made up of nanostructured biosilica feature a low RI with an exceptionally low absorption coefficient in a visible array. This property makes them appealing to optical studies. The diatom frustules show some new properties pointing to light waveguiding, focusing, scattering, diffraction, and photoluminescence. These features are observed when in non-aquatic conditions to improve structural index contrast and distribution of spectrum. The valve of the frustules is consisting of three parts – cribrum, foramen, and the wall which connects them is the areola. The three parts are aligned together by a hexagonal array of circular pores (Mcheik et al. 2018). The biosilica contributes to optimizing the light organization in leaf and engross to defend them from predators, infections, harmful radiation like UV rays and broad wavelength intensities (Schaller et al. 2013, Pierantoni et al. 2017), ensured metabolic exchange, maintain cellular integrity, and interacts with incident light. If the light wavelength and the holes are in the same order, then it allows interaction. The porosity and the periodicity enable light–frustule interaction as particle–particle dealings diffract and scatter. But the frustules are very rigid to provide mechanical potency under aqueous surroundings (Aitken et al. 2016). The exact role is investigated *in vivo* as it is difficult to measure the living organisms' light intensities or optical properties due to their organelles like chloroplasts that interfere in light and fluorescence gesture analyses (Mcheik et al. 2018). Therefore, the diatom frustule functions as an optical filter (Figure 8.3).

8.2.5 Electronic Properties

In diatoms, the frequency-dependent dielectric loss and dielectric constant are comparatively low as porous silica shows promising low-K-dielectric components having a large surface area. The consistency of the features depends on the increasing level of porosity. The dielectric properties of a diatom are calculated with an impedance

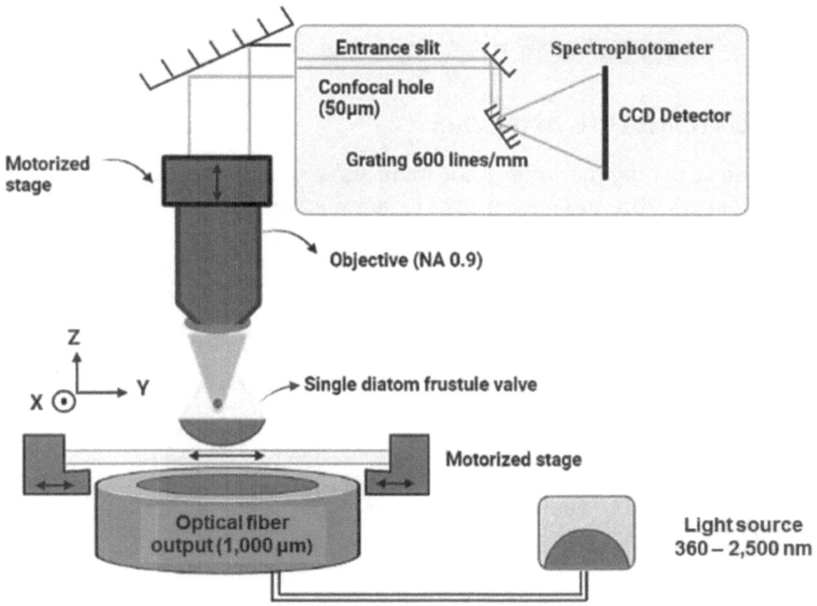

FIGURE 8.3 The diatom biosilica reflects the light, and its porosity enables light–frustule interaction to diffract and scatter, which is captured by a spectrophotometer.

analyser (HP 4192A LF, 10 MHz) and a network analyser (HP 8753A, 10 MHz-1GHz) (Jongwah et al. 2007).

8.2.6 Metallurgical Properties

The diatom plays a crucial role in the biological carbon cycle of the ocean, which influences the food chain leading to a better insight into the ecological factors selected for the species composition **(Smetacek 1999).** Diatom can dominate the blooms of the phytoplanktons as their rate of mortality is too low than any small algae with the same growth rate. Their silicified structure with diverse shapes, convoluted ornamentation, and material properties has an important contribution to succession and mechanical strength that protects them from predators (Milligan and Morel 2002, Hamm et al. 2003). The arm-race between predators and diatoms can influence the biochemical cycles and pelagic food web considerably. A 3D FEM (finite element model) was constructed to evaluate the stress values and distributions within the frustules. The external pressures reproduce a mandible bite to absorb the stress smoothly from the fragile areas where areolas are located mostly, the transversal ribs deflecting the stress concentration. To take out the ribs and to allocate an equivalent amount of material consistently during the frustule deformation and raise stress values more than 70%. Hence, more than 60% force is essential to break the ribless frustules. The material attributes of the diatom biosilica were examined to function as an FEM force of the deformation of an isolated pleura with a calibrated glass microneedle. An obvious deficiency of plastic deformation would point out that brittle failure is consistent in broken frustules. The organic matter is removed from the diatom in a culture medium that is monospecific to separate the purified siliceous components, to perform the load tests (Hamm et al. 2003).

8.3 SCIENTOMETRIC ANALYSIS

We start with searching the word "nanoarchitectonics" in the journal article title for the period January 2010 to January 2022 as summarized in Figure 8.4.

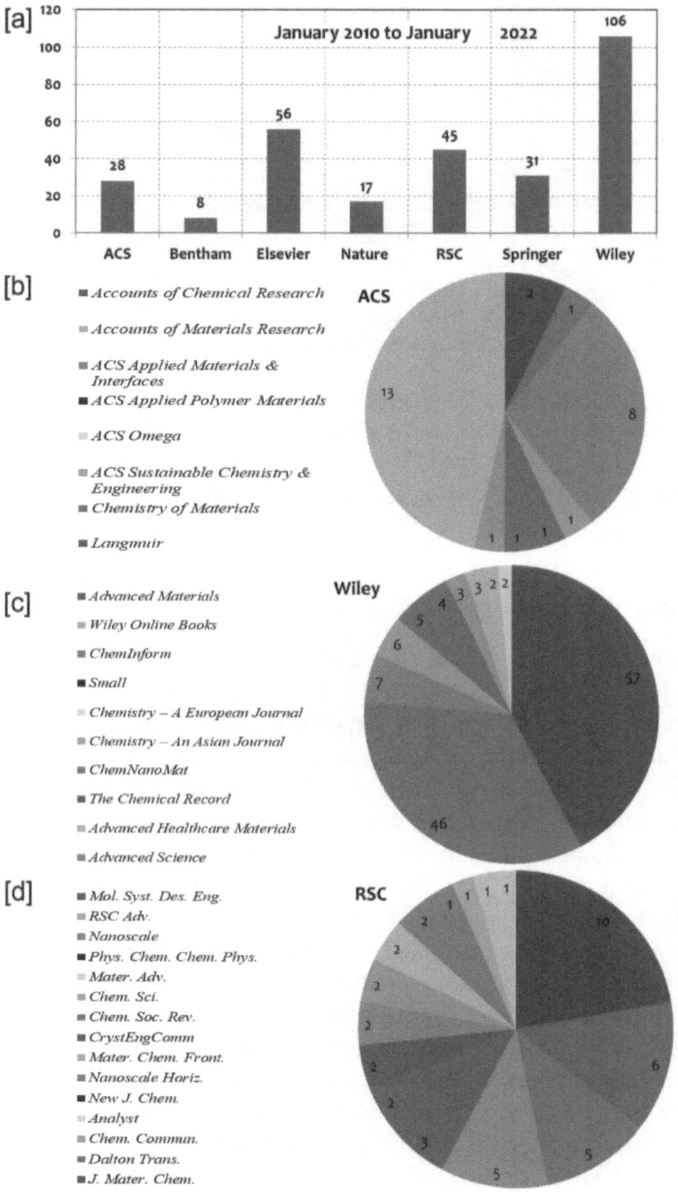

FIGURE 8.4 Scientometric analysis on diatom-based nanoarchitectonics.

8.4 NANOFABRICATION TECHNIQUES TO PREPARE HIERARCHICAL BIOSILICA MATRIX

8.4.1 Atomic Force Microscopy (AFM)

Over the last three decades, AFM has been highly instrumental in understanding the physico-chemical properties and structure of polymer surfaces with remarkable temporal and spatial resolution under diverse conditions. AFM generally provides a two-dimensional map (2D) of the electrical, mechanical, and topographical properties of the surface at a nanometer scale using a sharp tip attached to a force-sensing cantilever (Garcia and Herruzo 2012). AFM can also simultaneously provide both non-topographical and topographical information while operating in liquid conditions (Wang et al. 2018).

Suhtiwanich et al. (2020) visualized the nanophase arrangement and topographic, non-topographic, and mechanical properties of PMPC-b-PMPTSSi thin film using AFM. The FSMM-ASM revealed the nanophase arrangement in terms of adhesion force. AFM has also been used for understanding the *in-situ* adsorption of aqueous amphiphilic copolymer on hydrophobic silica substrate (Hamley et al. 2004). Huang et al. 2017 utilized intermodulation-AFM and tapping mode-AFM for characterizing the mechanical properties of poly(dimethylsiloxane)-hydrophobic silica nanocomposites. Recently, intermodulation-AFM and tapping mode-AFM have also been used to ascertain the surface properties of phospholipid coatings on spray-dried silica nanoparticles (NPs) at the nano-scale level (Forchheimer et al. 2020).

Li et al. (2021) confirmed the formation of vertical mesoporous silica on the nanoporous substrate, which has diverse applications, including enhanced bioactivity, drug delivery, and degradation capacity. Recently, Chen et al. (2021) demonstrated the characterization of advanced biominerals and potential functions of biomineralization in influencing elemental balance in plants supplemented with silica- and iron-enhanced biochar. In a recent study, AFM tips were altered using N-methyl aniline- and methylamine-terminated silane by Si–O–Si bonds to understand the interaction of amines on the graphite surface (Zhang et al. 2021).

8.4.2 Transmission Electron Microscopy (TEM)

TEM is a powerful analytical technique for the structural characterization of NPs (Rajeshkumar and Bharath 2017). Previously, the TEM technique has been utilized to capture the image of the specimen *ex-situ* by periodically terminating the kinetic reaction process (Liao et al. 2013). However, in recent years, liquid cell TEM has been used to understand the trajectories of NPs growth in real time.

Recently, the TEM analysis of poly(2APhS)/silicon dioxide nanocomposites demonstrated that silicon dioxide is entangled within the matrix of polymer, and samples are in the range of 70–120 nm (Benyakhou et al. 2017). Moreover, TEM analysis further suggested that the SiO_2 nanoparticle is uniformly distributed in a poly(2APhS) matrix. TEM was used for the characterization of microwave-synthesized silicon-dioxide-reinforced B–N-co-doped reduced graphene oxide nanocomposites (Xiong et al. 2021). TEM images of B–N–GO displayed the flocculated morphology of graphene oxide. Some researchers used 3D-TEM for the characterization of the 3D structure of silica in natural rubber (Kohjiya et al. 2005). The homogeneous arrangement of constituent phases in alumina–silica nano-hybrid oxide and soft sol-gel-processed alumina–titania catalysts was remarkably resolved by TEM analysis (Wunderlich et al. 2004).

In a recent study, the TEM analysis of PEG-grafted nano-silica composite revealed that spherical NPs are uniformly dispersed in aqueous solution and the polymer is effectively embedded onto the surface of silica NPs (Xu et al. 2018). Moradifar et al. (2018) investigated the thermal stability and chemical and microstructural transformation of highly ordered silica nano-opal meta-lattices at high temperatures. Recently, the structure of silica–polyalanine core-shell NPs was confirmed by TEM analysis (Jang et al. 2006). The outer shell of the nanoparticle demonstrated a remarkable escalation in brightness as compared to the dark inner core, and NPs were almost monodispersed and well distributed on the TEM grid.

8.4.3 X-Ray Photoelectron Spectroscopy (XPS)

XPS is a highly precise method for investigating the elemental composition and chemical state of a solid surface (Ding et al. 2021). XPS has a remarkable potential to investigate the chemical identity of the element due to the perfect fit of the size of particles to its probe length (Briggs 1990). XPS has been widely used for the identification of diverse core–shell-type nanostructures (Hoener et al. 1992). In a recent study, XPS has been used for the characterization of core-shell NPs with silica shells and gold core (Tunc et al. 2005). Recently, XPS was also used for the identification of epoxy soya bean oil (ESO)/nano-silica complex (Ding et al. 2021). XPS analysis revealed the presence of carbon, oxygen, and silica in the ESO complex. The chemical state of Ce, O, and Si elements in Ce-doped silicon dioxide nanocomposites was determined by XPS (El-Salamony et al. 2021). The Si peak at 103.3 eV was attributed to the signal of silicon dioxide. XPS technique has also been used to understand the distribution of Ti–O–Si species on the nanometric and micrometric support surface (Moreno et al. 2021).

Recently, XPS was used to understand the chemical nature of lithium silicate nanosheets and structural modifications during carbon dioxide desorption–adsorption steps (Belgamwar et al. 2021). XPS spectra displayed peaks at 54 eV (Li_4SiO_4 (major phase) and Li_2CO_3 (minor phase)) and 55.9 eV (Li_2SiO_3). The Si (2p) spectra displayed two Si signals at 101 eV (Li_4SiO_4) and 102 eV (Li_2SiO_3). XPS analysis also suggested a mixed-phase model of carbon dioxide desorption and adsorption by the lithium silicate nanosheets.

In a similar study by Soule et al. (2015), XPS spectra demonstrated that the stalk introduction in the silica nano-assemblies and grafting causes substitution of the Si 2p peak in lower energy as well as also showed the presence of CB6 on the nanoparticle surface. Alotaibi et al. (2020) used XPS to demonstrate that the uptake of cationic dyes was considerably enhanced by the novel cobalt–carbon silica adsorbent nanocomposites containing oxygen functional groups.

8.4.4 Surface-Enhanced Raman Scattering (SERS)

Raman spectroscopy is an ultrasensitive, highly specific, and non-invasive optical technique for bioanalytical and imaging applications (Pengphorm et al. 2021, Pham et al. 2017). It has opened the possibility for single-molecule spectroscopy (Nie and Emory 1997). SERS spectroscopy is a promising analytical technique in diverse detection fields (Hahm et al. 2018).

In a recent study, Zhou et al. (2020) demonstrated the formation of pomegranate-like silica-gold NPs by SERS-based analysis. Pham et al. (2017) have developed a silicon dioxide–gold–silver NP complex-based SERS probe for highly sensitive detection of trace amounts of chemicals in diverse fields. Recently, Pham et al. (2021) reported SERS-based detection of 4-mercaptophenyl boronic acid-immobilized Au–Ag core-shell assembled silica nanostructure for specific detection of glucose in neutral conditions. In a recent study, researchers used SiO_2–SO_3H spheres to make 3D silica/wax/Ag (Au) colloidosomes as 3D SERS substrates that have remarkable SERS improvement ability (Wang et al. 2019).

SERS has been used as a powerful tool for complex detection (Sanles-Sobrido et al. 2009). The metallic NPs such as silver NPs generated on silica backbone structure could considerably increase the SERS signal due to the easy-to-handle assembled nanostructure and generation of hot spots (Kim et al. 2018). They also reported a strong SERS signal from single aromatic asymmetric molecules assembly of magnetic and plasmonic NPs with fluorescent silica shell layer. In a recent study, Chang et al. (2017) demonstrated a positive correlation between the SERS activity and surface roughness of Ag-nanoshells by changing the silica core size. It is also supported by the fact that the SERS signal is relatively enhanced on the rough surface of the metallic particle due to the production of active sites with concentrated local electromagnetic fields (Schmidt et al. 2012).

8.4.5 Fourier-Transform Infrared Spectroscopy (FTIR)

FTIR spectroscopy is a powerful technique for understanding the functional groups and surface chemistry of NPs by measuring the infrared intensity versus the wavenumber (wavelength) of light (Vijayaraghavan and Ashokumar 2017). FTIR analysis has been used for the characterization of polyester-based polyurethane/nano-silica composites (Chen et al. 2003). The analysis revealed that certain chemical reactions were initiated between silica NPs and polyester molecules by *in situ* polymerization. Ranjbarzadeh et al. (2019) used FTIR analysis for understanding which functional groups are involved in the formation of silicate networks on the surface of silicon oxide NPs. The FTIR spectra depicted the presence of OH groups, Si–O and H-bonds, on the surface of the NPs. The authors concluded that the FTIR-based analysis can be potentially used to understand the stability of silica–water nanofluid.

FTIR was used for understanding the formation of spherical silver NPs (Bagenskiene et al. 2004). The FTIR spectra of NPs displayed absorption bands appearing from symmetric vibration of Si–O and asymmetric vibration of Si–O and Si–OH, respectively. The surface chemical structure of biomimetic core-shell silica NPs was analysed by FTIR (Tengjisi et al. 2021). FTIR analysis revealed that the SurSi peptide was integrated within the silica shell. FTIR spectra successfully demonstrated the copolymerization of poly(N-isopropyl acrylamide) and methacrylic acid on magnetic mesoporous silica NPs (Tian et al. 2018). FTIR analysis revealed the presence of macromolecular chains of poly(ethyleneimine) on the surface of silicon dioxide (Ghiyasi et al. 2018). The FTIR spectra confirmed the structure of silicon dioxide NPs by the presence of a strong absorption band at 1,000–1,100 cm^{-1} due to the stretching vibration of the Si–O–H and Si–O–Si groups.

8.4.6 X-Ray Powder Diffraction (XRD)

XRD is a useful technique to understand the arrangement of the atoms in the crystal by scattering X-rays by atoms, which results in a definite diffraction pattern (Chauhan and Chauhan 2014). Recently. the nano-β-silicon carbide powder–carbon compacts were used for fabricating silica-bonded porous silicon carbide ceramics (Malik et al. 2021). XRD pattern revealed that the raw nano-β-silicon carbide powders have stacking faults. It also showed polytypic β-to-α (3C→6H) phase transformations in silicon carbide–silicon dioxide ceramics at a minimal temperature (600°C). The structural properties of nano-SiO_2 local metakaolin-based-geopolymer were ascertained by XRD (Zidi et al. 2021). XRD analysis revealed that the nucleation site ofthe gel was substantially improved by the supplementation of 5% of nano-silica.

XRD analysis was used to discriminate the silica forms bedded in the siliceous geodes (Awadh and Yaseen 2019). The analysis showed that the geodes compromised several silica minerals from the outer (quartz, opaline) and the inner layer (calcite). The authors concluded that the XRD is a powerful tool for discriminating between the amorphous and microcrystalline silica forms. XRD analysis demonstrated that the inclusion of an optimum percentage of nanosilica in the cemented sandy soil improves its microstructure properties as confirmed by the decrease in the intensity of the calcium hydroxide peak and increase in the intensity of the calcium silicate hydrate (CSH) peak (Choobbasti and Kutanaei 2017). Recently, the XRD technique was used to understand the structural features of the silica–zirconia powder system (Musyarofah et al. 2019). The analysis demonstrated that the mixture contained a high zircon content. The authors concluded that the XRD data demonstrated the overall crystal structural characteristics of the silica–zirconia complex.

8.5 APPLICATION BASED ON DIATOMS SILICA NANOMATERIALS

8.5.1 Biotemplates

Biotemplating is an advanced process where nature-inspired materials are used as templates. Diatom is a unicellular, eukaryotic alga that has more than 1,00,000 different ubiquitous species, is highly porous, ordered, well ornamented, having broad surface area, and flexible due to its great surface modification making it suitable for undertaking as a material of biotemplating. It aims for designing drug incorporation in medical science due to its properties which have made it safe, risk-free, and low toxic (Chao et al. 2014). The diatom shells or frustules are permeated with SiO_2 which is amorphous. Its complexity and geometric ornamentation in micro and nanoscale give it a biomimetic approach. Photolithography technique is used to pattern the nanofabrics predominantly, but nowadays the nano imprinting lithography (NIL) is most favourable in comparison to the previous one as its reliability and compatibility to roll-to-roll processing are more cost-effective having high quality, resolution, and great speed. It is a smart robust tool that works directly or via a cheap detect-free template. It has two approaches: (i) the diatoms are natural templates that immobilize on a substrate by soft replication. Replication of a single diatom is okay, but the random assembly of diatoms due to replication seems problematic in the controlled area of fabrication. It needs a large surface to control. (ii) Lithographic fabrication makes a diatom-inspired model by 3D laser lithographical technique and maintains all structural qualities that are needed for optimized utilization in technical uses (Belegratis et al. 2013). The conjugation of DNA oligonucleotides to the frustules is used as a template in Au nanoparticles' sequence-specific multilayer assembly to make function it with complementary DNA strands (Losic et al. 2009). Though the diatom valve can be modified by a chemical named alkoxysilane compound, its reactive Si–OH group undergoes a condensation reaction to make the Si–O–Si bridge. The diatom frustules are also used to pattern biomolecules at the nanoscale which provides advantages to integrate easily in the conventional way of processing in the semiconductor industry (Dolatabadi and Guardia 2011).

8.5.2 Bioprinting

Bioprinting is a technology to control the demand for donated organs to be replaced as it includes synthetic polymerized tissues like heart tissue, blood vessels, and cartilage that are used in medical technology. In 1986, C.W. Hull first coined the term 3D bioprinting technology, which has been used in many fields like architecture, jewellery, textiles, machinery, military, aerospace, and especially in medical engineering. It is a regenerative technology in tissue and bone engineering combining rapid prototyping technology to biomanufacture and fabricate layer-by-layer precisely positioned 3D structures. This includes a bio-ink that contains cell suspensions placed in a cartridge controlled by computer patterning. It has many advantages like it saves time, with the highest resolution, great speed, high duplicating ability, and individuation.

There are three types of bioprinting, namely, inkjet, micro extrusion, and laser assisted. Features like cell viability, surface resolution, and material can affect the quality of the product. It is an *in-vitro* type of intricated 3D structure to fabricate organs, implantation of devices, drug screening, cell encapsulation, organ reconstruction, etc. Reconstructed organs can function normally, as it is a Computer-Aided Design (CAD) data-based technique, so it is also called stereolithography (Gu et al. 2016). Diatoms are a good source of biosilica and have broad surface and well-porous structure-intricated 3D assemblage suitable for 3D artificial scaffold matrices to deliver biomolecules. Though this way of diatomaceous biosilica-based approaches are under research or have not a wide way, yet diatomaceous biosilica-based approach regenerative capability has been utilized in bone tissue engineering. As an example, diatom frustules can easily be transformed into nanoparticles by combining with silk fibroin to form a bone tissue scaffold, resulting in increasing cell growth and osteogenesis-assisted key bone biomarker expression. The biocompatibility assessment with various functional groups shows non-toxicity and material suitability in artificial culture for bone repair and drug implementation (Lowe et al. 2019).

8.5.3 Biosensors

Diatoms have been used as bio-sensing platforms due to their distinct optical properties and large surface area. Biosilica derived from diatoms has been used to detect macromolecules by developing a biosensor by the immobilization of biological sensing elements such as enzymes, antibodies, DNA, cells, and nucleic acids onto the surface of the frustule. Silica beads can also conjugate with inorganic and organic dyes through lanthanide chelates and other biomolecules via well-known surface chemistry (Dolatabadi and Guardia 2011). The pore array of diatom silica greatly enhances the wavelength-specific-guiding modes of light absorption and acts as a photonic crystal. The diatom biosilica has electroluminescence and photoluminescence that can be fine-tuned by metabolically incorporating germanium to increase its luminescence properties and use in different sensor applications (Rorrer and Wang 2016). Table 8.2 represents the filtration ability and luminescent properties of diatoms, and the low manufacturing cost of pore arrays makes diatoms a suitable alternative for designing different biosensors (Losic et al. 2009; Ragni et al. 2017). Moreover, 7photoluminescence characteristics and large surface area of porous diatoms frustules make them highly efficient in optical sensing and selective immunocomplex biosensing (Mishra et al. 2017).

8.5.4 Biofiltration

The diatom frustule contains stacked 3D chambers which are hexagonal and are separated by silica plates. The pores that have the properties of providing mechanical defence, as well as mechanical and colloidal sorting and harvesting of light, are uniformly distributed in the plates, decreasing or increasing in size in a centripetal order. The analogy of the 3D chambers can create an asymmetric and constrained diffusion in the pore diameter which shows its potential in particle separation. Diatoms are unique organisms among the microbes to provide angularity, inertness, and rigidity through their biosilica which gives the capability of microfluidic sorting to the surface of the particle. They can sort the nanofluidic particles having the size of tens of nm and μm, 50 nm viruses, 1 mm bacteria, ammonium ions, etc. Diatoms have the process of sorting of the particles through three steps of control, separation, and sensing.

Though other microbes can sort 100 micrometres of sorting capacity, diatoms fix it to a sub-micrometre range with their rigidity. The control procedure is mainly based on the channel microtopography by which the colloidal particles are localized over the hexagonal ridges, then separated by the size deflection strength and sorted (Losic et al. 2009). The frustules also function as optical filters. They can protect living cells from excessive light intensities and harmful UV wavelengths. They use blue and red wavelengths to photosynthesize, and the excess blue light that produces

TABLE 8.2
Summary of Diatom Frustules Acting as Silica-Based Biosensors for Nanomaterials.

S. No.	Diatom	Molecules attached	Detection technique	Sensing materials	Reference
1.	*Coscinodiscus concinnus*	Antibody	Photoluminescence (quenching of biosilica)	Antigen	Dolatabadi and Guardia 2011
2.	*Coscinodiscus wailesii*	Antibody	Label-free electrochemical sensing	Cardiovascular biomarker proteins	Dolatabadi and Guardia 2011
3.	*Thalassiosira rotula*	-	Photoluminescence (quenching of biosilica)	NO_2	Dolatabadi and Guardia 2011
4.	*Cyclotella* sp.	Rabbit immune-globulin-G (I_gG)	Photoluminescence (quenching of biosilica)	Antigen	Gale et al. 2009
5.	*Pinnularia* sp.	Diatom photonic crystal biosilica with self-assembled Au NPs and in-situ growth Ag NPs	SERS	Food testing	Ren et al. 2013
6.	*Thalassiosira pseudonana*	Surface-immobilized cell-specific IgG	LiDSI method	Cancerous cells like B-lymphoma and neuroblastoma	Panwar and Dutta 2019
7.	*Pinnularia* sp.	Diatom photonic crystal biosilica with in-situ growth plasmonic NPs	SERs	Melamine for food safety	Kong et al. 2016
8.	*Pseudostaurosira trainorii*	Diatom biosilica integrated with Au NPs	SERS immunoassay	Interleukins in blood plasma	Kaminska et al. 2017
9.	*Phaeodactylum tricornutum*	Human immuneglobulin-G (IgG)	SERS immunoassay	Hepatitis B virus antigen (HBsAg) in human blood plasma	Gaëtan Vanier et al. 2015
10.	*Microshuttles*	Diatomite nanoparticles (DNP)	Photoluminescence	Medical applications	Sprynskyy et al. 2017
11.	Aulacoseira sp.	DOPA/Fe_3O_4	Magnetic guide, drug loading, and release	Indomethacin	Losic et al. 2010

ROS is inhibited by their ordered porous shells (Mcheik et al. 2018). Homogeneous permeability, fixed pore size, the capability of filtration and encapsulation, transport in a small number, and easy multiplication after that make the frustules a potential scaffold to fabricate a nanodevice for liquid filtration, DNA purification, and microbe's filtration, etc. (Terracciano et al. 2018).

8.5.5 Biocomposites

The main source of biosilica is cultivated from the fossilized mineral named diatomaceous earth (DE) which is a sediment of the dead frustules deposited at the bottom of the sea over million years. It is mined and sold for food, agriculture, beverage, and natural medicine at low cost. This intricate 3D structure of porous silica DE is considered as the most promising nanofabrication material. DE can also produce biofuels on a large scale, which contain high levels of fatty acid and oil. It is used in making more than 300 products in different industries like additives in construction stuff (e.g. CelaBrite®), toothpaste, coarse polishes, polymers, and as absorbents of air and water (Maher et al. 2018). The diatoms can develop waterproof and strong adhesives. The production of lubricating agents and self-healing-devising adhesives can develop the chemistries to indicate tribological problems like micro and nanoscaled frictional wear and tears. By analysing and mimicking this methodology, the bonding and lubrication of the electromechanical (micro or nano) system can be improved (Losic et al. 2009). The artificial lab-made biosilica is used as an electrode (−ve) in lithium-ion batteries. High thermostability, tough scattering in FTIR, periodic networking in pores, and opal-A-like amorphous silica composition are the most important features of biosilica. The precise capacity of the electrode substance is affected by the ratio of DB and CB (carbon black) in electrochemical experiments (Nowak et al. 2019).

8.5.6 Biomimetic Analogues

The generation of biosilica for nanostructured devices is of biochemical interest and so in material science. The reproducible nanostructure made up as the result of biomineralization shows that the process of biomimicking of this strategy can produce sophisticated materials at ambient temperature. These structures can fabricate custom-tailored nanostructures and have numerous applications in diverse fields. Diatom biosilica is composed of 97% inorganic compounds including pure silica with a small amount of ammonium and iron and a few percentages of purified glycoproteins as an organic compound. Freshly prepared silicic acid and oligo propylene amines are reacted together to form artificially customized cell wall silica sheets as a biomimetic model (Noll et al. 2002). At neutral pH, ambient temperature, and pressure "silaffins", phosphorylated polypeptides are formed by a modification to develop silica both in vivo and in vitro, in biosilicification. In biomimetic silica support, enzyme immobilization has been shown using the advantages of the properties of biosilicification under gentle conditions (Ariga et al. 2021, Ariga et al. 2013). The main organic components of biosilica, oligo-N-methyl-propyleneamines and long chains of polyamines, are modified to form silaffin proteins post-transcriptionally. This protein and different lengths of free propyleneamines force the precipitation of nanoparticles rapidly in the silicic acid solution in vitro. Tripropylenetetramine reacts with the aqueous silicic acid in $CHCl_3$ resulting in a thin nanostructured silica layer so similar to the granular nanostructured diatom cell wall. It is an experimental biomimetic approach (Noll et al. 2002).

8.5.7 Biomanufacturing Technology

The functionality and intricacy of naturally structured nano- or microparticles are better than the artificial devices making it a more suitable matter for biotechnological biomanufacturing. The great ornamentation, 3D shapes, large surface area, exclusive optical properties, and multi-level nanopores make diatoms a potential biomanufacturing component. Some experimented manufactured devices are drug delivery devices, biosensor biocarrier, solar cells, battery electrodes, and electroluminescent display devices, etc. In the use of original frustules, diatoms can limit the function due to a certain size, and arranging bonding techniques limits the detection stability. To improve the device functionality, the bonding pattern, modification, and assembly play an important role. As biogenic silica is compatible, it can fix itself well to a creature's body without any allergic reaction, in the field of tissue engineering and biodetection. These bio-manufactured items are preassembled to avoid assembling problems. As an example, we can address fuel cells with multiple channels; incorporated circuits; logical operation performing light processors and diatom cloths with multifunctions like air-conditioning, humidity adjustment, and light warming (De Yuan et al. 2012).

8.6 CHALLENGES ENCOUNTERED IN DIATOM-INSPIRED NANOSTRUCTURE TECHNOLOGIES

8.6.1 Photonic Nanotechnology

Diatoms are photosynthetic marine alga with a pair of valves in their skeletal structure, which are made up of hydrated amorphous silica. These are called frustules and possess hierarchical micro- or nano-scaled photon crystal-like pores (Kong et al. 2016). They can absorb or intake volatile substances like hydrocarbons and solvents, etc., which is an important characteristic of an optical transducer. They should be with a large surface where the gases and vapours can be induced to condense. The chemical vapour in the silicon shell and the organic molecular air as a substitution cause an average refractive index to increase because of capillary condensation. It is detected by several optical mechanisms (Spanier and Herman 2000, De Stefano et al. 2009). It is observed that the silicon shell of diatom, which is cleaned with concentrated sulphuric acid solution and wet deposited on silicon water that is intrinsic and single polished, contributes a negligible amount of photoluminescence signal at a considered wavelength. A minimal amount of NO_2 is used as a gaseous texture for observing the quenching effect, but it becomes reversible in the presence of air. The wavelength peak can be shifted due to the gaseous absorption through the diatomaceous nanopores that can increase the average refractive index of a silicon optical device (De Stefano et al. 2009, De Stefano et al. 2004).

A newly invented bioenabled nanoplasmonic sensor has been demonstrated to be mediating on biosilica photonic crystals with *in-situ* nanoparticles of silver to show the chemical and biological sensing (label-free) based on SERS (surface-enhanced Raman scattering) from the complex sample. The ratio of the biosilica and large surface area may accumulate more molecules to analyse in the SERS surface to detect biomolecules unable to absorb metallic particles (Kong et al. 2016). A ubiquitous material like diatom frustules shows a possibility to obtain dual parameters of an optical sensor suggested by the substances has an exposure to register the effect of luminescent intensity and band peak position. Due to high symmetrical 3D ornamentation and pores, it shows peculiar optical features as photonic crystals. Also, it is very cost-effective which makes it a good material for nanotechnology, once the controlled structural manipulation is optimized (Massera et al. 2004).

8.6.2 Bioreactor Nanotechnology

It is known that bioreactors are a type of medium where raw materials transform into by-products through whole cells or cell-free enzymes. In nanotechnology, there are a few examples of bioreactors in the form of living organisms. In an experiment, it has been shown that the diatom frustules are used as photobioreactors in a two-stage bioreactor cultivation process to insert titanium into the patterned diatomaceous biosilica as it is known that diatom can accumulate trace levels of titanium, whereas other marine organisms can contain 0.01 to 0.13 wt % titanium in silica. Here, the diatom named *Pinnularia* sp. has been used as its frustule's possess two-scale periodic order, a rectangular lattice at submission scale and a fine-featured concentric array that lines the base of each pore. A type of bubble column photobioreactor has been used in liquid suspension (Jeffryes et al. 2008b), ambient temperature, and neutral pH without any raspy chemicals or any advanced mechanisms. First, a silicon-starved cell suspension is made by growing the diatoms in dissolved silicon. When it is achieved, the desirable suspension of the titanium and silicon is fed to the starved suspension continuously in a controllable way up to 10 hours. The titanium addition does not harm the cell growth and the shell morphology. The co-feeding of titanium and silicon is a requirement for the complete intercellular uptake of titanium. Titanium has been deposited on the baselining of each pore of the diatom as nanocrystalline anatase TiO_2 by thermal annealing (720°C in the air) with an average size of 32-nm crystal (Jeffryes et al. 2008c). The intact frustules are then isolated with the SDS/EDTA treatment and studied with TEM and STEM-EDS. TiO_2 is a semiconductor that is fabricated by bottom-up self-assembly of a living organism reported first (Figure 8.5). The devices made with this semiconductor material can include dye-sensitized solar cells for extended light-capturing efficiency and structured photocatalysts for a massive breakdown of noxious chemicals (Mihi et al. 2008, Carbonell et al. 2008, Wang et al. 2020). This is a unique invention in bioreactor nanotechnology to date, and further research study about this is undergoing.

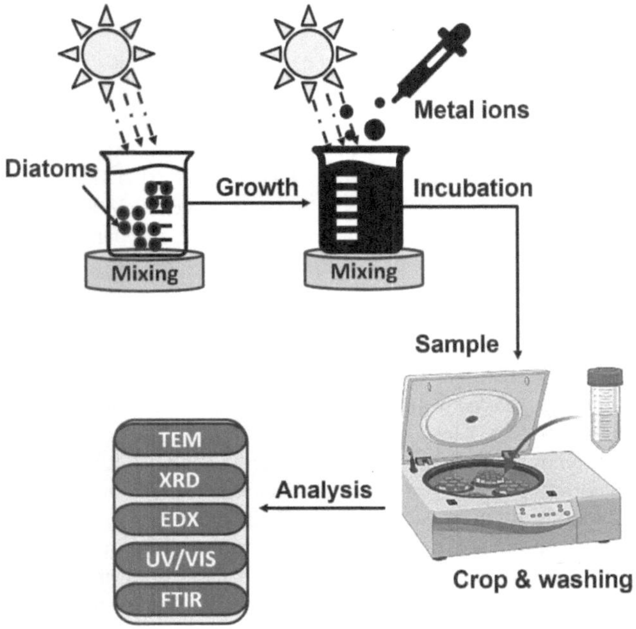

FIGURE 8.5 Diatom fabrication with metal ions by EDTA treatment and the study of growth analysis using TEM, XRD, EDX, UV/VIS spectrophotometry, and FTIR.

8.7 CONCLUSION

Diatoms are a promising and cost-effective biological source for the industrial production of silica-based biomaterials because of their higher growth rate and biomass productivity. Diatom-derived silica has numerous applications in diverse fields due to its high biocompatibility, low cost, eco-friendliness, high surface area, good thermal stability, and other physicochemical properties. Technological advancement has unleashed new insights into the morphology and structure of diatoms frustules. However, there are still some issues pertaining to the toxicity, low degradation rate, and purity of silica derived from diatoms. Moreover, it has not been approved as GRAS material yet by FDA and other government agencies. Therefore, future studies need to focus on sustainable biotechnological production of high-purity silica-based biomaterials from diatoms in environment friendly and cost-effective manner. Moreover, genetic engineering and recent advancement in upscale technologies may further increase the industrial applications of diatoms.

AUTHORSHIP CONTRIBUTION

Sahil Kapoor has contributed towards conceptualization, methodology, and data curation; **Meenakshi Singh** to supervision, writing–reviewing, and editing; **Sanchita Paul, Surojit Kar, and Trisha Bagchi** to writing – original draft preparation, revisions, and editing; and **Murthy Chavali** and **K. Chandrasekhar** have contributed towards supervision, software, and validation.

REFERENCES

A. Beganskienė, V. Sirutkaitis, M. Kurtinaitienė, R. Juškėnas, A. Kareiva, FTIR, TEM and NMR investigations of Stöber silica nanoparticles, *The Journal of Materials Science* 10 (2004) 287–290.

A. Chauhan, P. Chauhan, Powder XRD technique and its applications in science and technology, *Journal of Analytical & Bioanalytical Techniues* 5 (2014) 1–5.

A. Hamm, R. Merkel, O. Springer, P. Jurkojc, C. Maier, K. Prechtel, V. Smetacek, Architecture and material properties of diatom shells provide effective mechanical protection, *Nature* 421 (2003) 841–843. https://doi.org/10.1038/nature01416.

A. Kaminska, M. Sprynskyy, K. Winkler, T. Szymborski, Ultrasensitive SERS immunoassay based on diatom biosilica for detection of interleukins in blood plasma, *Analytical and Bioanalytical Chemistry* 409 (2017) 6337–6347.

A. Leynaert, C. Fradel, B. Beker, C. Soler, G. Delebecq, A. Lemercier, P. Pondaven, K. Heggarty, Diatom frustules nanostructure in pelagic and benthic environments, silicon. 10 (2018) 2701–2709.

A. Mcheik, S. Cassaignon, J. Livage, A. Gibaud, S. Berthier and P. J. Lopez, Optical properties of nanostructured silica structures from marine organisms, *Frontiers in Marine Science* 5 (2018) 123. https://doi.org/10.3389/fmars.2018.00123.

A. P. Garcia. Hierarchical and size-dependent mechanical properties of silica and silicon nanostructures inspired by diatom algae. Dissertation paper in fulfilment of the requirement for the degree of Master of Science, Department of Civil and Environmental Engineering, Massachusetts Institute of Technology, 2010.

A. P. Nowak, M. Sprynskyy, W. Brzozowska, A. Lisowska-Oleksiak, Electrochemical behavior of a composite material containing 3D-structured diatom biosilica, *Algal Research* 41 (2019) 101538, ISSN 2211–9264. https://doi.org/10.1016/j.algal.2019.101538

A. J. Choobbasti, S. S. Kutanaei, Microstructure characteristics of cement-stabilized sandy soil using nanosilica, *Journal of Rock Mechanics and Geotechnical Engineering* 9 (2017) 981–988. https://doi.org/10.1016/j.jrmge.2017.03.015.

A. J. Milligan, F. M. M. Morel, A proton buffering role for silica in diatoms, *Science* 297 (2002) 1848–1850.

B. Lowe, F. Guastaldi, M. L. Müller, F. Gootkind, M. J. Troulis, Q. Ye, *Nanobiomaterials for Bone Tissue Engineering*. Springer Nature Singapore Pte Ltd, 2019.

A. H. Choi, B. Ben-Nissan (eds.), Marine-derived biomaterials for tissue engineering applications, *Springer Series in Biomaterials Science and Engineering* 14. https://doi.org/10.1007/978-981-13-8855-2_4

B. Wang, Z. Wang, T. Chen, X. Zhou, Development of novel bioreactor control systems based on smart sensors and actuators, *Frontiers in Bioengineering and Biotechnology* 8 (2020) 1–7. https://doi.org/10.3389/fbioe.2020.00007

C. F. Hoener, K. A. Allan, A. J. Bard, A. Campion, M. A. Fox, T. E. Mallouk, S. E. Webber, J. M. White, Demonstration of a shell-core structure in layered cadmium selenide-zinc selenide small particles by x-ray photoelectron and Auger spectroscopies, *The Journal of Physical Chemistry* 96 (1992) 3812–3817. https://doi.org/10.1021/j100188a045.

C. Hamm, V. Smetacek, *Evolution of Primary Producers in the Sea*. Elsevier, 2007.

C. Jeffryes, T. Gutu, J. Jiao, G. L. Rorrer, Metabolic insertion of nanostructured germanium into the silica microstructure of the diatom pinnularia sp. *Materials Science and Engineering: C* 28 (2008a) 107–118.

C. Jeffryes, T. Gutu, J. Jiao, G. L. Rorrer, Two-stage photobioreactor process for the two-stage photobioreactor process for the metabolic insertion of nanostructured germanium into the silica microstructure of the diatom pinnularia sp, *Materials Science and Engineering: C* 28 (2008b) 107–118.

C. Jeffryes, T. Gutu, J. Jiao, G. L. Rorrer, metabolic insertion of nanostructured TiO_2 into the patterned biosilica of the diatom pinnularia sp. By a two-stage bioreactor cultivation process, *ACS Nano* 2 (2008c) 2103–2112. https://doi.org/10.1021/nn800470x

C. Tramontano, G. Chianese, M. Terracciano et al., Nanostructured biosilica of diatoms: From water world to biomedical applications. *Applied Sciences* 10(19) (2020): 6811. https://doi.org/10.3390/app10196811

D. A. Jadhav, A. D. Chendake, D. Ghosal, A. Singh Mathuriya, S. S. Kumar, S. Pandit, Advanced microbial fuel cell for biosensor applications to detect quality parameters of pollutants, In *Bioremediation, Nutrients, and Other valuable product Recovery*. Elsevier, 2021, 125–139. https://doi.org/10.1016/B978-0-12-821729-0.00003-8.

D. Briggs, Practical surface analysis, Auger X-Ray photoelect, *Spectroscopy* 1 (1990) 151–152.

D. Losic, J. G. Mitchell, N. H. Voelcker, Diatomaceous lessons in nanotechnology and advanced materials, *Advanced Materials* 21 (2009) 1–12. https://doi.org/10.1002/adma.200803778

D. Losic, Y. Yu, M. S. Aw, S. Simovic, B. Thierry, J. Addai-Mensah, Surface functionalisation of diatoms with dopamine modified iron-oxide nanoparticles: Toward magnetically guided drug microcarriers with biologically derived morphologies, *Chemical Communications* 46 (2010) 6323–6325. https://doi.org/10.1039/c0cc01305f.

D. Sethi, T. O. Buttler, F. Shuhaili, S. Vaidyanathan, Diatoms for carbon sequestration and Bio-based manufacturing, *Biology* 9 (8) (2020) 217 ISSN 2079–7737. https://doi.org/10.3390/biology9080217

D. Wang, T. P. Russell, Advances in atomic force microscopy for probing polymer structure and properties, *Macromolecules* 51 (2018) 3–24.

E. Carbonell, F. Ramiro-Manzano, I. Rodriguez, A. Corma, F. Meseguer, H. Garcia, Enhancement of TiO2 photocatalytic activity by structuring the photocatalyst film as a photonic sponge, photochem, *Photochemical & Photobiological Sciences* 7 (2008) 931–935.

E. De Tommasi, R. Congestri, P. Dardano, A. C. De Luca, S. Managò, I. Rea, M. De Stefano, UV-shielding and wavelength conversion by centric diatom nanopatterned frustules, *Scientific Reports* 8 (2018) 1–14.

E. De Tommasi, Light manipulation by single cells: The case of diatoms, *Journal of Spectroscopy* (2016) 2490128.

E. Hahm, M. G. Cha, E. J. Kang, X.-H. Pham, S. H. Lee, H.-M. Kim, D.-E. Kim, Y.-S. Lee, D.-H. Jeong, B.-H. Jun, Multilayer ag-embedded silica nanostructure as a surface-enhanced raman scattering-based chemical sensor with dual-function internal standards, *ACS Applied Materials & Interfaces* 10 (2018) 40748–40755. https://doi.org/10.1021/acsami.8b12640.

E. Massera, I. Nasti, L. Quercia, I. Rea, G. Di Francia, improvement of stability and recovery time in porous-silicon-based NO2 sensor, *Sensors & Actuators, B: Chemical* 102 (2004) 195–197. https://doi.org/10.1016/j.snb.2004.04.018.

F. Noll, M. Sumper, N. Hampp, Nanostructure of diatom silica surfaces and biomimetic analogues, *Nano Letters* 2 (2002). https://doi.org/10.1021/nl015581k

F. Ren, J. Campbell, X. Wang, G. L. Rorrer, A. X. Wang, Enhancing surface plasmon resonances of metallic nanoparticles by diatom biosilica, *Optics Express* 21 (2013) 15308–15313.

G. Chen, S. Taherymoosavi, S. Cheong, Y. Yin, R. Akter, C. E. Marjo, A. M. Rich, D. R. G. Mitchell, X. Fan, J. Chew, G. Pan, L. Li, R. Bian, J. Horvat, M. Mohammed, P. Munroe, S. Joseph, Advanced characterization of biomineralization at plaque layer and inside rice roots amended with iron- and silica-enhanced biochar, *Scientific Reports* 11 (2021) 159. https://doi.org/10.1038/s41598-020-80377-z.

G. Vanier, F. Hempel, P. Chan, M. Rodamer, D. Vaudry, U. G. Maier, P. Lerouge, M. Bardor. Biochemical characterization of human anti-hepatitis B Monoclonal Antibody Produced in the Microalgae Phaeodactylum tricornutum. https://doi.org/10.1371/journal.pone.0139282 October 5, 2015

G. Debra, T. Gutu, J. Jiao, C.-H. Chang, G. Rorrer, Photoluminescence detection of biomolecules by antibody-functionalized diatom biosilica, *Advanced Functional Materials* 19 (2009) 926–933. https://doi.org/10.1002/adfm.200801137.

G. Ding, X. Yu, J. Si, J. Mei, J. Wang, B. Chen, Influence of epoxy soybean oil-modified nano-silica on the compatibility of cold-mixed epoxy asphalt, *Materials and Structures* 54 (2021) 16. https://doi.org/10.1617/s11527-020-01611-8.

G. L. Rorrer and A. X. Wang, Nanostructured diatom frustule immune sensors, *Frontiers in Nanoscience and Nanotechnology* 2 (2016) 128–130. https://doi.org/10.15761/FNN.1000122

H. C. Schroder, X. Wang, W. Tremel, H. Ushijima and W. E. Muller, Bio-fabrication of biosilica-glass by living organisms, *Natural Product Reports* 25 (2008) 455–474.

H. Chang, E. Ko, H. Kang, M. G. Cha, Y.-S. Lee, D. H. Jeong, Synthesis of optically tunable bumpy silver nanoshells by changing the silica core size and their SERS activities, *RSC Advances* 7 (2017) 40255–40261. https://doi.org/10.1039/C7RA06170F.

H. E. Townley, K. L. Woon, F. P. Payne, H. White-Cooper, A. R. Parker, Modification of the physical and optical properties of the frustule of the diatom Coscinodiscus wailesii by nickel sulfate, *Nanotechnology* 18 (2007).

H. G. Liao, K. Niu, H. Zheng, Observation of growth of metal nanoparticles, *Chemical Communications* 49 (2013) 11720–11727. https://doi.org/10.1039/C3CC47473A.

H. Huang, I. Dobryden, P.-A. Thorén, L. Ejenstam, J. Pan, M. L. Fielden, D. B. Haviland, P. M. Claesson, Local surface mechanical properties of PDMS-silica nanocomposite probed with Intermodulation AFM, *Composites Science and Technology* 150 (2017) 111–119. https://doi.org/10.1016/j.compscitech.2017.07.013.

H. M. Kim, D.-M. Kim, C. Jeong, S. Y. Park, M. G. Cha, Y. Ha, D. Jang, S. Kyeong, X.-H. Pham, E. Hahm, S. H. Lee, D. H. Jeong, Y.-S. Lee, D.-E. Kim, B.-H. Jun, Assembly of plasmonic and magnetic nanoparticles with fluorescent silica shell layer for tri-functional SERS-magnetic-fluorescence probes and its bioapplications, *Scientific Reports* 8 (2018) 13938. https://doi.org/10.1038/s41598-018-32044-7.

I. Mihi, M. E. Calvo, J. A. Anta, H. Miguez, Spectral response of optal-based dye-sensitized solar cells, *The Journal of Physical Chemistry C* 112 (2008) 13–17.

I. Rogato, E. De Tommasi, Physical, chemical, and genetic techniques for diatom frustule modification: Applications in Nanotechnology, *Applied Sciences* 10 (2020) 8738. https://doi.org/10.3390/app10238738.

I. Tunc, S. Suzer, M. A. Correa-Duarte, L. M. Liz-Marzán, XPS characterization of Au (Core)/SiO2 (Shell) nanoparticles, *The Journal of Physical Chemistry B* 109 (2005) 7597–7600. https://doi.org/10.1021/jp050767j.

I. W. Hamley, S. D. Connell, S. Collins, In situ atomic force microscopy imaging of adsorbed block copolymer micelles, *Macromolecules* 37 (2004) 5337–5351. https://doi.org/10.1021/ma049396f.

J. An, D. Forchheimer, J. Sävmarker, M. Brülls, G. Frenning, Nanoscale characterization of PEGylated phospholipid coatings formed by spray drying on silica microparticles, *The Journal of Colloid and Interface Science* 577 (2020) 92–100.

J. E. N. Dolatabadi, M. de la Guardia, Applications of diatoms and silica nanotechnology in biosensing, drug and gene delivery, and formation of complex metal nanostructures, *Trends in Analytical Chemistry* 30 (9) (2011). https://doi.org/10.1016/j.trac.2011.04.015

J. E. Spanier, I. P. Herman, Use of hybrid phenomenological and statistical effective-medium theories of dielectric functions to model the infrared reflectance of porous SiC films, *Physical Review B* 61 (2000) 10437–10450. https://doi.org/10.1103/PhysRevB.61.10437.

J. Jang, J. Ha, B. Lim, Synthesis and characterization of monodisperse silica–polyaniline core-shell nanoparticles, *Chemical Communications* (2006) 1622–1624. https://doi.org/10.1039/B600167J.

J. Schaller, C. Brackhage, E. Bäucker, E. G. Dudel, UV-screening of Grasses by plant silica layer? *The Journal of Biosciences* 38 (2013) 413–416. https://doi.org/10.1007/s12038-013-9303-1.

J. T. Chao, M. J. P. Biggs, A. S. Pandit, diatoms: A biotemplating approach to fabricating drug delivery reservoirs, *Expert Opinion on Drug Delivery* 11 (2014) 1687–1695. https://doi.org/10.1517/17425247.2014.935336

J. W. Gössling, Biophotonics of diatoms: Linking frustule structure to photobiology. Ph.D. Thesis, Department of Biology, Faculty of Science, University of Copenhagen, Copenhagen, Denmark, 2017.

J. W. Jongwah, K. S. Butcher, M. Wintrebert-Fouquet, J. M. Dawes, J. M. Ferris, Optical and electronic properties of diatoms. In *Frontiers in Optics 2007/Laser Science XXIII/ Organic Materials and Devices for Displays and Energy Conversion*. Optical Society of America (OSA), 1–1. https://doi.org/10.1364/FIO.2007.JSuA43.

J. Xu, Z. Qiu, X. Zhao, H. Zhong, G. Li, W. Huang, Synthesis and characterization of shale stabilizer based on polyethylene glycol grafted nano-silica composite in water-based drilling fluids, *Journal of Petroleum Science and Engineering* 163 (2018) 371–377.

K. Ariga, R Fakhrullin, Nano architectonics on living cells, *RSC Advances* 11 (2021).

K. Ariga, K. Tsai, L. K. Shrestha, S. Hsu, Life science nanoarchitectonics at interfaces, *Materials Chemistry Frontiers* 5(3), (2021) 1018–1032.

K. Ariga, N. Horiz, Nano architectonics: What's coming next after nanotechnology? 6 (2021) 364. https://doi.org/10.1039/d0nh00680g.

K. Ariga, Q. Ji, T. Mori, M. Naito, Y. Yamauchi, A. Hideki, J. P. Hillab, Enzyme nano architectonics: organization and device application, *Chemical Society Reviews* 42 (2013) 6322–6345.

K. Ariga, X. Jia, J. Song, J. P. Hill, D. T. Leong, Y. Jia, J. Li, Nanoarchitectonics beyond self-assembly: Challenges to create bio-like hierarchic organization. *Angewandte Chemie International Edition* 59(36) (2020) 15424–15446. https://doi.org/10.1002/ANIE.202000802

Mondarte, M. Takai, T. Hayashi, Imaging the nanophase-separated structure of block copolymer thin film by atomic force microscopy in aqueous solution, *Chemistry Letters* 49 (2020) 641–644.

K. Vijayaraghavan, T. Ashokkumar, Plant-mediated biosynthesis of metallic nanoparticles: A review of literature, factors affecting synthesis, characterization techniques and applications, *The Journal of Environmental Chemical Engineering* 5 (2017) 4866–4883.

K. Wang, L. Feng, X. Li, W. Wang, Fabrication of 3D wax/silica/Ag(Au) colloidosomes as surface-enhanced Raman spectroscopy substrates based on Pickering emulsion and seed-mediated growth method of noble metal nanoparticles, *Journal of Materials Research* 34 (2019) 2137–2145.

L. De Stefano, I. Rendina, L. Moretti, A. M. Rossi, S. Tundo, Smart optical sensors for chemical substances based on porous silicon technology, *Applied Optics* 43, 167–172 (2004).

L. De Stefano, P. Maddalena, L. Moretti, I. Rea, I. Rendina, E. De tommasi, V. Mocella, M. De Stefano, Nano-biosilica from marine diatoms: A brand new material for photonic applications, *Superlattices and Microstructures* 46 (2009) 84–89.

M. Dobrosielska, R. Dobrucka, M. Gloc, D. Brzakalski, M. Szymanski, K. J. Kurzydlowski, R. E. Przekop, A new method of diatomaceous earth fractionation – A bio-raw material source for Epoxy-based composites, *Materials* 14 (2021) 1663. https://doi.org/10.3390/ma14071663

M. Mishra, A. P. Arukha, T. Bashir, D. Yadav, GBKS Prasad, All new faces of diatoms: Potential source of nanomaterials and beyond, *Frontiers in Microbiology* 81 (2017). https://doi.org/10.3389/fmicb.2017.01239

M. Pierantoni, R. Tenne, V. Brumfeld, V. Kiss, D. Oron, L. Addadi, et al., Plants and light manipulation: The integrated mineral system in okra leaves, *Advanced Science* 4 (2017).

M. R. Belegratis, V. Schmidt, V. Nees, B. Stadlober, Hartmann, Diatom- inspired templates for 3D replication: Natural diatoms versus laser written artificial diatoms, *Bioinspiration & Biomimetics* 9 (2013).

M. Sanles-Sobrido, W. Exner, L. Rodríguez-Lorenzo, B. Rodríguez-González, M. A. Correa-Duarte, R. A. Álvarez-Puebla, L. M. Liz-Marzán, Design of SERS-encoded, submicron, hollow particles through confined growth of encapsulated metal nanoparticles, *Journal of the American Chemical Society* 131 (2009) 2699–2705.

M. Sprynskyy, P. Pomastowski, M. Hornowska, A. Król, K. Rafi ́nska, B. Buszewski, Naturally organicfunctionalized 3D biosilicafrom diatom microalgae, *Materials & Design* 132 (2017) 22–29.

M. Terracciano, Diatoms green nanotechnology for biosilica- based drug delivery systems, *Pharmaceutics* 20 (2018).

M. Terracciano, L. De Stefano, I. Rea, Diatoms green nanotechnology for biosilica-based drug delivery systems, *Pharmaceutics* 10 (2018) 242.

M. S. Schmidt, J. Hübner, A. Boisen, Large area fabrication of leaning silicon nanopillars for surface-enhanced Raman spectroscopy, *Advanced Materials* 24 (2012) OP11–OP18.

Musyarofah, S. Soontaranon, W. Limphirat, Triwikantoro, S. Pratapa, XRD, WAXS, FTIR, and XANES studies of silica-zirconia systems, *Ceramics International* 45 (2019) 15660–15670.

N. Alotaibi, H. H. Hammud, N. Al Otaibi, S. G. Hussain, T. Prakasam, Novel cobalt–carbon@silica adsorbent, *Scientific Reports* 10 (2020) 18652. https://doi.org/10.1038/s41598-020-75367-0.

N. Bhalla, P. Jolly, N. Formisano, P. Estrela, Introduction to biosensors, *Essays in Biochemistry* 60 (1) (2016) 1–8. https://doi.org/10.1042/EBC20150001.

N. Kroger, Prescribing diatom morphology: Towards genetic engineering of biological nanomaterials, *Current Opinion in Chemical Biology* 11(2007) 662–669.

N. Sharma, D. P. Simon, A. M. Diaz-Garza, E. Fantino, A. Messaabi, F. Meddeb-Mouelhi, H. Germain, Desgagné-Penix, Diatoms biotechnology: Various industrial applications for a greener tomorrow, *Frontiers in Marine Science* 8 (2021) 636613. https://doi.org/10.3389/fmars.2021.636613

P. Moradifar, J. L. Russell, T. E. Mallouk, N. Alem, In-situ TEM study of formation of an ordered hollow structure metalattice from silica nano-opals through high-temperature annealing, *Microscopy and Microanalysis* 24 (2018) 320–321.

P. Pengphorm, P. Nuchuay, N. Boonrod, S. Boonsit, P. Srisamran, S. Thongrom, P. Pewkum, P. Kalasuwan, P. van Dommelen, C. Daengngam, Fabrication of 3D surface-enhanced Raman scattering, *Journal of Physics: Conference Series* 1719 (2021) 12082.

Q. Gu, H. Zhu, J. Li, X. Li, J. Hao, G. G. Wallace, Q. Zhou, Three-dimensional bioprinting speeds up smart regenerative medicine, *National Science Review* 3 (2016) 331–344, 2016.

R. Belgamwar, A. Maity, T. Das, S. Chakraborty, C. P. Vinod, V. Polshettiwar, Lithium silicate nanosheets with excellent capture capacity and kinetics with unprecedented stability for high-temperature CO_2 capture, *Chemical Science* 12 (2021) 4825–4835.

R. Garcia, E. T. Herruzo, The emergence of multifrequency force microscopy, *Nature Nanotechnology* 7 (2012) 217–226.

R. Malik, Y. W. Kim, W. H. Nam, Intrinsic microstructures of silica-bonded porous nano-SiC ceramics, *The Journal of the American Ceramic Society* 104 (2021) 706–710.

R. Ragni, S. Cicco, D. Vona, G. Leone, & G. M. R. Farinola, Biosilica from diatoms microalgae: Smart materials from biomedicine to photonics. *Journal of Materials Research* 32 (2017) 279–291. https://doi.org/10.1557/JMR.2016.459/METRICS

R. Ranjbarzadeh, A. Moradikazerouni, R. Bakhtiari, A. Asadi, M. Afrand, An experimental study on stability and thermal conductivity of water/silica nanofluid: Eco-friendly production of nanoparticles, *The Journal of Cleaner Production* 206 (2019) 1089–1100.

R. Sivangi Suryanarayana. Biofiltration- A versatile technique in pollution control. Conference: National Level Conference On Novel Products and Processes In Chemical Engineering. At: St. Joseph's College of Engineering, Chennai. 2008.

R. A. El-Salamony, E. Amdeha, A. M. El Shafey, A. M. Al Sabagh, Preparation and characterisation of Ce-doped SiO_2 nano-materials as effective photo-catalyst under visible light, *International Journal of Environmental Analytical Chemistry* (2021).

S. Benyakhou, A. Belmokhtar, A. Zehhaf, A. Benyoucef, Development of novel hybrid materials based on poly(2-aminophenyl disulfide)/silica gel: Preparation, characterization and electrochemical studies, *The* Journal *of* Molecular Structure 1150 (2017) 580–585.

S. Ghiyasi, M. G. Sari, M. Shabanian, M. Hajibeygi, P. Zarrintaj, M. Rallini, L. Torre, D. Puglia, H. Vahabi, M. Jouyandeh, F. Laoutid, S. M. R. Paran, M. R. Saeb, Hyperbranched poly(ethyleneimine) physically attached to silica nanoparticles to facilitate curing of epoxy nanocomposite coatings, *Progress in Organic Coatings* 120 (2018) 100–109.

S. Kohjiya, A. Katoh, J. Shimanuki, T. Hasegawa, Y. Ikeda, Three-dimensional nano-structure of in situ silica in natural rubber as revealed by 3D-TEM/electron tomography, *Polymer* (Guildf) 46 (2005) 4440–4446.

S. Maher, T. Kumeria, M. S. Aw, and D. Losic, Diatom silica for biomedical applications: Recent progress and advances, *Advanced Healthcare Materials* 7 (2018). https://doi.org/10.1002/adhm.201800552

S. Nie, S. R. Emory, Probing single molecules and single nanoparticles by surface-enhanced raman scattering, *Science* 275 (1997) 1102–1106.

S. Rajeshkumar, L. V. Bharath, Mechanism of plant-mediated synthesis of silver nanoparticles – A review on biomolecules involved, characterisation and antibacterial activity, *Chemico-Biological Interactions* 273 (2017) 219–227.

S. Soulé, J. Allouche, J.-C. Dupin, C. Courrèges, F. Plantier, W.-S. Ojo, Y. Coppel, C. Nayral, F. Delpech, H. Martinez, Thermoresponsive gold nanoshell@mesoporous silica nano-assemblies: An XPS/NMR survey, *Physical Chemistry Chemical Physics* 17 (2015) 28719–28728.

S. Xiong, B. Zhang, S. Luo, H. Wu, Z. Zhang, Preparation, characterization, and tribological properties of silica-nanoparticle-reinforced B-N-co-doped reduced graphene oxide as a multifunctional additive for enhanced lubrication, *Friction* 9 (2021) 239–249.

S. Zhou, M. Maeda, E. Tanabe, M. Kubo, M. Shimada, Bioinspired one-step synthesis of pomegranate-like Silica@Gold Nanoparticles with surface-enhanced raman scattering activity, *Langmuir* 36 (2020) 2553–2562.

S. M. Awadh, Z. M. Yaseen, Investigation of silica polymorphs stratified in siliceous geode using FTIR and XRD methods, Materials Chemistry *and* Physics 228 (2019) 45–50.

Tengjisi, Y. Hui, G. Yang, C. Fu, Y. Liu, C.-X. Zhao, Biomimetic core-shell silica nanoparticles using a dual-functional peptide, *The Journal of Colloid and Interface Science* 581 (2021) 185–194.

V. Panwar, T. Dutta, Diatom biogenic silica as a felicitous platform for biochemical engineering: Expanding frontiers, *ACS Applied Bio Materials* 2 (2019) 2295–2316.

V. Singh, S. K. Chakarvarti, Biotemplates and their uses in nanomaterials synthesis: A review, *American Journal of Bioengineering and Biotechnology* 2 (2016) 1–14. https://doi.org/10.7726/ajbebt.2016.1001.

V. Smetacek, Diatoms and the ocean carbon cycle, *Protist* 150 (1999) 25–32.

W. Wunderlich, P. Padmaja, K. G. K. Warrier, TEM characterization of sol-gel-processed alumina-silica and alumina–titania nano-hybrid oxide catalysts, *The Journal of the European Ceramic Society* 24 (2004) 313–317.

X. Chen, L. Wu, S. Zhou, B. You, In situ polymerization and characterization of polyester-based polyurethane/nano-silica composites, *Polymer International* 52 (2003) 993–998.

X. Kong, K. Squire, E. Li, P. LeDuff, G. L. Rorrer, S. Tang, B. Chen, C. P. McKay, R. Navarro-Gonazalez, A. X. Wang, Chemical and biological sensing using diatom photonic crystal biosilica with in-situ growth plasmonic nanoparticles, *IEEE Transactions on NanoBioscience* 15 (2016) 828–834.

X. Xiao, X. Zhang, H. Su, S. Chen, Z. He, C. Zhao, S. Yang, A visible-NIR responsive Dye-sensitized solar cell based on diatom frustules and cosensitization of photopigments from diatom and purple bacteria, *Journal of Chemistry* 2020 1710989.

X.-H. Pham, B. Seong, E. Hahm, K.-H. Huynh, Y.-H. Kim, J. Kim, S. H. Lee, B.-H. Jun, Glucose detection of 4-Mercaptophenylboronic acid-immobilized gold-silver core-shell assembled silica nanostructure by surface-enhanced raman scattering, *Nanomater* 11 (2021). https://doi.org/10.3390/nano11040948.

X.-H. Pham, M. Lee, S. Shim, S. Jeong, H.-M. Kim, E. Hahm, S. H. Lee, Y.-S. Lee, D. H. Jeong, B.-H. Jun, highly sensitive and reliable SERS probes based on nanogap control of an Au–Ag alloy on silica nanoparticles, *RSC Advances* 7 (2017) 7015–7021.

Y. Zhang, X. Zhu, X. Li, B. Chen, In situ quantitative determination of the intermolecular attraction between amines and a graphene surface using atomic force microscopy, *The Journal of Colloid and Interface Science* 581 (2021) 385–395.

Y. P. Moreno, W. L. da Silva, F. C. Stedile, C. Radtke, J. H. Z. dos Santos, Micro and nanodomains on structured silica/titania photocatalysts surface evaluated in RhB degradation: Effect of structural properties on catalytic efficiency, *Applied Surface Science Advances* 3 (2021) 100055.

Z. De Yuan, W. Yu, C. Jun, P. JunFeng, J. Xing Gang, J. Yong Gang, Bio-manufacturing technology based on diatom micro-and nanostructure, *Chinese Science Bulletin* 57 (2012).

Z. H. Aitken, S. Luo, S. N. Reynolds, C. Thaulow, J. R. Greer, Microstructure provides insights into evolutionary design and resilience of Coscinodiscus sp. frustules, *Proceedings of the National Academy of Sciences of the United States of America* 113 (2016) 2017–2022.

Z. Li, Y. He, L. H. Klausen, N. Yan, J. Liu, F. Chen, W. Song, M. Dong, Y. Zhang, Growing vertical aligned mesoporous silica thin film on the nanoporous substrate for enhanced degradation, drug delivery and bioactivity, *Bioactive Materials* 6 (2021) 1452–1463.

Z. Tian, X. Yu, Z. Ruan, M. Zhu, Y. Zhu, N. Hanagata, Magnetic mesoporous silica nanoparticles coated with thermo-responsive copolymer for potential chemo- and magnetic hyperthermia therapy, *Microporous and Mesoporous Materials* 256 (2018) 1–9.

Z. Zidi, M. Ltifi, I. Zafar, Synthesis and attributes of nano-SiO_2 local metakaolin based-geopolymer, *The Journal of Building Engineering* 33 (2021).

9 Algal Biofuel
A Promising Source of Green Energy

Prachi Agrawal, Kushal Kant Pant, Madan Sonkar and Vikas Chandra

9.1 INTRODUCTION

The world's population is expanding quickly, which is always driving up demand for fuel-based energy. Being unsustainable as well as nonrenewable, extensive utilization of fossil fuels around the world causes their depletion. As a result, the opportunity for biofuels to be used in place of fossil fuels is expanding globally. At commercial level, some wealthy countries are already generating biofuels. Over the past few decades, biofuels and other bioproducts have been produced using biomass, which is sourced from a biological precursor (Raheem et al. 2018). There are first-to-fourth generations of biofuels depending upon the type of biomass used (Cerri et al. 2017; Ho et al. 2010). Various generations of biofuels are depicted in Figure 9.1.

The first generation of biofuels, known as agrofuel, were produced using specific cultivated plants as feedstocks. Bioethanol is created by yeast fermentation of plant sugars or starches, while biodiesel is created by extracting plant oils (Lü et al. 2011). Unfortunately, food as well as water industries are adversely affected by such production processes (Rosenthal 2007). The second generation of biofuels was reliant on inedible crop components and plants that are not consumed by human (Brown and Brown 2013). Algal biofuels evolved as third generation of biofuels in order to eliminate excessive usage of water, land, and toxic pesticides (Raheem et al. 2018). The fourth generation of biofuel is centred on metabolic engineering of microalgae (by altering genomic sequences) to enhance productivity and/or reduce production costs (Dutta et al. 2014; Lü et al. 2011). Biodiesel, bioethanol, biomethanol, biohydrogen, bioether, and biogases are examples of biofuels. Among these, biodiesel and bioethanol are primary and most commonly used biofuels (Biofuel Basics n.d.). Compared to the fossil fuels (the biggest contributors to global warming), biofuels made from biomass pose less or no adverse impact on environment (Paul Abishek et al. 2014). Biofuels are green, affordable, and sustainable fuels made from renewable resources. Recent research has indicated that the genetic engineering of microalgae can increase their ability to collect carbon dioxide, produce biomass, and improve accumulation of lipids (Beacham et al. 2017; Dutta et al. 2014; Levitan et al. 2014; Shuba and Kifle 2018). Fuels made from microalgae have a high

DOI: 10.1201/9781003455950-9

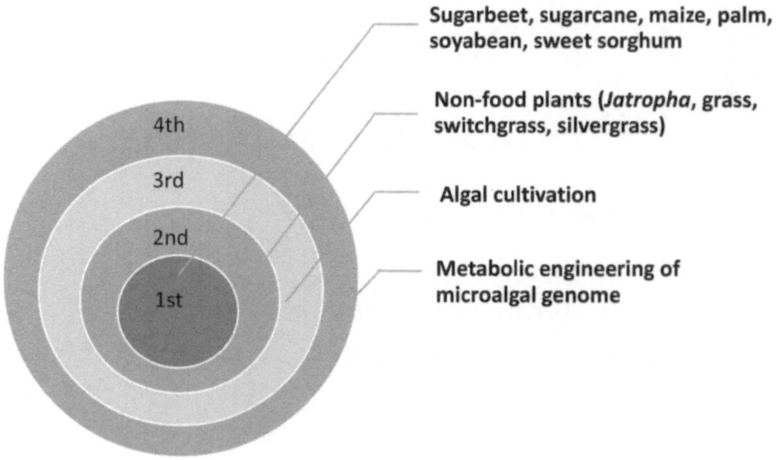

FIGURE 9.1 Various generations of biofuels.

potential to reduce global CO_2 emissions and are nontoxic and environmentally beneficial. According to studies, 1 kilogram of algal biomass can fix 1.83 kilogram of CO_2 (Gendy and El-Temtamy 2013). Some species also utilize SO_x and NO_x as nutritional sources in addition to CO_2. Up to 50% of the dry weight of algal biomass can comprise carbon dioxide. For regulating and optimizing energy structure and cost, the selection of biomass is critical for the production of biofuels (Cerri et al. 2017). In order to offset steadily rising need of production of biofuels and other biochemicals, the microalgal biomass is very promising (Ho et al. 2010; Paul Abishek et al. 2014). Apart from the production of bioethanol, biodiesel, and biogases, algae can also be used for photobiological production of biohydrogen and syngases. Algal biomasses can be converted into energy sources by chemical, biochemical, and thermochemical means (Kraan 2013). Some examples of biofuels obtained from algal biomasses are shown in Figure 9.2.

9.2 ALGAE

Algae, which exhibit plant-like characteristics, are typically found in aquatic environments. Algae are eukaryotes. Algae can survive in diverse habitat including saline or fresh water, damp soil, moist rocks, etc. They can come in various sizes, and their organization can be unicellular or multicellular. Algae neither have vascular tissue nor roots, stems, leaves, or flowers unlike angiosperms and higher plants. Being one of the primary producers in aquatic ecosystem, algae provide food to a variety of marine species. Thus, algae are main sources of sustenance of marine animals. In case of single-celled organisms, the varieties that reproduce asexually divide spontaneously and release spores that may or may not be motile. Sexually reproducing algae are typically encouraged to create gametes. The seven main varieties of algae are depicted in Table 9.1.

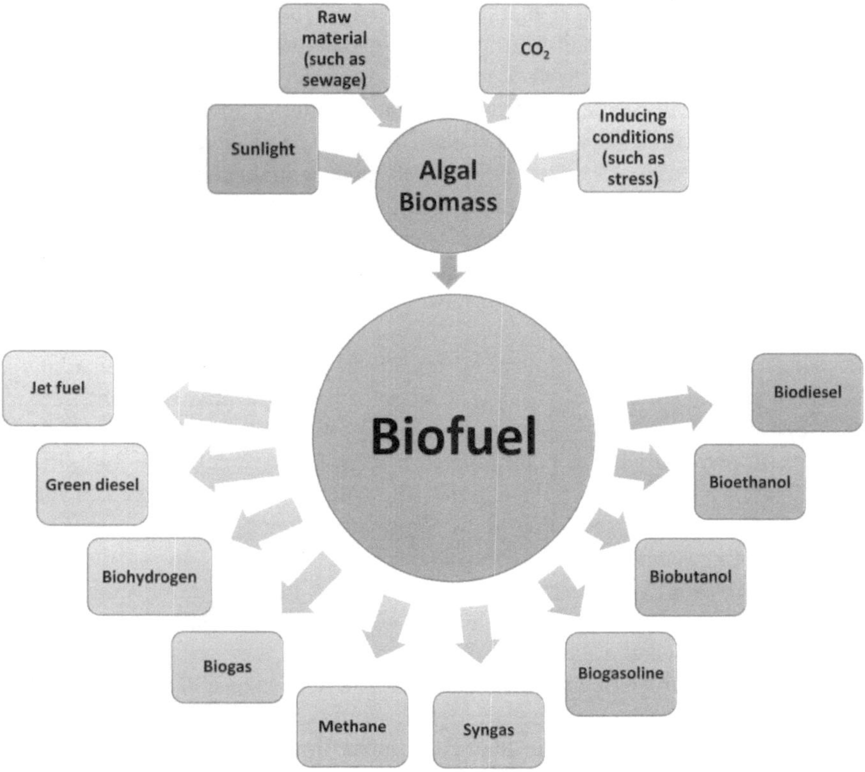

FIGURE 9.2 Different types of biofuels produced from algal biomasses.

TABLE 9.1
Differentiation of Algal Groups Based on Different Characteristics (Gamal Saad et al. 2019).

Algae	Storage material	Chlorophyll type	Cell wall
Phaeophyceae	Laminarin	a, c	Cellulose
Rhodophyceae	Floridean starch	a, d	Cellulose/mucilaginous material
Chlorophyceae	Starch	a, b	Cellulose/pectose
Xanthophyceae	Oil	A	Cellulose and hemicellulose
Pyrrophyceae	Starch	a, c	Stiff cellulose plates
Chrysophyceae	Leucosin and oil drops	a, c	Cellulose with silicate frustules
Euglenophyceae	Paramylon	a, b	No cell wall

a. **Phaeophyceae:** It is also known as brown algae, which include several types of seaweed and kelp found in marine habitats and are among the broadest algae species. These protists' life cycles entail the changeover of generations.

b. **Rhodophyceae:** Tropical marine environments frequently have red algae. These eukaryotic cells differ from other algae in that they lack centrioles and flagella. In addition to tropical reefs, red algae can also grow on or in association with other types of algae. They have cellulose and a wide variety of other carbohydrates in their cell walls. These algae grow and reproduce by the germination of monospores. They are also engaged in sexual reproduction and generational alternation. Several species of seaweed belong to red algae.
c. **Chlorophyceae:** While some species can be discovered in the ocean, freshwater conditions are where green algae thrive. Some kinds of green algae contain one or two flagella and have cellulosic cell walls. Chlorophyceae carry out photosynthesis and have chloroplasts. These algae come in thousands of different multicellular and unicellular types. Typically, multicellular members form colonies with a size range of four to several thousand cells. Reproduction occurs by aplanospores (non-motile) and zoospores (motile).
d. **Xanthophyceae:** Algae belonging to this group have cellulose and silica cell walls and one or two flagella that allow them to move. Frequently, they evolve within minuscule colonies comprising only a handful of cells. They are most frequently seen in freshwater.
e. **Pyrrophyceae:** Unicellular fire algae, which use flagella for movement, are typically found in oceans and in some freshwater sources. Dinoflagellates and cryptomonads are the two classes into which they are divided. A red tide is a phenomenon when the ocean appears red because of the abundance of dinoflagellates. Some species of Pyrrophyta are bioluminescent, similar to some fungus. They make the ocean appear to be on fire at night. Dinoflagellates are also dangerous because they release a neurotoxin that can impair human and animal muscle function. Similar to dinoflagellates, cryptomonads are capable of producing hazardous algal blooms that turn water red or dark brown.
f. **Chrysophyceae:** The most prevalent unicellular algae belonging to Chrysophyceae are diatoms and golden-brown algae.
g. **Euglenophyceae:** Euglenoids are protists found in both marine and freshwater environments. Instead of having a cell wall, they are protected by a pellicle, which is a layer rich in proteins.

The carbon molecules found in algae can be used to make biofuels, dietary supplements, medicines, and cosmetics (Raheem et al. 2018). Microalgae can also be utilized for cleaning wastewater and reducing atmospheric CO_2. To increase the potential of algae as a source of renewable bioproducts, growth-enhancing methods and gene-manipulating techniques may be employed (Shuba and Kifle 2018). Algae possess high quantities of lipids and carbohydrates, including starch and cellulose, making them highly valuable for the production of biofuels such as biodiesel and bioethanol. Algae adapt well to their habitats and develop rather quickly; under ideal growth conditions, the population of some species can double in a matter of hours. Microalgae have a very short harvesting cycle (less than ten days). Thus, microalgae can be harvested multiple times in a year (Razeghifard 2013; Schenk et al. 2008).

Microalgae are a broad group of single-celled organisms that may provide a range of options for meeting our needs of transportation fuel in several ways. Carbon dioxide emission is one of the major catastrophes leading to global warming. Nevertheless, nearly 40% of the global carbon can be utilized by algae (Falkowski et al. 1998; Parker et al. 2008). Algae can grow biomass very quickly; many species show two doublings per day, and other species can double in as little as six hours (Huesemann et al. 2009; Sheehan et al. 1998). Algal biomass has received a lot of attention due to its potential use in the biofuel industry. Figure 9.3 illustrates some of the benefits of algal biomass for the generation of biofuels.

Algal biomass feedstock has few drawbacks, including a higher production cost than traditional crops. Similar to other energy-intensive processes, the harvesting of algae accounts for up to 30% of the total production cost. For gathering biomasses, several methods including centrifugation, flocculation, floatation, sedimentation, and filtration are frequently used. In order to create an algal biorefinery, conversion technologies for obtaining useful bioproducts/bioenergy can be classified into three different categories (biochemical, chemical, and thermochemical) as depicted in Figure 9.4.

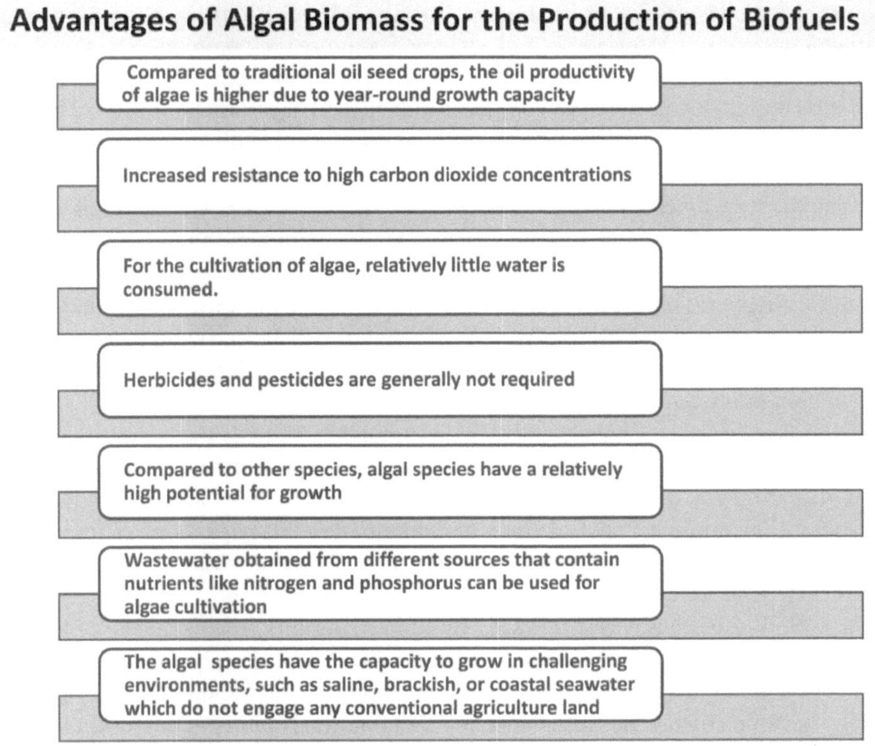

FIGURE 9.3 Advantages of algal biomass for the production of biofuels.

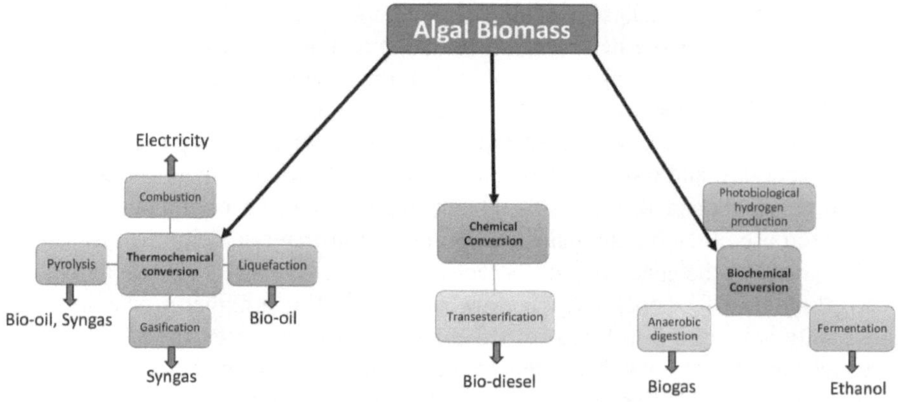

FIGURE 9.4 Different processes used for the conversion of algal biomass into useful products/energy.

9.3 CULTIVATION OF MICROALGAE

Algae have rapid growth rates, high levels of photosynthetic activity, and high effectiveness in sequestering carbon dioxide. Additionally, they utilize phosphorus and nitrogen present in wastewater of industries, agriculture, and municipality, thus lessening the nutrient load in the wastewater. They have a considerable portion of lipids that can be used for the synthesis of biodegradable and harmless biofuels. Given that algae require only a few basic conditions to develop, algal cultivation for the generation of biofuel seems rather straightforward. Algal growth and development and lipid accumulation in algae depend on a number of variables such as type and quantity of nutrients (micro and macro), carbon dioxide, pH, temperature, intensity of light, and photoperiod. Algae respond differently to these factors, particularly light and temperature. As physiological adaptation strategy, the amount of unsaturated fatty acids reduces as the temperature rises (Van Wagenen et al. 2012). Additionally, the type of fatty acid affects a number of biodiesel qualities, including lubricity, oxidative stability, melting point, and heating point. Palmitic- (C16:0), palmitoleic- (C16:1), stearic- (C18:0), oleic- (C18:1), linoleic- (C18:2), and linolenic- (C18:3) acids are the typical precursors converted into biodiesel (Zheng et al. 2012). Under restrictive conditions, certain microalgae alter their lipid production pathway to favour the synthesis of substantial quantities of neutral lipids, predominantly triacylglycerol. These lipids are then accumulated within cytosolic lipid bodies (Hu et al. 2008).

An extensive screening must be done for choosing the algal species with the highest biofuel productivities followed by optimizing all circumstances to achieve maximum production for these species. To increase a species' biomass and biofuel production, hundreds of experiments can be run on it. Microfluidics or on-chip technology became crucial due to the laborious and time-consuming nature of traditional lab methods. High-throughput single-cell analysis with microfluidics saves time and labour. This method also enables quick testing of multiple factors. Despite several flaws, this method was able to outweigh all the drawbacks of the earlier methods

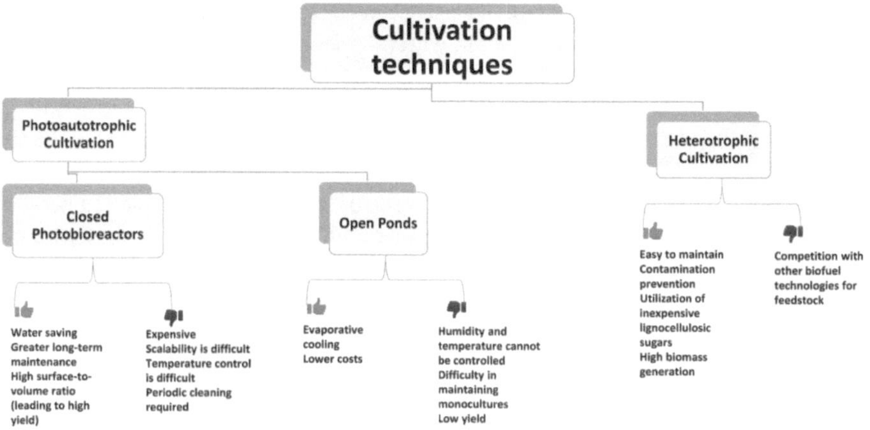

FIGURE 9.5 Different algal cultivation techniques for biofuel production.

(Love et al. 2013). Ecotoxicology screening (Zheng et al. 2012), cell identification (Hashemi et al. 2011), lipid analysis (Lim et al. 2014), and quantifying self-secreted macromolecules like ethanol (Abalde-Cela et al. 2015) and lactate (Hammar et al. 2015) are just a few applications where microfluidic chips have been used.

It is important to choose the appropriate cultivation system for algae cultivation since it influences the yield of biofuel and the effectiveness of phytoremediation. Different types of algal cultivation methods used for biofuel production are listed in Figure 9.5 along with pros and cons of each method.

9.3.1 Closed System

Closed systems, such as photobioreactors (PBRs), are extensively regulated, high-yielding systems designed for achieving optimal stirring and high light accessibility (Lee and Lee 2016; Zhang et al. 2018). PBRs can be made in a variety of shapes, with the tubular design being the most popular (Brennan and Owende 2010). They can be built as towers, tanks, or plastic or glass bags. According to reports, microalgal biomass was reported to be significantly greater in bubble columns and airlift photobioreactors. An additional tank is typically added to segregate it since too much oxygen can harm algae development (Rawat et al. 2013). Despite the elimination of contamination (Huang et al. 2010), controlled facilities suffer from a significant drawback in terms of their costliness (Brennan and Owende 2010).

9.3.2 Open System

The presence of a revolving arm in open systems is an important feature that ensures continuous stirring of the culture, while the location of these systems plays a crucial role in determining the amount of available sunlight, which is essential for their capacity to exploit atmospheric CO_2 (Brennan and Owende 2010). The main problem with these systems is contamination from bacteria or even from other microalgae (Suganya et al. 2016). However, variations in light, water, temperature, and evaporation have a greater impact on this system. The cost and productivity per unit area are the two key criteria used to compare various cultivation methods. Due to controlled growth conditions that guarantee strong light penetration, photobioreactors are more productive than open ponds. However, compared to closed systems that need expensive infrastructure, operations, and maintenance, open ponds are more affordable (Leite et al. 2013). Utilizing photoautotrophic systems has a number of benefits, one of which is use of CO_2 as a C source. Additionally, contamination is less severe when employing photoautotrophic systems compared to other farming techniques (Mata et al. 2010).

9.3.3 Hybrid System

A hybrid system is an arrangement that combines an open system and a closed system. Outstanding biomass productivity can be achieved by hybrid systems (Razzak et al. 2017). They are designed to provide higher yield in a cost-effective manner compared to closed systems. To increase yield, microalgae are first cultivated in closed PBRs before being moved to open system (Schenk et al. 2008). Hybrid system is ideal for cultivating algae at a large scale (Rawat et al. 2013).

Algal metabolism can be classified as autotrophic, heterotrophic, mixotrophic, or photoheterotrophic (Daliry et al. 2017). According to Durbin and Chen (1994), autotrophic pathways utilize light to transform inorganic carbon into organic energy through photosynthesis. In order to grow in the dark, heterotrophic pathways require organic carbon (Huppe and Turpin 1994), whereas mixotrophic pathways allow cells to grow either autotrophically or heterotrophically, depending on the available food sources (Andrews 1968). The photoheterotrophic process operates in the presence of light and organic carbon (Durbin and Chen 1994). Growth occurs more quickly under heterotrophic metabolism than under autotrophic metabolism (Martinez et al. 1991). Mixotrophic metabolism is said to be the most effective approach to achieve the highest levels of biomass and lipid productivities (Scarsella et al. 2010).

9.4 HARVESTING OF MICROALGAE

The cells grown in culture are collected by extracting them from their nutrient solution while preserving their water content. A number of techniques including flocculation, filtration, flotation, sonication, centrifugation, and precipitation, etc., are used for harvesting biomass of algae. In rare circumstances, combining more than one harvesting techniques may increase biomass yield. Dehydration may proceed after harvesting in varying situations. It involves taking the water out of cells to obtain dried material (Chen et al. 2015).

9.5 ALGAL BIOFUELS

Algae have a significant potential for biofuel production as the third-generation feedstock because of their rapid growth, outstanding biomass output, and accumulation of lipid and carbohydrate in huge amounts. Among the useful biofuels created by algae are biodiesel, biogas, bioethanol, and biomethane. The leftover biomass can be utilized to make medicines, biocontrol agents, fertilisers, and animal feed. Anaerobic digestion of organic material results in the production of biogas or biomethane (Ward et al. 2014). The steps of anaerobic digestion are as follows:

1) Bacterial hydrolysis of biopolymers to generate monosaccharides;
2) Fermentation of the monosaccharides to acids;
3) Synthesis of acetate by the acetogenic bacteria; and
4) Synthesis of CH_4 and CO_2 by the action of methanogenic bacteria (Oncel et al. 2015).

Kerosene, diesel, and gasoline can be produced by the conversion of microalgal hydrocarbons. For instance *Botryococcus braunii* generates hydrocarbons outside of the cell that can be used for oil production and are easier to extract (Ranga Rao et al. 2007). Biomass gasification generates bio-syngas, which contains methane, hydrogen, carbon monoxide, water, and ashes (Buxy et al. 2013).

The minimum temperature needed for the gasification process is in the range of 800–1200°C. A water content of less than 20% is preferred for biomass (Ghasemi et al. 2012). Microalgae can directly synthesize hydrogen in the absence of O_2 (Tiwari and Pandey 2012). Fermentation of algal carbohydrates by yeast yields bioethanol. A starch concentration of more than 50% has been found in certain microalgae (Chen et al. 2013; Markou et al. 2012). It is also possible to convert microalgal cellulose and hemicellulose into sugars and ethanol (Hamelinck et al. 2005).

9.5.1 Biodiesel Production

Fatty acid methyl esters (FAME), produced from various renewable lipid feedstocks, are combinations of monoalkyl esters of long-chain fatty acids that are used to make biodiesel. Different diesel engines can utilize them directly (Demirbas 2009). In the middle of the 1980s, research began to be focused at microalgae as a potential substrate for making liquid fuels. During World War II, diatoms were used to extract lipids to help with the energy crisis (Cohen et al. 1995). Algal biomass can be economically converted into biodiesel using a variety of technologies since it produces more oil than oil seed crops.

Growth rate and quantity of biomass obtained from a particular species of algae can affect the oil productivity. *Chlamydocapsa bacillus*, *Ankistrodesmus fusiformis*, *Kirchneriella lunaris*, and *Ankistrodesmus falcatus* are a few examples of the microalgae species that are typically selected for the manufacture of biodiesel due to high amounts of polyunsaturated FAME (Nascimento et al. 2012). During exponential growth, they frequently double their biomass. As much as 80% of the dry biomass of microalgae is found to be made up of oil, which indicates very high oil content. Algal biomass has the potential to yield between 5,000 to 15,000 gal of biodiesel/acre/year (Chisti 2007; Spolaore et al. 2006).

9.5.1.1 Harvesting and Drying of Algal Biomass

Macroalgae are harvested with nets, whereas microalgae are harvested using conventional techniques, such as filtration, flocculation, centrifugation, foam fractionation, and sedimentation (Csordas and Wang 2004; Liu et al. 2013; Prochazkova et al. 2013; Rossignol et al. 1999). The type of algal species determines which harvesting technique is used. Before storing algal biomass, drying is a crucial technique to increase its shelf life and prevent deterioration (Munir et al. 2013). Microalgal biomass has been dried using the majority of effective drying techniques, including spray drying, drum drying, freeze drying or lyophilization, and sun drying (Leach et al. 1998; Richmond 2004; Williams and Laurens 2010). Due to high water content in the biomass, sun drying is not regarded as a highly efficient approach (Mata et al. 2010). Widjaja et al. (2009) demonstrated the significance of drying temperature in the lipid extraction process from algal biomass, following the example set by Prakash et al. (1997), who utilized conventional solar drying equipment to dry *Spirulina* and *Scenedesmus* with 90% moisture content. Drying temperature influences lipid yield and triglyceride content.

9.5.1.2 Extraction of Oil from Algal Biomass

Various methods for lipid extraction from microalgae have been documented in scientific literature, such as mechanical and solvent extraction, but these techniques are known to be costly and demand substantial energy inputs (Li et al. 2014).

9.5.1.2.1 Mechanical Oil Extraction

The mechanical extraction technique for obtaining oil from algae involves the use of mechanical presses or expellers, which require the algal biomass to be dried beforehand. Applying pressure with these devices breaks down the cells and allows for oil extraction. This method is relatively simple and can recover around 75% of the oil (Munir et al. 2013). Various methods, such as screw expellers, mechanical pressing apparatus like pistons, and exposure to osmotic shocks, have been used for extracting oil from algal cells (Topare et al. 2011). However, compared to other approaches, the mechanical extraction technique is considered less favorable and less efficient due to its longer extraction time (Popoola and Yangomodou 2006).

9.5.1.2.2 Solvent-Based Oil Extraction

Less than 1% of the oil is left in biomass after solvent oil extraction, which typically recovers almost all of the oil. So rather than mechanically extracting oil and fats, this method is more appropriate (Topare et al. 2011). Two-stage solvent extraction methods have been used for removing lipids from microalgae. The solvent that is employed largely determines how much lipid is obtained from the biomass and the subsequent output of the biodiesel. Acetone, chloroform, cyclohexane, hexane, and benzene are just a few of the organic solvents that are utilized individually or in combination (Afify et al. 2010).

9.5.1.3 Transesterification

The conversion of algal oil into biodiesel involves multiple stages of interaction between triglycerides or fatty acids and alcohol. Various alcohols, including ethanol, butanol, methanol, propanol, and amyl alcohol, can be used for this reaction. However, due to their affordability and favorable physical and chemical characteristics, ethanol and methanol are frequently employed for commercial purposes (Bisen et al. 2010; Surendhiran and Vijay 2012). Enzymes (such as lipases) and catalysts (acids and alkalies such as sodium and potassium hydroxides) can be used to speed up the reaction. According to this approach, three moles of alcohol and one mole of glycerol are produced for every mole of triglyceride used (Chisti 2007; Meher et al. 2006; Sharma and Singh 2009; Stergiou et al. 2013; Surendhiran and Vijay 2012). Glycerol can be occasionally or continually withdrawn during transesterification process because it is denser than biodiesel. The presence of methanol, a co-solvent used to retain soap and glycerol in oil, has been linked to engine failure (Munir et al. 2013). To salvage the biodiesel, the removal of glycerol and methanol is accomplished by subjecting it to multiple rounds of water washing (Chisti 2007).

9.5.2 Bioethanol Production

According to a number of experts, certain species of algae that generate significant amounts of carbohydrates as reserve polymers can also make bioethanol. Biomass of algae has been deemed more suited for the manufacture of bioethanol because it contains less lignin and hemicellulose than lignocellulosic biomass (Chen et al. 2013). Recently, efforts have been undertaken to produce bioethanol by fermenting algae as the feedstocks (Nguyen and Vu 2012; Singh et al. 2011). For these algae to create large amounts of polysaccharides like starch and cellulose, they typically need light, nutrients, and carbon dioxide. These polysaccharides may be hydrolysed into fermentable sugars, followed by fermentation to produce bioethanol, and then they can be separated using distillation.

Recent reports indicate that the cell walls of various green algae species, such as *Spirogyra* and *Chlorococcum*, contain notable quantities of polysaccharides. With a 65% conversion efficiency, microalgae like *C. vulgaris*, the dry weight of which has a starch content of roughly 37%, are the greatest source of bioethanol (Eshaq and Ali 2010; Lam and Lee 2012). Before fermentation, polysaccharide-based biomass typically requires additional processing, including pre-treatment and saccharification (Harun et al. 2010). An alternative approach involves utilizing a strain that naturally produces the enzyme amylase, allowing for the simultaneous saccharification and fermentation processes to occur, resulting in the production of ethanol in a single step. Pre-treatment is a crucial step that makes biomass more accessible to the enzymes required to produce the monosaccharides. Acid pre-treatment is frequently employed to transform cell wall polymers into simple forms. The pre-treatment procedure uses relatively little energy and is a productive one (Harun and Danquah 2011; Yazdani et al. 2011).

Like starch, various polymers such as fucoidan, mannitol, and alginate are found in the cell walls of different types of algae. However, before these polymers can be used for fermentation, they require additional processing through pre-treatment and saccharification. Laminarin, which may be hydrolysed by 1,3-glucanases or laminarinases, is an additional type of storage carbohydrate present in a variety of brown seaweeds as well as microalgae (Kumagai and Ojima 2010). When pretreated and saccharified biomass of algae is fermented in anaerobic environment to synthesize bioethanol, a variety of microbes, including yeast, bacteria, and fungus, can be employed (Nguyen and Vu 2012). The microorganisms that ferment bioethanol are now thought to be *Saccharomyces cerevisiae* and *Zymomonas mobilis*. These well-known microbes, however, cannot ferment mannitol anaerobically. Instead, mannitol fermentation needs oxygen to proceed, and only *Zymobacter palmae* can do this (Horn et al. 2000). Agar, a polymer extracted from red algae, can be implied in bioethanol production (Yoon et al. 2010).

9.5.3 Biogas Production

The high polysaccharide content found in algae has garnered a significant interest in the generation of biogas through anaerobic digestion. The most common opinion is that seaweeds provide great feedstock for biogas generation. The generation of biogas from a variety of algae species, including *Scenedesmus, Spirulina, Euglena*, and *Ulva*, has been proven by several researchers (Ras et al. 2011; Samson and Leduy 1982; Saqib et al. 2013; Yen and Brune 2007; Zhong et al. 2012).

Algae contain organic material that can be converted anaerobically to produce biogas. During this process, various gases such as CH_4, CO_2, and traces of H_2S are emitted. The conversion occurs in four stages. Initially, with the help of an enzyme produced by specific obligate anaerobic bacteria like *Clostridia* and *Streptococci*, insoluble organic material and larger biomolecules, including lipids, carbohydrates, and proteins, are broken down into soluble organic material in the first stage. The next stage, known as acidogenesis, involves the conversion of soluble organic materials into volatile fatty acids (VFAs) and alcohols, facilitated by enzymes secreted by acidogenic bacteria. Furthermore, these VFAs and alcohols undergo acetogenesis, where acetogenic bacteria transform them into acetic acid and hydrogen. Finally, methanogens break down the acetic acid and hydrogen into methane and carbon dioxide (Brennan and Owende 2010; Cantrell et al. 2007; Romagnoli et al. 2010; Vergara-Fernández et al. 2008).

9.5.4 Biohydrogen Production

Algae can be exploited to produce energy or gaseous fuels. Different methods, including biophotolysis and photo fermentation, can be used to make biohydrogen (Sharma et al. 2013). Some microalgae have glycogen in their cells rather than starch. Hydrogen is produced in anaerobic environment, where ferrodoxin is oxidized by hydrogenase. Numerous researchers have focused their efforts on identifying enzyme activities that engage with ferredoxin and other metabolic processes to facilitate the production of microalgal photobiohydrogen. In order to increase the generation of biohydrogen, the researchers are also involved in altering underlying genes (Gavrilescu and Chisti 2005; Hankamer et al. 2007; Rajkumar et al. 2013; Wecker et al. 2011).

9.5.5 Bio-Oil and Syngas Production

Algal biomass in the liquid phase is used to produce bio-oil by applying high temperature in anaerobic condition. Bio-oil undergoes pyrolysis for processing, and composition of the oil changes depending on the feedstocks and circumstances (Iliopoulou et al. 2007; Li et al. 2008). The productivity of bio-oil is influenced by several factors, including water, ash content, composition of the biomass, pyrolysis temperature, and vapour residence duration (Fahmi et al. 2008). To enhance the quality and usefulness of bio-oil, certain procedures are necessary (Bae et al. 2011). Bio-oils can be effectively utilized to generate electricity through various means such as external combustion engines, organic rankine cycles, stirling engines, and steam. Additionally, internal combustion engines like diesel and gas turbines can also be employed for electricity production (Chiaramonti et al. 2007). While there is limited research on algal pyrolysis compared to lignocellulosic biomass, the literature indicates that fluidized bed rapid pyrolysis technologies yield a significant amount of bio-oil. To address the challenges associated with high carrier gas flow and energy inputs, different pyrolysis processes have been developed (Oyedun et al. 2012).

Gasification is a high-temperature process (800–1000°C) that converts biomass into a combustible gas mixture called syngas or producer gas. This method involves a partial oxidation of biomass using oxygen and steam, resulting in the production of syngas comprising several gases such as CO, CO_2, CH_4, H_2, and N_2. Syngas generated through the gasification of woody biomass can be utilized as a fuel or in gas turbines, despite its relatively low calorific value. Moreover, gasification offers the advantage of being able to utilize various types of biomass feedstocks for energy generation (Carvalho et al. 2006; Lv et al. 2007; Prins et al. 2006).

9.6 CURRENT STATUS AND BOTTLENECKS

Finding new alternative resources that can be used in place of fossil fuels has recently proven to be difficult. Algal biofuels have been regarded as the ideal resource to replace liquid petroleum fuel due to their abundance of benefits, including minimal land requirements for cultivation and high oil content. Nevertheless, one of its barriers to industrial manufacturing is the poor biomass yield. Another drawback is the collection of biomasses, which requires a lot of energy. A low-energy input, cost-effective, and energy-efficient harvesting technique is needed. Better biomass collection techniques and high oil productivity via genetic alteration are also needed to produce low-cost microalgal biofuels. The conventional algae harvesting method, the biorefinery idea, and the improvements in photobioreactor design will all help to further lower the cost of producing biofuels using algae.

Bioprospecting for microalgae is required to find out species with very high lipid content (Bošnjaković and Sinaga 2020). Algal biofuels often have low-to-no detrimental environmental effects (Adeniyi et al. 2018). In practice, the production of biofuels is associated with diverse environmental applications, including bioremediation for wastewater treatment, generating electricity and heat, bio-fixation for CO_2 removal, creating biofertilisers, producing animal feed, and contributing to healthcare and food production (Adeniyi et al. 2018). Algae represent the most environmentally friendly fuel source that can aid in reducing greenhouse gas emissions (Suganya et al. 2016). The development of affordable commercialization technologies is essential for the success of algae-based biofuels. The most enticing alternative, nevertheless, is to employ genetically modified algae with fast growth rate that are able to over produce precursors at high rate. An open pond is a good place to introduce these species to the ecosystem. This open pond may serve as the initial stage of water filtration in a wastewater treatment facility. The biomass slurry of such ponds may be utilized to make biofuels, and the used biomass may further be utilized as fertiliser or animal feed. Although the manufacturing of algae-based biofuels requires a lot of water, nitrogen, phosphorus, and CO_2 and is expensive, they are ecologically friendly and do not compete with other sources of energy for land or water, contrary to popular belief. For the manufacture of biofuels at large scale, lot of supplies, machinery, electricity, water, and fertilisers are required (Ferrell and Sarisky-Reed 2010). Wastewater is used to supply part of the nitrogen and phosphorus while also preventing competition with crops grown for food and feed. The presence of heavy metals, algae pathogens and predators, and other toxins in wastewater is a significant downside. Furthermore, it is impractical, expensive, and energy-intensive to transport poor-quality water over long distances because the majority of algae production sites are located distant from wastewater treatment facilities. Additionally, water won't be recycled entirely (Ferrell and Sarisky-Reed 2010). Combining wastewater treatment and biofuel generation is the most optimistic plan (Cabanelas et al. 2013). Lead, mercury, arsenic, and cadmium are only a few of the heavy metals found in industrial wastes. Algae are particularly good at absorbing such harmful substances. However, leftover by-products cannot be utilized as animal feed.

The selection of algal strains presents another difficulty. Natural microalgal species typically need extensive time and laborious procedures for site selection, isolation, purification, and identification. Media and environment used for cultivation of the target algal species must be optimized in order to achieve optimal microalgal proliferation. Both phenotypic and genomic characteristics are frequently involved in species identification. To make the most of the in vitro research, well-optimized environmental factors at the sample location must be noted. The isolated microalgal strain is then cultivated for the generation of lipids and biomass in sophisticated systems (Mutanda et al. 2011). Even bio-engineered algae, which are more susceptible to viruses and predators, may not survive in the original environment while doing well in laboratory settings. Adding enough inoculum to the organisms' original environments can help remedy this issue (Leite et al. 2013). Ten times more energy is required for the extraction of algal oil than for the extraction of soya bean oil. These challenges need to be addressed for commercialization of algal biofuels.

9.7 CONCLUSION

Algae make desirable feedstock for the generation of biofuels. Algae may be used to produce valuable products such as biodiesel, biogas, bioethanol, medicines, nutraceuticals, and many more. Biofuels are environment friendly since they are renewable and biodegradable. The quick growth and high lipid content of algae are the two of many advantageous characteristics they have, making them suitable for biofuel production. The majority of algae, known as chlorophytes, are used in bioremediation, water treatment, production of food and medicines, and energy generation. We discussed several methods of cultivation, harvesting, and processing techniques in this book chapter. The requirement of high infrastructure and high operation, maintenance expenses, the selection of high lipid-containing algae strains, commercial-scale harvesting, and problems with water evaporation appear to be the primary hurdles for using algae as a source of biofuel. To make algae biofuel production more appealing, new and effective methods must be used. Increased biofuel production through algae will aid in protecting conventional resources of fuels and, as a result, the environment.

COMPETING INTEREST

The authors declare no conflict of interest.

REFERENCES

Abalde-Cela, S., Gould, A., Liu, X., Kazamia, E., Smith, A. G., & Abell, C. (2015). High-throughput detection of ethanol-producing cyanobacteria in a microdroplet platform. *Journal of the Royal Society, Interface*, *12*(106), 20150216. https://doi.org/10.1098/rsif.2015.0216

Adeniyi, O. M., Azimov, U., & Burluka, A. (2018). Algae biofuel: Current status and future applications. *Renewable and Sustainable Energy Reviews*, *90*, 316–335. https://doi.org/10.1016/j.rser.2018.03.067

Afify, A. E.-M. M. R., Shalaby, E. A., & Shanab, S. M. M. (2010). Enhancement of biodiesel production from different species of algae. *Grasas y Aceites*, *61*(4), Article 4. https://doi.org/10.3989/gya.021610

Andrews, J. F. (1968). A mathematical model for the continuous culture of microorganisms utilizing inhibitory substrates. *Biotechnology and Bioengineering*, *10*(6), 707–723. https://doi.org/10.1002/bit.260100602

Bae, Y. J., Ryu, C., Jeon, J.-K., Park, J., Suh, D. J., Suh, Y.-W., Chang, D., & Park, Y.-K. (2011). The characteristics of bio-oil produced from the pyrolysis of three marine macroalgae. *Bioresource Technology*, *102*(3), 3512–3520. https://doi.org/10.1016/j.biortech.2010.11.023

Beacham, T. A., Sweet, J. B., & Allen, M. J. (2017). Large scale cultivation of genetically modified microalgae: A new era for environmental risk assessment. *Algal Research*, *25*, 90–100. https://doi.org/10.1016/j.algal.2017.04.028

Biofuel Basics. (n.d.). *Energy.Gov*. www.energy.gov/eere/bioenergy/biofuel-basics

Bisen, P. S., Sanodiya, B. S., Thakur, G. S., Baghel, R. K., & Prasad, G. B. K. S. (2010). Biodiesel production with special emphasis on lipase-catalyzed transesterification. *Biotechnology Letters*, *32*(8), 1019–1030. https://doi.org/10.1007/s10529-010-0275-z

Bošnjaković, M., & Sinaga, N. (2020). The perspective of large-scale production of algae biodiesel. *Applied Sciences*, *10*(22), Article 22. https://doi.org/10.3390/app10228181

Brennan, L., & Owende, P. (2010). Biofuels from microalgae—A review of technologies for production, processing, and extractions of biofuels and co-products. *Renewable and Sustainable Energy Reviews*, *14*(2), 557–577. https://doi.org/10.1016/j.rser.2009.10.009

Brown, T. R., & Brown, R. C. (2013). A review of cellulosic biofuel commercial-scale projects in the United States. *Biofuels, Bioproducts and Biorefining*, *7*(3), 235–245. https://doi.org/10.1002/bbb.1387

Buxy, S., Diltz, R., & Pullammanappallil, P. (2013). *Biogasification of Marine Algae Nannochloropsis Oculata* (G. Wicks, J. Simon, R. Zidan, R. Brigmon, G. Fischman, S. Arepalli, A. Norris, & M. McCluer, Eds.; pp. 59–67). John Wiley & Sons, Inc. https://doi.org/10.1002/9781118585160.ch6

Cabanelas, I. T. D., Ruiz, J., Arbib, Z., Chinalia, F. A., Garrido-Pérez, C., Rogalla, F., Nascimento, I. A., & Perales, J. A. (2013). Comparing the use of different domestic wastewaters for coupling microalgal production and nutrient removal. *Bioresource Technology*, *131*, 429–436. https://doi.org/10.1016/j.biortech.2012.12.152

Cantrell, K., Ro, K., Mahajan, D., Anjom, M., & Hunt, P. G. (2007). Role of thermochemical conversion in livestock waste-to-energy treatments: Obstacles and opportunities. *Industrial & Engineering Chemistry Research*, *46*(26), 8918–8927. https://doi.org/10.1021/ie0616895

Carvalho, A. P., Meireles, L. A., & Malcata, F. X. (2006). Microalgal reactors: A review of enclosed system designs and performances. *Biotechnology Progress*, *22*(6), 1490–1506. https://doi.org/10.1021/bp060065r

Cerri, C. E. P., You, X., Cherubin, M. R., Moreira, C. S., Raucci, G. S., Castigioni, B. de A., Alves, P. A., Cerri, D. G. P., Mello, F. F. de C., & Cerri, C. C. (2017). Assessing the greenhouse gas emissions of Brazilian soybean biodiesel production. *PLoS One*, *12*(5), e0176948. https://doi.org/10.1371/journal.pone.0176948

Chen, C.-L., Chang, J.-S., & Lee, D.-J. (2015). Dewatering and drying methods for microalgae. *Drying Technology*, *33*(4), 443–454. https://doi.org/10.1080/07373937.2014.997881

Chen, C.-Y., Zhao, X.-Q., Yen, H.-W., Ho, S.-H., Cheng, C.-L., Lee, D.-J., Bai, F.-W., & Chang, J.-S. (2013). Microalgae-based carbohydrates for biofuel production. *Biochemical Engineering Journal*, *78*, 1–10. https://doi.org/10.1016/j.bej.2013.03.006

Chiaramonti, D., Oasmaa, A., & Solantausta, Y. (2007). Power generation using fast pyrolysis liquids from biomass. *Renewable and Sustainable Energy Reviews*, *11*(6), 1056–1086. https://doi.org/10.1016/j.rser.2005.07.008

Chisti, Y. (2007). Biodiesel from microalgae. *Biotechnology Advances*, *25*(3), 294–306. https://doi.org/10.1016/j.biotechadv.2007.02.001

Cohen, Z., Norman, H. A., & Heimer, Y. M. (1995). Microalgae as a source of omega 3 fatty acids. *World Review of Nutrition and Dietetics*, *77*, 1–31.

Csordas, A., & Wang, J.-K. (2004). An integrated photobioreactor and foam fractionation unit for the growth and harvest of Chaetoceros spp. in open systems. *Aquacultural Engineering*, *1–2*(30), 15–30. https://doi.org/10.1016/j.aquaeng.2003.07.001

Daliry, S., Hallajisani, A., Mohammadi Roshandeh, J., Nouri, H., & Golzary, A. (2017). Investigation of optimal condition for *Chlorella vulgaris* microalgae growth. *Global Journal of Environmental Science and Management*, *3*(2), 217–230. https://doi.org/10.22034/gjesm.2017.03.02.010

Demirbas, A. (2009). Progress and recent trends in biodiesel fuels. *Energy Conversion and Management*, *50*(1), 14–34. https://doi.org/10.1016/j.enconman.2008.09.001

Durbin, E. G., & Chen, C. Y. (1994). Effects of pH on the growth and carbon uptake of marine phytoplankton. *Marine Ecology Progress Series*. https://doi.org/10.3354/meps109083

Dutta, K., Daverey, A., & Lin, J.-G. (2014). Evolution retrospective for alternative fuels: First to fourth generation. *Renewable Energy*, *69*, 114–122. https://doi.org/10.1016/j.renene.2014.02.044

Eshaq, F. S., & Ali, M. N. (2010). *Spirogyra Biomass a Renewable Source for Biofuel (Bioethanol) Production*. www.semanticscholar.org/paper/Spirogyra-biomass-a-renewable-source-for-biofuel-Eshaq-Ali/d4a6293466b1b4b2471dffa7d5137dce98f15a6e

Fahmi, R., Bridgwater, A. V., Donnison, I., Yates, N., & Jones, J. M. (2008). The effect of lignin and inorganic species in biomass on pyrolysis oil yields, quality and stability. *Fuel*, *87*(7), 1230–1240. https://doi.org/10.1016/j.fuel.2007.07.026

Falkowski, P. G., Barber, R. T., & Smetacek, V. (1998). Biogeochemical controls and feedbacks on ocean primary production. *Science (New York, N.Y.)*, *281*(5374), 200–207. https://doi.org/10.1126/science.281.5374.200

Ferrell, J., & Sarisky-Reed, V. (2010). *National Algal Biofuels Technology Roadmap* (DOE/EE-0332). EERE Publication and Product Library, Washington, D.C. (United States). https://doi.org/10.2172/1218560

Gamal Saad, M., Dosoky, N., Zoromba, M., & Shafik, H. (2019). Algal biofuels: Current status and key challenges. *Energies*, *12*, 1920. https://doi.org/10.3390/en12101920

Gavrilescu, M., & Chisti, Y. (2005). Biotechnology-a sustainable alternative for chemical industry. *Biotechnology Advances*, *23*(7–8), 471–499. https://doi.org/10.1016/j.biotechadv.2005.03.004

Gendy, T. S., & El-Temtamy, S. A. (2013). Commercialization potential aspects of microalgae for biofuel production: An overview. *Egyptian Journal of Petroleum*, *22*(1), 43–51. https://doi.org/10.1016/j.ejpe.2012.07.001

Ghasemi, Y., Rasoul-Amini, S., Naseri, A. T., Montazeri-Najafabady, N., Mobasher, M. A., & Dabbagh, F. (2012). Microalgae biofuel potentials (Review). *Applied Biochemistry and Microbiology*, *48*(2), 126–144. https://doi.org/10.1134/S0003683812020068

Hamelinck, C. N., Hooijdonk, G. van, & Faaij, A. P. (2005). Ethanol from lignocellulosic biomass: Techno-economic performance in short-, middle- and long-term. *Biomass and Bioenergy*, *28*(4), 384–410. https://doi.org/10.1016/j.biombioe.2004.09.002

Hammar, P., Angermayr, S. A., Sjostrom, S. L., van der Meer, J., Hellingwerf, K. J., Hudson, E. P., & Joensson, H. N. (2015). Single-cell screening of photosynthetic growth and lactate production by cyanobacteria. *Biotechnology for Biofuels*, *8*, 193. https://doi.org/10.1186/s13068-015-0380-2

Hankamer, B., Lehr, F., Rupprecht, J., Mussgnug, J. H., Posten, C., & Kruse, O. (2007). Photosynthetic biomass and H_2 production by green algae: From bioengineering to bioreactor scale-up. *Physiologia Plantarum*, *131*(1), 10–21. https://doi.org/10.1111/j.1399-3054.2007.00924.x

Harun, R., & Danquah, M. K. (2011). Influence of acid pre-treatment on microalgal biomass for bioethanol production. *Process Biochemistry*, *46*(1), 304–309.

Harun, R., Danquah, M. K., & Forde, G. M. (2010). Microalgal biomass as a fermentation feedstock for bioethanol production. *Journal of Chemical Technology & Biotechnology*, *85*(2), 199–203. https://doi.org/10.1002/jctb.2287

Hashemi, J., Worrall, C., Vasilcanu, D., Fryknäs, M., Sulaiman, L., Karimi, M., Weng, W.-H., Lui, W.-O., Rudduck, C., Axelson, M., Jernberg-Wiklund, H., Girnita, L., Larsson, O., & Larsson, C. (2011). Molecular characterization of acquired tolerance of tumor cells to picropodophyllin (PPP). *PLoS One*, *6*(3), e14757. https://doi.org/10.1371/journal.pone.0014757

Ho, S.-H., Chen, W.-M., & Chang, J.-S. (2010). Scenedesmus obliquus CNW-N as a potential candidate for CO_2 mitigation and biodiesel production. *Bioresource Technology*, *101*(22), 8725–8730. https://doi.org/10.1016/j.biortech.2010.06.112

Horn, S. J., Aasen, I. M., & Østgaard, K. (2000). Production of ethanol from mannitol by Zymobacter palmae. *Journal of Industrial Microbiology and Biotechnology*, *24*, 51–57.

Hu, Q., Sommerfeld, M., Jarvis, E., Ghirardi, M., Posewitz, M., Seibert, M., & Darzins, A. (2008). Microalgal triacylglycerols as feedstocks for biofuel production: Perspectives and advances. *The Plant Journal: For Cell and Molecular Biology*, *54*(4), 621–639. https://doi.org/10.1111/j.1365-313X.2008.03492.x

Huang, G., Chen, F., Wei, D., Zhang, X., & Chen, G. (2010). Biodiesel production by microalgal biotechnology. *Applied Energy*, *87*(1), 38–46. https://doi.org/10.1016/j.apenergy.2009.06.016

Huesemann, M. H., Hausmann, T. S., Bartha, R., Aksoy, M., Weissman, J. C., & Benemann, J. R. (2009). Biomass productivities in wild type and pigment mutant of Cyclotella sp. (Diatom). *Applied Biochemistry and Biotechnology*, *157*(3), 507–526. https://doi.org/10.1007/s12010-008-8298-9

Huppe, H. C., & Turpin, D. H. (1994). Integration of carbon and nitrogen metabolism in plant and algal cells. *Annual Review of Plant Physiology and Plant Molecular Biology*, *45*(1), 577–607. https://doi.org/10.1146/annurev.pp. 45.060194.003045

Iliopoulou, E. F., Antonakou, E. V., Karakoulia, S. A., Vasalos, I. A., Lappas, A. A., & Triantafyllidis, K. S. (2007). Catalytic conversion of biomass pyrolysis products by mesoporous materials: Effect of steam stability and acidity of Al-MCM-41 catalysts. *Chemical Engineering Journal*, *134*(1), 51–57. https://doi.org/10.1016/j.cej.2007.03.066

Kraan, S. (2013). Mass-cultivation of carbohydrate rich macroalgae, a possible solution for sustainable biofuel production. *Mitigation and Adaptation Strategies for Global Change*, *18*(1), 27–46. https://doi.org/10.1007/s11027-010-9275-5

Kumagai, Y., & Ojima, T. (2010). Isolation and characterization of two types of beta-1,3-glucanases from the common sea hare Aplysia kurodai. *Comparative Biochemistry and Physiology. Part B, Biochemistry & Molecular Biology*, *155*(2), 138–144. https://doi.org/10.1016/j.cbpb.2009.10.013

Lam, M. K., & Lee, K. T. (2012). Microalgae biofuels: A critical review of issues, problems and the way forward. *Biotechnology Advances*, *30*(3), 673–690. https://doi.org/10.1016/j.biotechadv.2011.11.008

Leach, G., Oliveira, G., & Morais, R. (1998). Spray-drying of *Dunaliella salina* to produce a β-carotene rich powder. *Journal of Industrial Microbiology and Biotechnology*, *20*(2), 82–85. https://doi.org/10.1038/sj.jim.2900485

Lee, O. K., & Lee, E. Y. (2016). Sustainable production of bioethanol from renewable brown algae biomass. *Biomass and Bioenergy*, *92*, 70–75. https://doi.org/10.1016/j.biombioe.2016.03.038

Leite, G. B., Abdelaziz, A. E. M., & Hallenbeck, P. C. (2013). Algal biofuels: Challenges and opportunities. *Bioresource Technology*, *145*, 134–141. https://doi.org/10.1016/j.biortech.2013.02.007

Levitan, O., Dinamarca, J., Hochman, G., & Falkowski, P. G. (2014). Diatoms: A fossil fuel of the future. *Trends in Biotechnology*, *32*(3), 117–124. https://doi.org/10.1016/j.tibtech.2014.01.004

Li, Y., Ghasemi Naghdi, F., Garg, S., Adarme-Vega, T. C., Thurecht, K. J., Ghafor, W. A., Tannock, S., & Schenk, P. M. (2014). A comparative study: The impact of different lipid extraction methods on current microalgal lipid research. *Microbial Cell Factories*, *13*(1), 14. https://doi.org/10.1186/1475-2859-13-14

Li, Y., Horsman, M., Wu, N., Lan, C. Q., & Dubois-Calero, N. (2008). Biofuels from microalgae. *Biotechnology Progress*, *24*(4), 815–820. https://doi.org/10.1021/bp070371k

Lim, H. S., Kim, J. Y. H., Kwak, H. S., & Sim, S. J. (2014). Integrated microfluidic platform for multiple processes from microalgal culture to lipid extraction. *Analytical Chemistry*, *86*(17), 8585–8592. https://doi.org/10.1021/ac502324c

Liu, J., Zhu, Y., Tao, Y., Zhang, Y., Li, A., Li, T., Sang, M., & Zhang, C. (2013). Freshwater microalgae harvested via flocculation induced by pH decrease. *Biotechnology for Biofuels*, *6*(1), 98. https://doi.org/10.1186/1754-6834-6-98

Love, K. R., Bagh, S., Choi, J., & Love, J. C. (2013). Microtools for single-cell analysis in biopharmaceutical development and manufacturing. *Trends in Biotechnology*, *31*(5), 280–286. https://doi.org/10.1016/j.tibtech.2013.03.001

Lü, J., Sheahan, C., & Fu, P. (2011). Metabolic engineering of algae for fourth generation biofuels production. *Energy & Environmental Science*, *4*(7), 2451–2466. https://doi.org/10.1039/C0EE00593B

Lv, P., Yuan, Z., Wu, C., Ma, L., Chen, Y., & Tsubaki, N. (2007). Bio-syngas production from biomass catalytic gasification. *Energy Conversion and Management*, *48*(4), 1132–1139. https://doi.org/10.1016/j.enconman.2006.10.014

Markou, G., Angelidaki, I., & Georgakakis, D. (2012). Microalgal carbohydrates: An overview of the factors influencing carbohydrates production, and of main bioconversion technologies for production of biofuels. *Applied Microbiology and Biotechnology*, *96*(3), 631–645. https://doi.org/10.1007/s00253-012-4398-0

Martinez, F., Ascaso, C., & Orus, M. I. (1991). Morphometric and stereologic analysis of Chlorella vulgaris under heterotrophic growth conditions. *Annals of Botany*, *67*(3), 239–245.

Mata, T. M., Martins, A. A., & Caetano, Nidia. S. (2010). Microalgae for biodiesel production and other applications: A review. *Renewable and Sustainable Energy Reviews*, *14*(1), 217–232. https://doi.org/10.1016/j.rser.2009.07.020

Meher, L. C., Vidya Sagar, D., & Naik, S. N. (2006). Technical aspects of biodiesel production by transesterification—A review. *Renewable and Sustainable Energy Reviews*, *10*(3), 248–268. https://doi.org/10.1016/j.rser.2004.09.002

Munir, M., Sharif, N., Naz, S., Saleem, F., & Manzoor, F. (2013). Harvesting and processing of microalgae biomass fractions for biodiesel production. *Science Technology and Development*, *32*(3), 235–243.

Mutanda, T., Ramesh, D., Karthikeyan, S., Kumari, S., Anandraj, A., & Bux, F. (2011). Bioprospecting for hyper-lipid producing microalgal strains for sustainable biofuel production. *Bioresource Technology*, *102*(1), 57–70. https://doi.org/10.1016/j.biortech.2010.06.077

Nascimento, I. A., Marques, S. S. I., Cabanelas, I. T. D., Pereira, S. A., Druzian, J. I., de Souza, C. O., Vich, D. V., de Carvalho, G. C., & Nascimento, M. A. (2012). Screening microalgae strains for biodiesel production: Lipid productivity and estimation of fuel quality based on fatty acids profiles as selective criteria. *BioEnergy Research*, *6*(1), 1–13. https://doi.org/10.1007/s12155-012-9222-2

Nguyen, T. H. M., & Vu, V. H. (2012). Bioethanol production from marine algae biomass: Prospect and troubles. *Journal of Vietnamese Environment*, *3*(1), 25–29. https://doi.org/10.13141/jve.vol3.no1.pp25-29

Oncel, S. S., Kose, A., Faraloni, C., Imamoglu, E., Elibol, M., Torzillo, G., & Vardar Sukan, F. (2015). Biohydrogen production from model microalgae *Chlamydomonas reinhardtii*: A simulation of environmental conditions for outdoor experiments. *International Journal of Hydrogen Energy*, *40*(24), 7502–7510. https://doi.org/10.1016/j.ijhydene.2014.12.121

Oyedun, A. O., Lam, K.-L., Gebreegziabher, T., Lee, H. K. M., & Hui, C.-W. (2012). Optimisation of operating parameters in multi-stage pyrolysis. *Chemical Engineering Transactions*, *29*, 655–660. https://doi.org/10.3303/CET1229110

Parker, M. S., Mock, T., & Armbrust, E. V. (2008). Genomic insights into marine microalgae. *Annual Review of Genetics*, *42*, 619–645. https://doi.org/10.1146/annurev.genet.42.110807.091417

Paul Abishek, M., Patel, J., & Prem Rajan, A. (2014). Algae oil: A sustainable renewable fuel of future. *Biotechnology Research International*, *2014*, e272814. https://doi.org/10.1155/2014/272814

Popoola, T. O. S., & O. D. Yangomodou. (2006). Extraction, properties and utilization potentials of cassava seed oil. *Biotechnology*, *5*(1), 38–41. https://doi.org/10.3923/biotech.2006.38.41

Prakash, J., Pushparaj, B., Carlozzi, P., Torzillo, G., Montaini, E., & Materassi, R. (1997). Microalgal biomass drying by a simple solar device. *International Journal of Solar Energy*, *18*(4), 303–311.

Prins, M. J., Ptasinski, K. J., & Janssen, F. J. J. G. (2006). More efficient biomass gasification via torrefaction. *Energy*, *31*(15), 3458–3470.

Prochazkova, G., Safarik, I., & Branyik, T. (2013). Harvesting microalgae with microwave synthesized magnetic microparticles. *Bioresource Technology*, *130*, 472–477. https://doi.org/10.1016/j.biortech.2012.12.060

Raheem, A., Prinsen, P., Vuppaladadiyam, A. K., Zhao, M., & Luque, R. (2018). A review on sustainable microalgae based biofuel and bioenergy production: Recent developments. *Journal of Cleaner Production*, *181*, 42–59. https://doi.org/10.1016/j.jclepro.2018.01.125

Rajkumar, R., Yaakob, Z., & Takriff, M. S. (2013). Potential of micro and macro algae for biofuel production: A Brief Review. *BioResources*, *9*(1), 1606–1633. https://doi.org/10.15376/biores.9.1.1606-1633

Ranga Rao, A., Sarada, R., & Ravishankar, G. A. (2007). Influence of CO_2 on growth and hydrocarbon production in *Botryococcus braunii*. *Journal of Microbiology and Biotechnology*, *17*(3), 414–419.

Ras, M., Lardon, L., Bruno, S., Bernet, N., & Steyer, J.-P. (2011). Experimental study on a coupled process of production and anaerobic digestion of *Chlorella vulgaris*. *Bioresource Technology*, *102*(1), 200–206. https://doi.org/10.1016/j.biortech.2010.06.146

Rawat, I., Ranjith Kumar, R., Mutanda, T., & Bux, F. (2013). Biodiesel from microalgae: A critical evaluation from laboratory to large scale production. *Applied Energy*, *103*, 444–467. https://doi.org/10.1016/j.apenergy.2012.10.004

Razeghifard, R. (2013). Algal biofuels. *Photosynthesis Research*, *117*. https://doi.org/10.1007/s11120-013-9828-z

Razzak, S. A., Ali, S. A. M., Hossain, M. M., & deLasa, H. (2017). Biological CO_2 fixation with production of microalgae in wastewater – A review. *Renewable and Sustainable Energy Reviews*, *76*(C), 379–390.

Richmond, A. (2004). Handbook of microalgal culture: Biotechnology and applied phycology. *Handbook of Microalgal Culture: Biotechnology and Applied Phycology*. www.cabdirect.org/cabdirect/abstract/20053041910

Romagnoli, F., Blumberga, D., & Gigli, E. (2010). Biogas from marine macroalgae: a new environmental technology—Life cycle inventory for a further LCA. *Environmental and Climate Technologies*, *4*(1), 97–108. https://doi.org/10.2478/v10145-010-0024-5

Rosenthal, E. (2007, November 17). U.N. Report describes risks of inaction on climate change. *The New York Times*. www.nytimes.com/2007/11/17/science/earth/17climate.html

Rossignol, N., Vandanjon, L., Jaouen, P., & Quemeneur, F. (1999). Membrane technology for the continuous separation of microalgae culture medium: Compared performances of cross-flow microfiltration and ultrafiltration. *Aquacultural Engineering*. https://scholar.google.com/scholar_lookup?title=Membrane+technology+for+the+continuous+separation+microalgae+culture+medium%3A+compared+performances+of+cross-flow+microfiltration+and+ultrafiltration&author=Rossignol%2C+N.&publication_year=1999

Samson, R., & Leduy, A. (1982). Biogas production from anaerobic digestion of *Spirulina maxima* algal biomass. *Biotechnology and Bioengineering*, *24*(8), 1919–1924. https://doi.org/10.1002/bit.260240822

Saqib, A., Tabbssum, M. R., Rashid, U., Ibrahim, M., Gill, S. S., & Mehmood, M. A. (2013). Marine macroalgae Ulva: a potential feed-stock for bioethanol and biogas production. *Asian Journal of Agriculture and Biology*, *1*(3), 155–163.

Scarsella, M., Belotti, G., De Filippis, P., & Bravi, M. (2010). Study on the optimal growing conditions of Chlorella vulgaris in bubble column photobioreactors. *Chemical Engineering*, *20*, 85–90.

Schenk, P. M., Thomas-Hall, S. R., Stephens, E., Marx, U. C., Mussgnug, J. H., Posten, C., Kruse, O., & Hankamer, B. (2008). Second generation biofuels: high-efficiency microalgae for biodiesel production. *BioEnergy Research*, *1*(1), 20–43. https://doi.org/10.1007/s12155-008-9008-8

Sharma, S., Narayan, S., & Tripathi, S. (2013). Biohydrogen from algae: Fuel of the future. *International Research Journal of Environmental Sciences*, *2*, 44–47.

Sharma, Y. C., & Singh, B. (2009). Development of biodiesel: Current scenario. *Renewable and Sustainable Energy Reviews*, *13*(6–7), 1646–1651.

Sheehan, J., Dunahay, T., Benemann, J., & Roessler, P. (1998). *Look Back at the U.S. Department of Energy's Aquatic Species Program: Biodiesel from Algae; Close-Out Report* (NREL/TP-580–24190). National Renewable Energy Lab. (NREL), Golden, CO (United States). https://doi.org/10.2172/15003040

Shuba, E. S., & Kifle, D. (2018). Microalgae to biofuels: 'Promising' alternative and renewable energy, review. *Renewable and Sustainable Energy Reviews*, *81*, 743–755. https://doi.org/10.1016/j.rser.2017.08.042

Singh, A., Nigam, P. S., & Murphy, J. D. (2011). Mechanism and challenges in commercialisation of algal biofuels. *Bioresource Technology*, *102*(1), 26–34. https://doi.org/10.1016/j.biortech.2010.06.057

Spolaore, P., Joannis-Cassan, C., Duran, E., & Isambert, A. (2006). Commercial applications of microalgae. *Journal of Bioscience and Bioengineering*, *101*(2), 87–96. https://doi.org/10.1263/jbb.101.87

Stergiou, P.-Y., Foukis, A., Filippou, M., Koukouritaki, M., Parapouli, M., Theodorou, L. G., Hatziloukas, E., Afendra, A., Pandey, A., & Papamichael, E. M. (2013). Advances in lipase-catalyzed esterification reactions. *Biotechnology Advances*, *31*(8), 1846–1859. https://doi.org/10.1016/j.biotechadv.2013.08.006

Suganya, T., Varman, M., Masjuki, H. H., & Renganathan, S. (2016). Macroalgae and microalgae as a potential source for commercial applications along with biofuels production: A biorefinery approach. *Renewable and Sustainable Energy Reviews*, *55*(C), 909–941.

Surendhiran, D., & Vijay, M. (2012). Microalgal biodiesel—A comprehensive review on the potential and alternative biofuel. *Research Journal of Chemical Sciences*, *2*(11), 71–82.

Tiwari, A., & Pandey, A. (2012). Cyanobacterial hydrogen production – A step towards clean environment. *International Journal of Hydrogen Energy*, *1*(37), 139–150. https://doi.org/10.1016/j.ijhydene.2011.09.100

Topare, N. S., Raut, S., Renge, V., Khedkar, S. V., Chavan, Y., & Bhagat, S. (2011). Extraction of oil from algae by solvent extraction and oil expeller method. *International Journal of Chemical Sciences*. www.semanticscholar.org/paper/Extraction-of-Oil-from-Algae-by-Solvent-Extraction-Topare-Raut/8bdf4e3a316402f8b84a45a7d398a31b17e0a0a7

Van Wagenen, J., Miller, T. W., Hobbs, S., Hook, P., Crowe, B., & Huesemann, M. (2012). Effects of light and temperature on fatty acid production in nannochloropsis salina. *Energies*, *5*(3), Article 3. https://doi.org/10.3390/en5030731

Vergara-Fernández, A., Vargas, G., Alarcón, N., & Velasco, A. (2008). Evaluation of marine algae as a source of biogas in a two-stage anaerobic reactor system. *Biomass and Bioenergy*, *32*(4), 338–344. https://doi.org/10.1016/j.biombioe.2007.10.005

Ward, A. J., Lewis, D. M., & Green, F. B. (2014). Anaerobic digestion of algae biomass: A review. *Algal Research*, *5*, 204–214. https://doi.org/10.1016/j.algal.2014.02.001

Wecker, M. S. A., Meuser, J. E., Posewitz, M. C., & Ghirardi, M. L. (2011). Design of a new biosensor for algal H_2 production based on the H_2-sensing system of *Rhodobacter capsulatus*. *International Journal of Hydrogen Energy*, *36*(17), 11229–11237. https://doi.org/10.1016/j.ijhydene.2011.05.121

Widjaja, A., Chien, C.-C., & Ju, Y.-H. (2009). Study of increasing lipid production from fresh water microalgae *Chlorella vulgaris*. *Journal of the Taiwan Institute of Chemical Engineers*, *40*(1), 13–20. https://doi.org/10.1016/j.jtice.2008.07.007

Williams, P. J. le B., & Laurens, L. M. L. (2010). Microalgae as biodiesel & biomass feedstocks: Review & analysis of the biochemistry, energetics & economics. *Energy & Environmental Science*, *3*(5), 554–590. https://doi.org/10.1039/B924978H

Yazdani, P., Karimi, K., & Taherzadeh, M. J. (2011). *Improvement of Enzymatic Hydrolysis of A Marine Macro-Alga by Dilute Acid Hydrolysis Pretreatment*. https://ep.liu.se/en/conference-article.aspx?series=ecp&issue=57&volume=1&Article_No=25

Yen, H.-W., & Brune, D. E. (2007). Anaerobic co-digestion of algal sludge and waste paper to produce methane. *Bioresource Technology*, *98*(1), 130–134. https://doi.org/10.1016/j.biortech.2005.11.010

Yoon, J. J., Kim, Y. J., Kim, S. H., Ryu, H. J., Choi, J. Y., Kim, G. S., & Shin, M. K. (2010). Production of polysaccharides and corresponding sugars from red seaweed: International Conference on Functionalized and Sensing Materials, FuSeM 2009. *Functionalized and Sensing Materials*, 463–466. https://doi.org/10.4028/www.scientific.net/AMR.93-94.463

Zhang, L., Zhang, B., Zhu, X., Chang, H., Ou, S., & Wang, H. (2018). Role of bioreactors in microbial biomass and energy conversion. *Bioreactors for Microbial Biomass and Energy Conversion*, 39–78.

Zheng, G., Wang, Y., & Qin, J. (2012). Microalgal motility measurement microfluidic chip for toxicity assessment of heavy metals. *Analytical and Bioanalytical Chemistry*, *404*(10), 3061–3069. https://doi.org/10.1007/s00216-012-6408-6

Zhong, W., Zhang, Z., Luo, Y., Qiao, W., Xiao, M., & Zhang, M. (2012). Biogas productivity by co-digesting Taihu blue algae with corn straw as an external carbon source. *Bioresource Technology*, *114*, 281–286. https://doi.org/10.1016/j.biortech.2012.02.111

10 Life Cycle Assessment (LCA), Techno-Economic Analysis (TEA) and Environmental Impact Assessment (EIA) of Algal Biorefinery

Bikash Kumar, Tonmoy Ghosh, Sukhvinder Singh Purewal and Kiran Bala

10.1 INTRODUCTION

Energy consumption and the ability to produce energy are one of the most significant parameters to judge a nation's socioeconomic status (Hassan et al. 2019). The fast-paced growth of human civilization has been controlled by the need to consume more energy and technology. By 2030–2035, the energy demand is estimated to rise 40–50% above the 2010 levels (Moazeni and Khazaei 2020; Shuba and Kifle 2018). The methodology used to attain this energy demand contributes to climate change (Sharifi 2020). Still, fossil fuels are a major energy resource to fulfill transportation, industrial, and domestic application (Dominković et al. 2018). Fossil fuels plagued with many problems contribute to environmental pollution and climate change. Fossil fuels are nonrenewable, and reserve sites are positioned in politically unstable regions of the world compromising their continuous availability (Asomaning et al. 2018). One of the most recent examples is the Russia–Ukraine conflict which is causing global fossil fuel price fluctuations (Osiichuk and Shepotylo 2020). Therefore, the current need is to search for potential alternatives which can help in fulfilling energy needs via environment-friendly resources. Fossil fuels are a source not only of energy but also of numerable chemicals which are used for the survival of human civilization. Therefore, while we search for energy sources, we also need to search for greener alternatives to these fossil-based chemicals. Biomass and microbial systems are being exploited to develop technologies and products that can help in fulfilling energy and chemical needs and mitigating carbon emissions (Ahmad et al. 2022). The energy generated from these biological sources called biofuels, results in net-zero carbon emission as the carbon content is sequestered from the atmosphere. Several countries have come up

DOI: 10.1201/9781003455950-10

with greener technologies for commercial production of biofuels and chemicals to meet their needs (Chen et al. 2021; Khan et al. 2018a; San Juan et al. 2019).

Biofuels are classified into different generations based on the type of feedstock used. The first-generation biofuels come from edible food crops and thus have an ethical concern in food-fuel conflicts. Thus, lignocellulosic biomass or waste generated during the production of edible food crops is used for the production of second-generation biofuels (Kumar et al. 2020a). The major substrate for the third generation of biofuel or algal-based biorefinery concept are the algae (Agrawal et al. 2020). Genetically modified plants and microorganisms used for biofuels and biochemical production are classified as fourth-generation (Kumar et al., 2020b; Mat Aron et al. 2020). Large amounts of microalgal biomass can be grown in small arable land/barren land at a very fast rate (Goswami et al. 2022). Algae can even grow on wastewater utilizing carbon dioxide and natural sunlight, helping in simultaneous waste-water treatment and energy generation (Chiu et al. 2015; Goswami et al. 2020; Kumar et al. 2019; Pittman et al. 2011). The biodiesel generated from algae is emission-free as it generates less or no NO_x, SO_x, and particulate matter when compared with fossil-based diesel. Thus, biodiesel from algae is not only a greener source but also helps in mitigating pollution. However, the unideal net energy gains and enormous investment required to run algal-based biofuels production shifted the focus from single product to multiproduct microalgal biorefinery to utilize potential synergies during downstream/upstream process integrations (De Bhowmick et al. 2019; Khoo et al. 2020).

The algae are a group of photosynthetic prokaryotes or eukaryotes. The algal biomass consists of cell walls rich in carbohydrates and cellular protein and lipids. Several biofuels such as biodiesel, bioethanol, biobutanol, biochar, biohydrogen syngas, and fuel additives have been reported to be produced from algae (Goswami et al. 2021a, 2021b; Goswami et al. 2022a, 2022b). Also, chemicals such as antibiotics, antioxidants, bioplastics, dietary fibers, glycerol, nutraceuticals, pharmaceuticals polyols, and animal feeds steroids are reported from algal sources (D'Este 2017; Dong et al. 2016; Mehariya et al. 2021a). The algae are also capable of generating pigments such as anthocyanins, astaxanthin, β-carotene, lutein, and phycoerythrin phycobilins (Barsanti and Gualtieri 2014; Linares et al. 2017; Molino et al. 2020; Saini et al. 2021; Timmis et al. 2010) (Figure 10.1).

In the late 1950s first algal biomass-based biofuel production was demonstrated and in the 1970s first large-scale microalgae cultivation at larger scale was demonstrated for sustainable biofuel production (Iwamoto 2004; Paul Abishek et al. 2014; Pulz and Gross 2004). The Solar Energy Research Institute in Golden, Colorado (USA) researched improving the microalgal biomass yield for enhanced biofuel production. They attempted to control environmental factors such as light illumination, temperature, and nutrient supply (Aguirre et al. 2013; Chisti 2007; Linares et al. 2017; Schenk et al. 2008; Sing et al. 2013).

The selection of algal strain and growth conditions such as nutrients, light, and temperature is often regulated by the choice of product (Aratboni et al. 2019; Khan et al. 2018b). For example, under stress conditions of nutrient deficiency such as lack of nitrogen in the medium, *Scenedesmus* resulted in a higher accumulation of polyhydroxyalkanoates/ bioplastics (García et al. 2021). Similarly, under cold stress, the microalgae *Chlamydomonas reinhardtii* resulted in higher pigment production (Supakorn et al. 2021). As elaborated earlier as well, the ability of algae to grow using the resources and

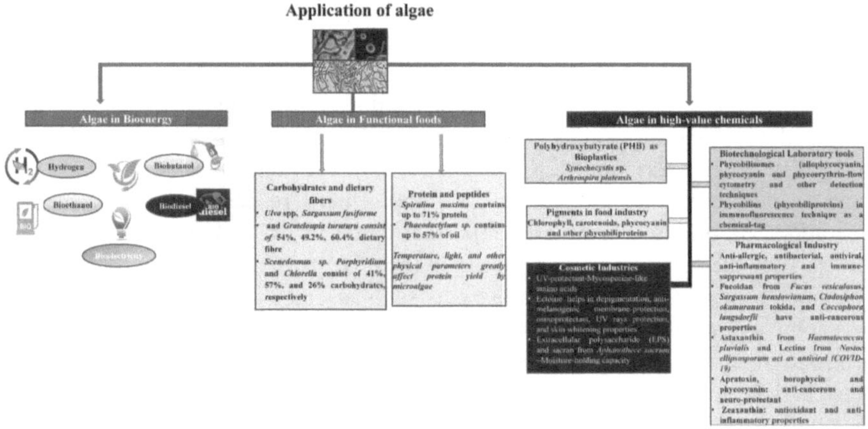

FIGURE 10.1 Potential application of algae.

nutrients accessible in the diverse wastewaters and CO_2 from the different industrial sources makes it a more potent microorganism for greener production of bioenergy and value-added compounds (Goswami et al. 2021a; Mehariya et al. 2021b; Molino et al. 2020). Thus, the CO_2 and wastewater from anthropogenic activities may be directed toward the development of multiproduct algal biorefinery fulfilling the broader purpose of bioremediation and resource recovery (biofuel, bioelectricity, biochemicals) simultaneously (Gajda et al. 2015; Goswami et al. 2021c; Ndayisenga et al. 2018).

The integrated biorefinery is developed with the primary objective of generating maximum productivity at the lowest energy and cost usage (Chandra et al. 2019). The algal biorefinery has several advantages and is key to achieving the objective of a circular bioeconomy, but several bottlenecks such as high infrastructure and running costs, suitable algal choice, and contamination still limit its commercialization and profitability (Gifuni et al. 2019). The imprudent application of fossil fuels without taking into concern its sustainability and several long-term negative impacts such as air pollution, global temperature rise, and uneven rains are already posing a danger to the existence of human civilization. Thus, before making a technology to be commercially available to the general masses, it is necessary to assess the economic and environmental feasibility and sustainability of the technology. The life cycle assessment (LCA), techno-economic assessment (TEA), and environmental sustainability assessment method offer an approach to measure and recognize the economics, environmental feasibility, and sustainability of the processes (Kumar and Verma 2021; Kumar et al. 2019; Kumar and Verma 2020).

10.2 GENERAL OVERVIEW OF LIFE CYCLE ASSESSMENT

LCA is a predictive tool that helps in the assessment of the impact of a process/product development during the entire cycle of its development on economic, environmental, health, and social parameters. The LCAs help in estimating the sustainability of process and product on different essential aspects before the application of technology or generation of product on a large scale (Luz et al. 2018). There are several studies on LCA for assessing the performance of biomass biorefineries in terms of energy output and different environmental performance indices (Righi 2019).

Gabrielle and Gagnaire (2008) demonstrated the LCA analysis for biorefineries utilizing food crops-based agricultural residues as starting material. Similarly, Liang et al. (2017), Raman and Gnansounou (2015), Soh et al. (2014), and Tonini and Astrup (2012) demonstrated LCA analysis for forest residues, non-food crops, algal biomass, and municipal solid-wastes-based biorefinery, respectively, for biofuel and value-added product recovery. The important facets of LCA analysis in biorefinery processes are the net energy required to run the process from start to end; net energy or economic output; environmental aspects such as greenhouse gases (GHG) emission, global warming, pollution generation, or mitigation ability (Jeswani et al. 2020). These parameters of any developed biorefinery are assessed to give techno-economic life cycle and environmental impact assessment. These impact assessments are key to evaluating the sustainability of the overall design of the operation for its execution aimed at the generation of biofuel and value-added products (Vance et al. 2022).

The LCA analysis framework as recommended by ISO 14040 and 14044 involves majorly four components, that is, goal and scope, definition, life cycle inventory, and life cycle impact assessment (ISO(14040) 2006; ISO(14044) n.d.; Neri et al. 2016) (Figure 10.2). LCA framework is not only limited to biorefinery but can also be applied to sectors such as the generation of synthetic and biological chemicals, waste management, and resource recovery (Davis et al. 2016; Thornley et al. 2015), (Cespi et al. 2015; Cespi et al. 2016; Cespi et al. 2015), (Passarini et al. 2014).

In terms of evaluating the life cycle assessment of biorefineries important framework and key methods are the definition of (a) goals and scope, (b) functional unit identification, (c) data inventory and allocation of key parameters, (d) uncertainties, (e) sensitivity analysis, (f) allocation and estimation of biorefinery outputs, and (g) impact categories such as land use and GHG emission patterns (Ahlgren et al. 2015).

Therefore, in the present study, LCA for algal-based biorefinery is examined with a specific emphasis on essential parameters and data inventory required for analysis. The methodology and interpretation of some case studies for algal-based biorefineries have also been addressed. The chapter will also discuss how LCA studies can help in the development of future integrated biorefineries to attain the objective of a circular bioeconomy. LCA studies also control the policy decision, thus a summary of integrated sustainability assessment (combining economic, environmental, and social factors) will also be presented for future policy decisions.

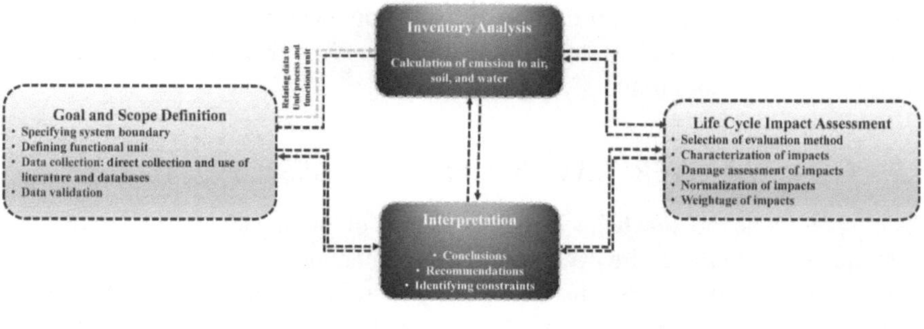

FIGURE 10.2 Life cycle assessment framework.

10.3 TOOLS USED FOR THE LCA AND IMPACT ASSESSMENT ANALYSIS

There are several tools available for the LCA analysis and are being used are summarized subsequently.

10.3.1 SimaPro

SimaPro is a powerful sustainability software solution suitable for product design and sustainability and used widely by decision-makers and sustainability experts. The tool provides insight into fact-based LCA to be enabled to take better decisions to make the overall process economical with minimal negative environmental footprints for different services and products. It is developed by PRé_Consultants (2019) and regularly updated. The key features of the current version as described by the company are given next.

- A systematic and transparent way to model and evaluate complex life cycles in an easy manner
- It can compute environmental impact across all life cycle stages of services and products.
- It has ability to recognize the hotspots at each stage of the supply chain such as raw materials' extraction, synthesis and manufacturing, supply, application by user, and waste disposal (SimaPro 2023).

10.3.2 openLCA

The software is developed by Jutta Hildenbrand, Michael Srocka, and Andreas Ciroth in the year 2005. It is developed on three concepts.

1. Its basic idea is utilizing open-source software and standard programming design for creating a sophisticated high-performance modular framework that is fast and reliable toward developing visually attractive and flexible life cycle modeling and sustainability assessment.
2. Engaging the programming community to contribute to the development of different models for open-source LCA.
3. Enable users to design their modules by making the availability of newly built modules by independent contributors (openLCA 2023).

Different available modules on the websites are software for automatic and graphical modeling, allocation and system expansion, contribution tree, data quality systems, developer tools, Monte Carlo simulation, parameters, regionalized impact assessment, and life cycle costing (openLCA Features 2023).

10.3.3 One Click LCA

It is a platform for easy and programmed life cycle assessment software that facilitates the calculation of environmental impacts of products. It enables the estimation of the sustainability of products, processes, services, and infrastructure. It has a wide range of software for design and construction, manufacturing, and enterprises (One Click LCA 2023).

10.3.4 GaBi

This is developed by Sphera(R) and provides productivity, safety, and sustainability analysis (GaBi 2023). It utilizes software and proprietary data for enabling stakeholders to highlight, manage, and mitigate risks in the area of product development, operation and management, environmental impact, health safety, and sustainability. The Sphera(R) provides Life cycle assessment (LCA_GaBi 2023) and product sustainability data and software (Sphera_PSSD 2023).

10.3.5 BEES (Building for Environmental and Economic Sustainability)

Developed by the Engineering Laboratory of NIST (National Institute for Standards and Technology), United States, BEES utilizes a life cycle assessment approach (ISO 14040) standards for measuring the environmental performance of building products. BEES analyses all stages of products such as the acquisition of raw materials, manufacturing and construction, transport of goods, installation, application, and recycling/waste management. Similarly, BEES measures economic performance utilizing the standard life cycle cost methods (ASTM) involving initial investment cost, operation, replacement maintenance, and repair. Further, the results obtained from LCA and economic performances are subjected to Multi-Attribute Decision Analysis for comprehensive economic and environmental impact assessment (Kneifel and Lavappa 2018).

10.3.6 esg.tech

esg.tech is a web portal providing a group of LCA, economic assessment, and environmental assessment tools, for example, carbon accounting, ESG portfolio management, ESG Reporting, Life Cycle Assessment, Product Stewardship, Supply Chain Sustainability, and Sustainability Reporting software (ESG.TECH_SS 2023). Some famous LCA software such as Ecochain (Mobius and Helix) and Earthshift Global are available on this portal (ESG.TECH_SS 2023).

10.3.7 Ecoinvent Database

ecoinvent is a Zurich (Switzerland)-based not-for-profit association that enables worldwide accessibility of high-quality data for sustainability assessments. With an aim to provide free data on sustainability, ecoinvent focuses on the compilation of LCI data, their linkage, and easy access/distribution. Ecoinvent association also takes part in global data management initiatives and projects and coordinates/collaborates at the international level for promotion and awareness for the creation of high-quality life cycle inventories and environmental assessments and making them affordably accessible worldwide. This helps several stakeholders such as designers, architects, scientists, entrepreneurs, NGOs, industrialists, academicians, and policymakers (Ecoinvent 2023).

Ecoinvent data has been used efficiently in leading software and applications such as eco-calculators, ecodesign tools, LCA tools, computer-aided design (CAD), and business and product lifecycle management tools. Some of the examples are Simapro, GiBA, opneLCA, Umberto, Altermaker, SAP, one-click LCA, eBalance, Carbonstop, EarthSmart, Places by The circulate initiative, eFootprint (Ecoinvent Software 2023).

These tools and databases help in identifying important economic and environmental factors and also provide inventories and data for further analysis. These benefit in predicting or calculating the economic and energy requirement for each stage of the process or product development. These life cycle assessment approaches play a key role in designing or choosing a sustainable process for industrial setup and are also adapted for policy development (Seidel 2016; Somé et al. 2018).

10.4 METHODS, FRAMEWORK, AND LCA AND LCIA OF THE ALGAL-BIOREFINERY

Several groups have analyzed the life cycle assessment of algal-based biorefineries (Gao et al. 2013; Gnansounou and Raman 2016; Mu et al. 2014; Passell et al. 2013) for the biofuel and value-added compounds generation. Ubando et al. (2022) demonstrated LCA of microalgal-based biorefinery summarizing challenges in the assessment of different aspects of algal biorefinery and also providing a comprehensive insight into design choices for future microalgal biorefineries. Similarly, Mobarezkhoo et al. (2022) evaluated the LCA of supply network of microalgae-based biorefinery supply network using a sustainable mixed integer linear programming model (MILP) for designing and planning a distribution network for microalgae biofuel. Further, Nilsson et al. (2022) demonstrated the LCA for seaweed-based biorefinery in Ireland for the production of sodium alginate from seaweed *Saccharina latissimi*. The work focused on evaluating 19 impact categories with special emphasis on climate change factors.

As discussed, different components and parameters for LCA analysis of algal biorefinery are as given here.

10.4.1 Component and Parameters for LCA of Algal Biorefinery

10.4.1.1 Goal and Scope Definition

The potential application of the product/service developed by the process is referred to as a goal. The scope set up the boundaries for the analysis consisting of functional units (Kylili et al. 2016).

10.4.1.2 Life Cycle Inventory for Algal Biorefinery

The life cycle inventory (LCI) flow involves the grouping of inputs and outputs associated with the targeted product functional unit. The functional units of the product are the focal points of all LCA studies and are expressed in terms of energy or cost.

Input and output cycle data at each step of the algal biorefinery: Different stages of the algal biorefinery are cultivation, harvesting, oil and protein extraction,

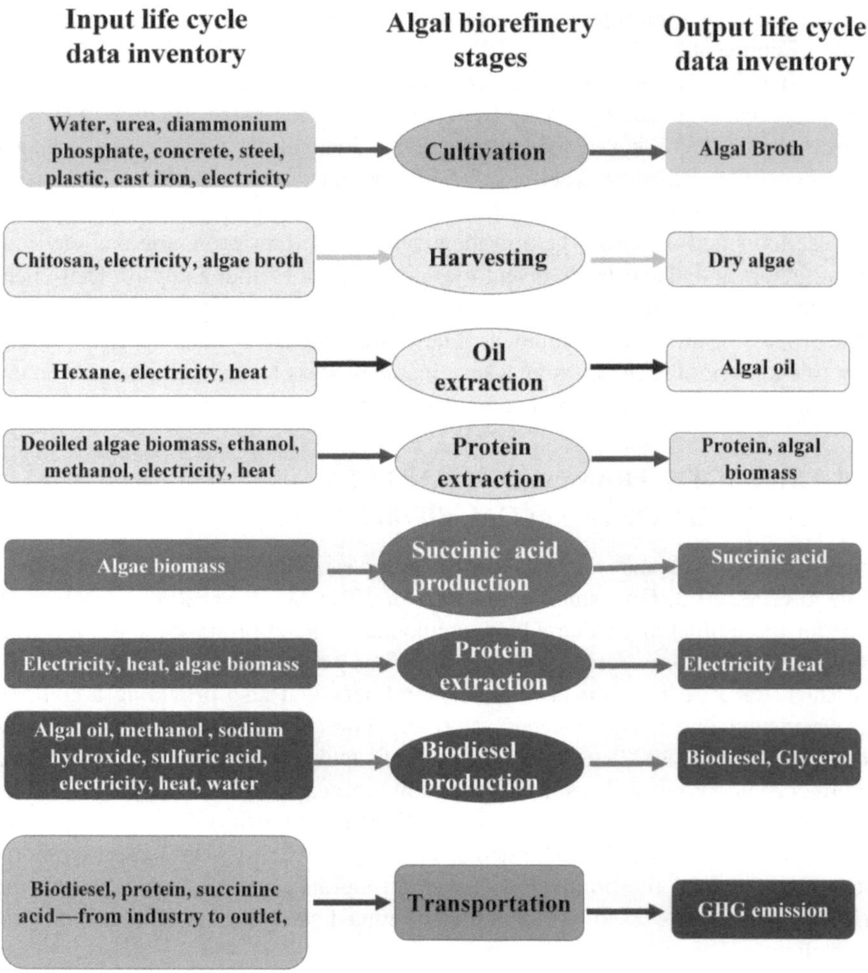

FIGURE 10.3 Life cycle data inventory for algal biorefinery (production of biodiesel, succinic acid, and algal protein).

value-added compound production (succinic acid, methane, biodiesel), co-product handling, and transportation. The output inventory involves algal broth, dry biomass, algal oil, protein and biomass, succinic acid, electricity, heat, biodiesel, and glycerol. The input and output inventory data classified for algal biorefinery is provided in Figure 10.3 (Gnansounou and Raman 2016, 2017)).

The energy and water requirement of the overall process along with the cost of repair of working equipment at each stage is included in LCI as well (Siqueira et al. 2018).

Similarly, Nilsson et al. (2022) demonstrated the LCA of seaweed biorefinery for alginate production using *S. latissimi* in Ireland. The important stages for the LCI are seaweed cultivation, fertile algae collection, preparation of spore and seeding, deployment of seed to farming site, growth, harvesting, preservation/drying post harvesting, and product extraction (sodium alginate) by conventional and ultrasound systems. The inventory data for each stage are different; for example, for the fertile algae collection stage, the inventory data involve the cost of plastic bags, cloth, and transposition to the hatchery. Similarly, inventory parameters such as nutrient components, illumination system, cooling and heating system, boat operation, and transportation cost are important inventory data for the seaweed-based biorefinery. The output inventory for each stage was considered as alginate, liquid residue from alginate extraction, cellulosic film, electricity, and fertilizer (NH_3).

10.4.1.3 Life Cycle Impact Assessment (LCIA)

The LCIA connects input and output categories to the respective impact categories related to techno-economic aspects, environmental, human health, or social impacts (Kumar and Verma 2020, 2021; Othoniel et al. 2016). The LCIA results are assessed keeping in mind the defined goal and scope within the system boundary, and decisions on process parameters are decided. Thus, it also acts as a tool for the economic and environmental assessment of divergent processes with analogous functions (Ubando et al. 2019).

The environmental impact assessment parameters such as fuel-specific consumption, GHG emission, net energy ratio, and energy balance calculation are given in the formula in Figure 10.4.

In the case of seaweed biorefinery life cycle impact assessment reported by Nilsson et al. (2022), they utilized SimaPro version 9.1 software using LCIA method EF 3.0 with 19 different impact categories recommended by the EU initiative Product Environmental Footprint available on the European platform of LCA (EPLCA 2023). During the study, data from Ecoinvent 3.6 database and Agribalyse 3 was used. The data for farming equipment and seaweed cultivation/post harvesting was retrieved from Ecoinvent database and farm data for the year 2021, respectively. The inventory data for seaweed cultivation and seedling step was obtained from Thomas et al. (2021).

Environmental Impact Assessment (EIA) Parameters	Formulas and Explanation
Fuel Specific consumption$_{biofuel}$ (FSC $_{biofuel}$)	$FSC\ biofuel = \frac{Biofuel\ blend\ factor \times Fuel\ specific\ consumption_{fuel\ blend}}{1-[\left(\frac{1-(Biofuel\ blend\ factor) \times Fuel\ specific\ consumption_{fuel\ blend}}{Fuel\ economuy_{fossil\ fuels}}\right)]}$ *Fuel specific consumption represents the amount of fuel consumed by the vehicle for a specific distance covered #Blend factor represent the percentage of biofuel mix in fossil fuel
Greenhouse gases'emission well-to-wheel ($GHG_{biofuel}$ WtW)	$GHG_{biofuel}\ WtW = GHG_{biofuel}\ WtT \times FSC_{biofuel}$ WtT: Well-to-tank
Net energy ratio (NER)	$NER = \sum E_{out} / \sum E_{in}$ The net energy ratio is energy produced by the system and energy used during the operation
Energy balance (EB) calculation	$EB = \sum input - \sum output$ In all the cases, energy output from biofuel is calculated in terms of the energy potential of microalgal-based biofuel (biodiesel), usually represented in MJ/kg

FIGURE 10.4 Formulas and explanation for different environmental impact assessment parameters.

The results of LCIA from the aforementioned study were presented in terms of alginate extraction and environmental factors. Alginate extraction yield in terms of mass percentage of dry *S. latissima* biomass increased with the ultrasound method as compared to conventional methods. During the study, climate change impact was assessed and it shows post-harvest drying of seaweed requires high electricity consumption (75% of overall climate impact) and are represented showing 6.12 kg CO_2 equivalent per kg dry seaweed. Whereas wet seaweed has a much lower climate impact of 0.16 kg CO_2 eq. per kg wet weight. The initial collection of algae and spore preparation has negligible climate impact whereas boat fuel also contributes significantly to climate hotspots, just second to drying and higher than deploying, growth, and maintenance step (Table 10.1) (Figure 10.5).

TABLE 10.1
Life Cycle Impacts Assessment Results for 19 Impact Categories for 1 kg Dry and 1 kg Wet Seaweed (Nilsson et al., 2022).

Impact category (Unit)	Amount per 1 kg dry *S. latissima* at post-harvest	Amount per 1 kg wet *S. latissima* at harvest
Acidification (mol H+ eq)	0.025	0.000
Climate change (kg CO_2 eq)	6.211	0.155
Climate change—biogenic (kg CO_2 eq)	0.012	0.000
Climate change—fossil (kg CO_2 eq)	6.194	0.155
Climate change—land use and LU change (kg CO_2 eq)	0.005	0.000
Ecotoxicity, freshwater (CTUe)	95.934	5.380
Eutrophication, marine (kg P eq)	0.004	0.000
Eutrophication, freshwater (kg P eq)	0.001	0.000
Eutrophication, terrestrial (mol N eq)	0.043	0.001
Human toxicity, cancer (CTUh)	0.000	0.000
Human toxicity, non-cancer (CTUh)	0.000	0.000
Ionising radiation (kBq U-235 eq)	0.403	0.013
Land use (Pt)	20.942	0.466
Ozone depletion (kg CFC11 eq)	0.000	0.000
Particulate matter (disease causing PM_{10}, $PM_{2.5}$)	0.000	0.000
Photochemical ozone formation (kg NMVOC eq)	0.012	0.000
Resource use, fossils (MJ)	96.383	2.937
Resource use, minerals, and metals (kg Sb eq)	0.000	0.000
Water use (m^3 deprivation)	0.661	0.016

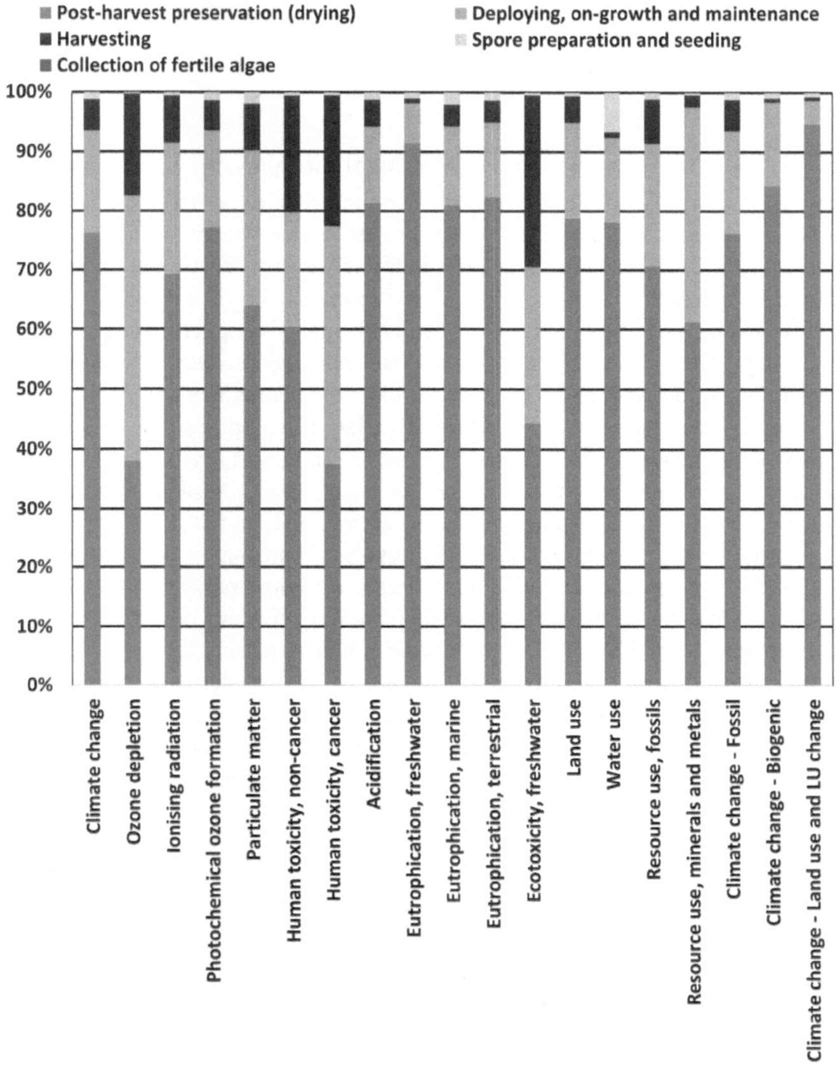

FIGURE 10.5 Environmental, health impact of 19 impact categories from five cultivation steps of dried seaweed (*S. latissima*) (as shared by Creative Common license (Nilsson et al. 2022)).

10.4.1.4 Uncertainties in LCIA

The trade-offs among several conflicting factors in LCA occasionally affect uncertainties in dominant factors. For example, in exchange for higher ecotoxicity allocation, it may be traded off with lower global warming impact (Guinée et al. 2017). To simplify the LCIA, some uncertainties of material or process that can impact significantly the result of impact categories may be omitted. These uncertainties may be a contribution from different impact categories, consumption, or environmental emissions. To overcome the uncertainties, selection of an accurate data set is a must. Monte Carlo simulation is an advanced, universally acceptable statistical method employed for studying the impact of alterations on the inputs and outputs (Michiels and Geeraerd 2020).

10.4.1.5 Sensitivity Analysis in LCIA of Algal Biorefinery

During LCA analysis, the lower and upper bound of a parameter is selected, and several permutations of parameter levels are then calculated for sensitivity analysis of a process. In addition to this, scenario analysis was also performed where it was measured if uncertain parameters have any impact on the overall result. This can help in providing a clear picture of the sensitivity analysis (Eggemann et al. 2020).

In the algal biorefinery, the variation in efficiency may be due to variations in the compositional analysis. This, in turn, can impact the economics and environmental impact of the overall system. This effect of variation in composition on economics and environmental impacts is considered under the sensitivity analysis. The data under study as per previous literature can be selected as a base/reference case for the sensitivity analysis. The data inventories are selected, and LCIA is performed in comparison with the base case or reference system (Gnansounou and Raman 2017).

The study by Nilsson et al. (2022) evaluated sensitivity analysis and the potential of product improvement. A small pilot scale plant has been administered for alginate extraction and extrapolated to industrial scale production, but, still, it is not able or represent/mimic the large scale system. Therefore, different alternative circumstances were assessed for production improvement and identification of weaknesses (sensitivity analysis) in the modeling. For example, i) evade algal biomass drying stage, ii) to increase the recovery rate, precipitation was performed using isopropanol instead of ethanol, and this increased the rate of recovery, iii) to stimulate the optimized condition for industrial scale processes, reduction in inputs by a factor of 3 may be done, iv) the application of alcohol for precipitation instead of $CaCl_2$ process for biomolecule recovery was described by McHugh and FAO UN (2003), v) the application of all improvements from (i) to (iv). This process optimization leads to a huge impact on the performance of the environmental indicators. The drying of algal biomass and precipitation during alginate purification have more energy requirement and have a huge impact on product yield. Thus, the reduction of drying and precipitation can have a positive impact on overall performance indicators. Also, even a reduction in input parameters can have positive effects, but it may be not feasible as it may give some false positive results statistically but the actual economic and environmental impact.

10.4.1.6 Allocation and Impact Categories

Within the system boundary, the partition of environmental burden across common pathways to various products or functions is defined as allocation. The allocation procedure can be categorized into either of these four procedures such as displacement (system expansion), energy variation, mass change, and economics (Kolosz et al. 2020). The allocation is subject to substantial operational parameters in LCAs, and no perfect allocation model can be singled out for different or similar studies (Kok et al. 2020). Acidification potential, climate change parameters, greenhouse gases' emission, global warming potential (GWP), etc., are some of the impact categories that are beyond the scope of emissions for LCA methodology for a technology or system (Amini Toosi et al. 2020; Kolosz et al. 2020)

10.5 COMPREHENSIVE REVIEWS OF LCA AND LCIA FOR DIFFERENT ALGAL BIOREFINERIES PROCESSES

Chowdhury and Franchetti (2017) demonstrated LCIA for four different scenarios (stand-alone or in combination) for algal-based biorefinery for the production of methane, biodiesel, bioethanol, and bio-oils using anaerobic-digestion (AD), biodiesel-production, enzymatic-hydrolysis, and pyrolysis, respectively. The combination of algae biodiesel using dairy waste as a substrate along with enzymatic hydrolysis resulted in maximum net energy value. The combination of enzymatic liquefaction and subsequent pyrolysis for leftover algal biomass resulted in 38% more net energy as compared to the stand-alone pyrolysis. Chowdhury and Franchetti (2017) also recommended that multi-product biorefinery is the future as standalone algal biofuel production is not economical.

Barlow et al. (2016) reported the LCA of an algal biorefinery for the production of biodiesel utilizing hydrothermal liquefaction. The study recommended that the feedstock cost is a rate-limiting factor in deciding the overall process economics and subsequently regulating the biodiesel final cost. It was observed that considering a system boundary, the NER and GWP potential of the system are 0.33, and −44 g CO_2-e MJ^{-1}, respectively. The study also suggested that the final selling price of biodiesel produced is dependent mainly on the cost of algal biomass production. Also, algal cultivation/production method cost is key to algal biorefinery both in terms of economics and environmental impacts. Further, the cultivation of algae requires a huge amount of water so substituting fresh water with marine or wastewater can not only decrease the cost but also help in simultaneously addressing two problems, that is, pollution mitigation and high-value production from waste. The wastewater often contains nutrients and carbon sources and thus can help in decreasing overall production costs and also minimize environmental impact as wastewater will not be dumped directly causing pollution. In addition, algae don't require arable land and can be set up in barren lands; also, the CO_2 from industries outlets can also be directed to algal biorefinery plants where it can be used by algae for their growth and thus making the overall process with negative carbon emission and net zero GHG (Barlow et al. 2016; Chowdhury and Franchetti 2017; Javed et al. 2019; Nagarajan et al. 2020).

Maga (2017) performed an LCA analysis of wastewater-based algal biorefinery for biomethane production. It was demonstrated that this biomethane is used as fuel for vehicles in place of natural gas. The study recommended that biomethane wastewater-based algal biorefinery can efficiently be used as an alternative to natural gas as it has lesser climate impact and can prevent ozone depletion. However, the LCIA of parameters such as eutrophication, overall efficiency, particulate matter generation, photochemical oxidation, and water limitation was not very motivating. A comprehensive summary of sensitivity analysis, functional unit, allocation technique, global warming potential, and impact category for LCA system boundary for algal biorefinery is presented in Table 10.2 (Ubando et al. (2022)

TABLE 10.2

Comprehensive Summary of Sensitivity Analysis, Functional Unit, Allocation Technique, Global Warming Potential, and Impact Category for LCA System Boundary.

Sensitivity analysis, functional unit, allocation technique and LCA system boundary per article					Global warming potential impact category	The other LCA impact categories	
References	**Functional Unit**	**System boundary***	**Software utilized**	**Sensitivity analysis**	**GHG* (kg CO_2eq/F.U.)**	**Impact categories**	**Findings**
Ferreira et al. (2013)	1 $MJ_{consumed}/MJ_{Biodiesel\ produced}$ and 1 $gCO_{2\ emission}/MJ_{Biodiesel}$ produced	*CTGa	SimaPro 7.1	No	**SE (12.471–14.994) **SFE (14.881–17.913)	Energy Cost	172–239 $MJ/MJ_{produced}$ 286 $MJ/MJ_{produced}$ 660.56 €/$kg_{algal\ oil}$ 365.42€/$kg_{algal\ oil}$
Beckstrom et al. (2020)	1 kg of bioplastic feedstock resulting/ kg CO_{2eq}/kg bioplastic feedstock	*WTW	–	Yes	**CF-PBR/Drying (–0.315) **ORP/ Fractionation (0.656)	Areal energy	17 and 28 kWh/ha/day
Rodríguez et al. (2018)	1 kg dry biomass	*CTGa	SuperPro Designer 9.5 GaBi 6.0	No	0.06	–	–
Khanum et al. (2020)	1 kg of biodiesel	*CTGr	Gabi 6 Eco-Invent Database	Yes	128.3	NPV cost NPV revenue EP AP ODP PCOP	US$ 372.5 million US$ 493.2 million 129 kg PO_{4-}^{3} eq (42 kg P_{eq}) 3 g SO_{2eq} 0.18 µgCFC_{eq} 0.22 g C_2H_{4eq}

Montazeri et al. (2016)	1 kg biodiesel	*CTGa	GREET TRACI	No	0.5	CED Eutrophication factors	1.4 MJ/kg $_{biodiesel}$ 22 g $_{Neq}$
Thomassen et al. (2018)	11 tons of β-carotene and 128 tons of fertilizer per year over a project lifetime of 10 years	*CTGr	Oracle Crystal Ball	Yes	(53×10^7) 1 Belgium (2×10^7) 2 Belgium (1×10^7) 3 Belgium (2×10^7) 1 India (13×10^7) 2 India (8×10^7) 3 India	Over a 10-year period: NPV ODP	India: 33, 40, and 2 (million €) Belgium: –7, 25, and –29 (million €) India: 2, 1, and 3 (102 $_{kg\ CFC\text{-}11\ eq}$) Belgium: 2, 1, 2 (102 kg $_{CFC\text{-}11\ eq}$)
Somers and Quinn (2019)	1 MJ fuel product	*WTW	–	No	0.0281	NER	0.13 MJ/MJ $_{biofuel}$
Espada et al. (2020)	1 kg of β-carotene	*CTGa	Gabi 6.0	No	***BR1-SBEP (525) ****BR2-SCEP (275)	Energy Cost	~16000 MJ/kg$_{\beta\text{-carotene}}$ ~7000 MJ/kg$_{\beta\text{-carotene}}$ 4.24 million €/year 9.84 million €/year

*System boundary legends: Cradle-to-grave (CTGr), cradle-to-consumption (CTC), cradle-to-gate (CTGa), well-to-tank (WTT), well-to-wheels (WTW).

**GHG legends: Soxhlet Extraction (SE), Supercritical Fluid Extraction (SFE), Cyclic Flow Photobioreactors (CF-PBR),

***Open Raceway Ponds (ORP), Biorefinery 1 which includes a solvent-based extraction process (BR1-SBEP).

****Biorefinery 2 which includes a supercritical extraction process (BR2-SCEP).

For algal biorefinery, considering 1 MJ of biodiesel produced as a functional unit, the GHG emission and net energy ratio reported are 35–140 gCO_{2eq} and 0.1–4.2, respectively (Campbell et al. 2011; Sills et al. 2013). Similarly, comparing the LCA of biodiesel and petroleum-based diesel, a mixed response is observed, for example, the study by Sheehan et al. (1998) recommended that blending of biodiesel with petroleum diesel has a positive impact and reduces the CH_4, SOx, HF, PM_{10}, and CO whereas the NO_x, NMHC, and HCl emission increased. Nanaki and Koroneos (2012) performed a comparative LCA for the application of biodiesel, diesel, and gasoline for transportation in Greece. The study showed that biodiesel is a better choice as it results in a reduction in GHG emissions, and lower well-to-wheel methane emissions as compared to petroleum-based diesel and gasoline. However, the application of biodiesel results in higher PM10, NO, and NO_x emissions and also causes the release of eutrophication agents. Xu et al. (2022) demonstrated the life cycle analysis of biodiesel generated from renewable resources such as oil crops in the United States and compared it with petroleum diesel. The study showed that renewable biodiesels can cause higher GHG reductions by up to 79%–86% than petroleum diesel. The process parameters such as hydro-processing, transesterification, pretreatment, and land usage add value to the impact generated during biodiesel production using different strategies. As discussed earlier, each LCA analysis is different based on the choice of functional unit, reference system, and system boundaries. These factors impact the overall LCA performance, and variations in results are observed due to variation in the choice of parameters, goal and scope design, and also inventory selection, etc. (Gnansounou et al. 2009; Ika Rinawati et al. 2022; Panesar et al. 2017; Sills et al. 2013).

10.6 LCA OF THE MICROALGAE-BASED BIOREFINERY SUPPLY NETWORK AND THE NEED FOR INTEGRATED BIOREFINERIES

Biofuels' supply chain such as transportation and distribution is key while performing LCA and LCIA. Mobarezkhoo et al. (2022) demonstrated the life cycle evaluation of the supply network of microalgal biorefinery in Iran. The study involved designing and planning for sustainable microalgae biofuel production and building a strong distribution network. They performed mixed integer linear programming (MILP) modeling minimizing networking cost and maximizing saving of GHG emission as the function for the LCIA. A two-stage approach is utilized in which the first stage involves the determination of suitable candidates for the designing and construction of microalgal biorefinery. The analysis approach used is Data Envelopment Analysis (DEA), Combinative Distance-Based Assessment (CODAS) approaches, and Best-Worst Method (BWM) for economic, social, and environmental assessment parameters. During stage 2, OpenLCA tool followed by the bi-objective linear programming (BoLP) model was used for assessing the environmental impact of the microalgal biofuel supply chain network. Further, the epsilon constraint method is used to solve the (BoLP) model. The results of the study showed that since Iran's economy is facing several sanctions, a lack of technology, technicians, and equipment push them to focus on saving energy, minimizing energy wastage, and accessibility to transportation, investment, and maintenance cost to establish biorefineries. GHG emission is 42.3% for both the functional unit was observed, that is, GHG saving and production cost minimization. A national-level case study was also performed for Iran involving the microalgae biofuel supply chain. This case study recommended that replacing fossil fuels with biofuel can help in saving GHG emissions by 99%.

Further, the sensitivity analysis for cost and conversions rate cost was performed to analyze the model. The results show that the total cost of the transportation network has the highest impact on biodiesel cost. This network cost is high because the microalgal cultivation facility is situated at a distance from the refinery. Therefore, integrated or establishment of biorefineries and microalgae production facilities in nearby locations can minimize the network cost thus decreasing the overall cost.

Besides this, for improved service to customers and reduction of cost of the supply chain, maintenance and repair of facility or transportation units can be done for improved efficiency. Further uncertainties in transportation due to climatic changes and natural calamities such as flood earthquakes also need to be considered. Also, uncertainties associated with the cost of biomass procurement, rate of biomass to bioenergy conversion, the loss incurred during microalgae holding, and transport need to be considered before the evaluation of LCIA. Also, the inclusion of social objective functional needs such as health hazards and employability also needs to be taken care of.

The study also strongly recommended multiproduct integrated biorefineries as the way forward. The Iranian case study provided evidence for the statement where they recommended the production of ethane to minimize the cost coming from intermediate products. Similarly, even the integration of different biomass-based biorefineries such as agricultural waste, food waste, lignocellulosic waste, and municipal and animal waste can be a way forward (Kumar and Verma 2021). Developing decentralized biorefineries at the small-town level can provide employment, developing entrepreneur opportunities for social upliftment along with waste management processes such as integrating wastewater coming from cities to microalgae units and using food waste and CO_2 from industries for the growth of microalgae. Thus, developing a multi-waste multi-product biorefinery reaches the target of a circular bioeconomy and keeps the environment clean and society healthy and prosperous.

10.7 ROLE OF LCA AND LCIA IN POLICY DECISIONS BASED ON ALGAL BIOREFINERIES

LCA and LCIA provide the overall economic, environmental, and social impact of a technology, product, process, and service. Taking into consideration an example of a seaweed-based study by Nilsson et al. (2022), we observe that the cost of seaweed production and extraction process contributes significantly to the cost of the overall process. The fuel used for cultivation and running of boats for regular maintenance of the cultivation unit for 6 months requires much higher cost and has a negative environmental impact as well. Looking after The LCA and LCIA analysis may help the policymakers while designing policies for the future. Providing subsidies to procure equipment associated with microalgal refineries, motivating the green industries by providing them carbon credits, etc., are incentives to bring bigger business players to invest in biorefinery. Several certification programs are also run to highlight businesses with net zero carbon emissions.

Further, LCIA and LCA studies can act as guiding lights to make changes in designs and technologies for making processes both economically and environmentally favorable.

Some of the examples of government policies that were developed to promote biorefineries area are given next.

1. For the promotion of biofuel production and consumption, the US government devised the Energy Policy Act of 2005 and the Energy Independence and Security Act, 2007, with the target to produce 36 billion gallons of biofuel by 2022 (Sorda et al. 2010; Yacobucci and Bracmort 2009). Similarly, through the "Biomass Program 2008", the US government set a benchmark to reduce 30% of fossil-based gasoline consumption by 2030 (Saravanan et al. 2018a; Sorda et al. 2010).
2. In the European Union, several acts, action plans, and resolutions were moved for blending and shifting to a substantial percentage of energy usage through greener mode (Czyrnek-Delêtre et al. 2017; Kimble et al. 2008; Kumar et al. 2013; Lima and Gupta 2014; Martinot et al. 2005; Saravanan et al. 2018b; Smyth et al. 2010; Sorda et al. 2010; Su et al. 2015).
3. For ethanol blending, the National Biodiesel Mission and National Hydrogen Plan by the Government of India is based on impact assessment studies and literature developed in 2000–2020.

Thus, LCA and LCIA are required as key literature for devising policy for any biorefinery development for a sustainable future.

10.8 CONCLUSIONS

Overall, in an algal-based biorefinery, the cost of fuel and the environmental impact of each stage of biorefinery depend on the choice of methods used for cultivation. The cost of biomass production is the most dominant factor economically, and GHG or climate change factors are key during the application stage contributing to environmental impact. Cultivation also contributes to environmental impacts based on the choice of methods used. Thus, the choice of technology used during the process development is key and must be selected on the basis of LCA and LCIA analysis for minimizing cost and maximizing saving for GHG emissions. LCA and LCIA are key to the policy decisions that can help in the promotion of the development of greener technologies such as algal biorefinery through pro-biorefinery policies, subsidies, and certifications. Thus, LCIA acts as a guiding light for further technological advancements in algal-based biorefinery.

COMPETING INTEREST

All the authors declare that they do not have any competing interest.

FUNDING & ACKNOWLEDGMENT

BK is thankful to DST-SERB NPDF (PDF/2022/001781) scheme for his post-doctoral research

REFERENCES

Agrawal, K., and Bhatt, A., and B. N. and K. B. and V. P. (2020). Algal biomass: Potential renewable feedstock for biofuels production – Part I. In M. and M. P. K. and G. V. K. Srivastava Neha and Srivastava (Ed.), *Biofuel Production Technologies: Critical Analysis for Sustainability* (pp. 203–237). Springer. https://doi.org/10.1007/978-981-13-8637-4_8

Aguirre, A.-M., Bassi, A., & Saxena, P. (2013). Engineering challenges in biodiesel production from microalgae. *Critical Reviews in Biotechnology*, *33*(3), 293–308.

Ahlgren, S., Björklund, A., Ekman, A., Karlsson, H., Berlin, J., Börjesson, P., Ekvall, T., Finnveden, G., Janssen, M., & Strid, I. (2015). Review of methodological choices in LCA of biorefinery systems-key issues and recommendations. *Biofuels, Bioproducts and Biorefining*, *9*(5), 606–619.

Ahmad, M., Yousaf, M., Wang, S., Cai, W., Sang, L., Li, Z., & Zhao, Z. P. (2022). Development of rapid CO_2 utilizing microbial ecosystem onto the novel & porous FPUF@nZVI@TAC@ASP hybrid for green coal desulphurization. *Chemical Engineering Journal*, *433*. https://doi.org/10.1016/j.cej.2021.134361

Amini Toosi, H., Lavagna, M., Leonforte, F., Del Pero, C., & Aste, N. (2020). Life cycle sustainability assessment in building energy retrofitting; A review. In *Sustainable Cities and Society* (vol. 60). Elsevier Ltd. https://doi.org/10.1016/j.scs.2020.102248

Aratboni, H. A., Rafiei, N., Garcia-Granados, R., Alemzadeh, A., & Morones-Ramirez, J. R. (2019). Biomass and lipid induction strategies in microalgae for biofuel production and other applications. *Microbial Cell Factories*, *18*(1), 178.

Asomaning, J., Haupt, S., Chae, M., & Bressler, D. C. (2018). Recent developments in microwave-assisted thermal conversion of biomass for fuels and chemicals. In *Renewable and Sustainable Energy Reviews* (vol. 92, pp. 642–657). Elsevier Ltd. https://doi.org/10.1016/j.rser.2018.04.084

Barlow, J., Sims, R. C., & Quinn, J. C. (2016). Techno-economic and life-cycle assessment of an attached growth algal biorefinery. *Bioresource Technology*, *220*, 360–368.

Barsanti, L., & Gualtieri, P. (2014). *Algae: Anatomy, Biochemistry, and Biotechnology*. CRC press.

Beckstrom, B. D., Wilson, M. H., Crocker, M., & Quinn, J. C. (2020). Bioplastic feedstock production from microalgae with fuel co-products: A techno-economic and life cycle impact assessment. *Algal Research*, *46*. https://doi.org/10.1016/j.algal.2019.101769

Campbell, P. K., Beer, T., & Batten, D. (2011). Life cycle assessment of biodiesel production from microalgae in ponds. *Bioresource Technology*, *102*(1), 50–56. https://doi.org/10.1016/j.biortech.2010.06.048

Cespi, D., Beach, E. S., Swarr, T. E., Passarini, F., Vassura, I., Dunn, P. J., & Anastas, P. T. (2015). Life cycle inventory improvement in the pharmaceutical sector: Assessment of the sustainability combining PMI and LCA tools. *Green Chemistry*, *17*(6), 3390–3400.

Cespi, D., Passarini, F., Mastragostino, G., Vassura, I., Larocca, S., Iaconi, A., Chieregato, A., Dubois, J.-L., & Cavani, F. (2015). Glycerol as feedstock in the synthesis of chemicals: A life cycle analysis for acrolein production. *Green Chemistry*, *17*(1), 343–355.

Cespi, D., Passarini, F., Vassura, I., & Cavani, F. (2016). Butadiene from biomass, a life cycle perspective to address sustainability in the chemical industry. *Green Chemistry*, *18*(6), 1625–1638.

Chandra, R., Iqbal, H. M. N., Vishal, G., Lee, H. S., & Nagra, S. (2019). Algal biorefinery: A sustainable approach to valorize algal-based biomass towards multiple product recovery. In *Bioresource Technology* (vol. 278, pp. 346–359). Elsevier Ltd. https://doi.org/10.1016/j.biortech.2019.01.104

Chen, W. H., Lin, B. J., Lin, Y. Y., Chu, Y. S., Ubando, A. T., Show, P. L., Ong, H. C., Chang, J. S., Ho, S. H., Culaba, A. B., Pétrissans, A., & Pétrissans, M. (2021). Progress in biomass torrefaction: Principles, applications and challenges. In *Progress in Energy and Combustion Science* (vol. 82). Elsevier Ltd. https://doi.org/10.1016/j.pecs.2020.100887

Chisti, Y. (2007). Biodiesel from microalgae. *Biotechnology Advances*, *25*(3), 294–306. https://doi.org/https://doi.org/10.1016/j.biotechadv.2007.02.001

Chiu, S.-Y., Kao, C.-Y., Chen, T.-Y., Chang, Y.-B., Kuo, C.-M., & Lin, C.-S. (2015). Cultivation of microalgal Chlorella for biomass and lipid production using wastewater as nutrient resource. *Bioresource Technology*, *184*, 179–189.

Chowdhury, R., & Franchetti, M. (2017). Life cycle energy demand from algal biofuel generated from nutrients present in the dairy waste. *Sustainable Production and Consumption*, *9*, 22–27.

Czyrnek-Delêtre, M. M., Smyth, B. M., & Murphy, J. D. (2017). Beyond carbon and energy: The challenge in setting guidelines for life cycle assessment of biofuel systems. In *Renewable Energy* (vol. 105, pp. 436–448). Elsevier Ltd. https://doi.org/10.1016/j.renene.2016.11.043

Davis, S. C., Kauneckis, D., Kruse, N. A., Miller, K. E., Zimmer, M., & Dabelko, G. D. (2016). Closing the loop: integrative systems management of waste in food, energy, and water systems. *Journal of Environmental Studies and Sciences*, *6*(1), 11–24.

De Bhowmick, G., Sarmah, A. K., & Sen, R. (2019). Zero-waste algal biorefinery for bioenergy and biochar: A green leap towards achieving energy and environmental sustainability. In *Science of the Total Environment* (vol. 650, pp. 2467–2482). Elsevier B.V. https://doi.org/10.1016/j.scitotenv.2018.10.002

D'Este, M. (2017). *Algal Biomass for Bioenergy and Bioproducts Production in Biorefinery Concepts*. Ph.D Thesis Department of Environmental Engineering, Technical University of Denmark (DTU). DTU Library, https://backend.orbit.dtu.dk/ws/portalfiles/portal/133576888/Thesis_online_version_Martina_D_Este.pdf

Dominković, D. F., Bačeković, I., Pedersen, A. S., & Krajačić, G. (2018). The future of transportation in sustainable energy systems: Opportunities and barriers in a clean energy transition. In *Renewable and Sustainable Energy Reviews* (vol. 82, pp. 1823–1838). Elsevier Ltd. https://doi.org/10.1016/j.rser.2017.06.117

Dong, T., Knoshaug, E. P., Davis, R., Laurens, L. M. L., Van Wychen, S., Pienkos, P. T., & Nagle, N. (2016). Combined algal processing: A novel integrated biorefinery process to produce algal biofuels and bioproducts. *Algal Research*, *19*, 316–323.

Ecoinvent. (2023). *About Ecoinvent Database*. https://ecoinvent.org/the-ecoinvent-association/

Ecoinvent Software. (2023). *Software Tools—Ecoinvent*. https://ecoinvent.org/the-ecoinvent-association/software-tools/

Eggemann, L., Escobar, N., Peters, R., Burauel, P., & Stolten, D. (2020). Life cycle assessment of a small-scale methanol production system: A power-to-fuel strategy for biogas plants. *Journal of Cleaner Production*, *271*. https://doi.org/10.1016/j.jclepro.2020.122476

EPLCA. (2023). *European Platform on LCA _ EPLCA*. https://eplca.jrc.ec.europa.eu/

ESG.TECH_SS. (2023). *Global ESG and Sustainability Software List*. https://esg.tech/directory/?_software_sorting=title_asc&_page=2

Espada, J. J., Pérez-Antolín, D., Vicente, G., Bautista, L. F., Morales, V., & Rodríguez, R. (2020). Environmental and techno-economic evaluation of β-carotene production from Dunaliella salina. A biorefinery approach. *Biofuels, Bioproducts and Biorefining*, *14*(1), 43–54. https://doi.org/https://doi.org/10.1002/bbb.2012

Ferreira, A. F., Ribeiro, L. A., Batista, A. P., Marques, P. A. S. S., Nobre, B. P., Palavra, A. M. F., da Silva, P. P., Gouveia, L., & Silva, C. (2013). A biorefinery from Nannochloropsis sp. microalga—Energy and CO_2 emission and economic analyses. *Bioresource Technology*, *138*, 235–244. https://doi.org/10.1016/j.biortech.2013.03.168

GaBi. (2023). *Search Life Cycle Assessment Datasets Search Life Cycle Assessment Datasets (GaBi)*. https://sphera.com/product-sustainability-gabi-data-search/

Gabrielle, B., & Gagnaire, N. (2008). Life-cycle assessment of straw use in bio-ethanol production: A case study based on biophysical modelling. *Biomass and Bioenergy*, *32*(5), 431–441.

Gajda, I., Greenman, J., Melhuish, C., & Ieropoulos, I. (2015). Self-sustainable electricity production from algae grown in a microbial fuel cell system. *Biomass and Bioenergy*, *82*, 87–93.

Gao, X., Yu, Y., & Wu, H. (2013). Life cycle energy and carbon footprints of microalgal biodiesel production in Western Australia: A comparison of byproducts utilization strategies. *ACS Sustainable Chemistry & Engineering*, *1*(11), 1371–1380.

García, G., Sosa-Hernández, J. E., Rodas-Zuluaga, L. I., Castillo-Zacarías, C., Iqbal, H., & Parra-Saldívar, R. (2021). Accumulation of pha in the microalgae scenedesmus Sp. Under nutrient-deficient conditions. *Polymers*, *13*(1), 1–23. https://doi.org/10.3390/polym13010131

Gifuni, I., Pollio, A., Safi, C., Marzocchella, A., & Olivieri, G. (2019). Current Bottlenecks and Challenges of the Microalgal Biorefinery. In *Trends in Biotechnology* (vol. 37, Issue 3, pp. 242–252). Elsevier Ltd. https://doi.org/10.1016/j.tibtech.2018.09.006

Gnansounou, E., Dauriat, A., Villegas, J., & Panichelli, L. (2009). Life cycle assessment of biofuels: energy and greenhouse gas balances. *Bioresource Technology*, *100*(21), 4919–4930.

Gnansounou, E., & Raman, J. K. (2016). Life cycle assessment of algae biodiesel and its co-products. *Applied Energy*, *161*, 300–308.

Gnansounou, E., & Raman, J. K. (2017). *Life cycle Assessment of Algal Biorefinery*. In *Life-Cycle Assessment of Biorefineries* 2017, Pages 199-219. Elsevier https://doi.org/10.1016/B978-0-444-63585-3.00007-3

Goswami, R. K., Agrawal, K., Mehariya, S., Molino, A., Musmarra, D., & Verma, P. (2020). Microalgae-based biorefinery for utilization of carbon dioxide for production of valuable bioproducts. In *Chemo-Biological Systems for CO_2 Utilization* (pp. 203–228). Taylor and Francis CRC Press https://doi.org/10.1201/9780429317187-11/microalgae-based-biorefinery-utilization-carbon-dioxide-production-valuable-bi

Goswami, R. K., Agrawal, K., Upadhyaya, H. M., Gupta, V. K., & Verma, P. (2022a). Microalgae conversion to alternative energy, operating environment and economic footprint: An influential approach towards energy conversion, and management. In *Energy Conversion and Management* (vol. 269). Elsevier Ltd. https://doi.org/10.1016/j.enconman.2022.116118

Goswami, R. K., Agrawal, K., & Verma, P. (2022b). Microalgae biomass biorefinery: a sustainable renewable energy feedstock of the future. In P. Verma (Ed.), *Micro-algae: Next-generation Feedstock for Biorefineries: Cultivation and Refining Processes* (pp. 1–29). Springer Nature. https://doi.org/10.1007/978-981-19-0793-7_1

Goswami, R. K., Mehariya, S., Obulisamy, P. K., & Verma, P. (2021a). Advanced microalgae-based renewable biohydrogen production systems: A review. In *Bioresource Technology* (vol. 320). Elsevier Ltd. https://doi.org/10.1016/j.biortech.2020.124301

Goswami, R. K., Mehariya, S., Verma, P., Lavecchia, R., & Zuorro, A. (2021b). Microalgae-based biorefineries for sustainable resource recovery from wastewater. *Journal of Water Process Engineering*, *40*. https://doi.org/10.1016/j.jwpe.2020.101747

Goswami, R. K., and Agrawal, K. and Verma, P. (2021c). Microalgae-based biofuel-integrated biorefinery approach as sustainable feedstock for resolving energy crisis. In N. and S. R. Srivastava Manish and Srivastava (Eds.), *Bioenergy Research: Commercial Opportunities & Challenges* (pp. 267–293). Springer. https://doi.org/10.1007/978-981-16-1190-2_9

Guinée, J. B., Heijungs, R., Vijver, M. G., & Peijnenburg, W. J. G. M. (2017). Setting the stage for debating the roles of risk assessment and life-cycle assessment of engineered nanomaterials. *Nature Nanotechnology*, *12*(8), 727–733. https://doi.org/10.1038/nnano.2017.135

Hassan, S. S., Williams, G. A., & Jaiswal, A. K. (2019). Moving towards the second generation of lignocellulosic biorefineries in the EU: Drivers, challenges, and opportunities. In *Renewable and Sustainable Energy Reviews* (vol. 101, pp. 590–599). Elsevier Ltd. https://doi.org/10.1016/j.rser.2018.11.041

Ika Rinawati, D., Ryota Keeley, A., Takeda, S., & Managi, S. (2022). Life-cycle assessment of hydrogen utilization in power generation: A systematic review of technological and methodological choices. *Frontiers in Sustainability*, *3*(920876), 1–19. https://doi.org/https://doi.org/10.3389/frsus.2022.920876

ISO(14040). (2006). *Environmental Management, Assessment of the Life Cycle, Principles and Framework, The International Organization for Standardization (ISO) 14040:2006 (en)* . Italian National Unification (UNI), Multidisciplinary Digital Publishing Institute.

ISO(14044). (n.d.). *Environmental Management, Assessment of the Life Cycle, Requirements and Guidelines; The International Organization for Standardization (ISO)* 14044:2006 (en). Springer, Italian National Unification (UNI).

Iwamoto, H. (2004). Industrial production of microalgal cell-mass and secondary products-major industrial species. *Handbook of Microalgal Culture: Biotechnology and Applied Phycology*, *255*, 263.

Javed, F., Aslam, M., Rashid, N., Shamair, Z., Khan, A. L., Yasin, M., Fazal, T., Hafeez, A., Rehman, F., Rehman, M. S. U., & others. (2019). Microalgae-based biofuels, resource recovery and wastewater treatment: A pathway towards sustainable biorefinery. *Fuel*, *255*, 115826.

Jeswani, H. K., Chilvers, A., & Azapagic, A. (2020). Environmental sustainability of biofuels: A review: Environmental sustainability of biofuels. In *Proceedings of the Royal Society A: Mathematical, Physical and Engineering Sciences* (vol. 476, Issue 2243). Royal Society Publishing. https://doi.org/10.1098/rspa.2020.0351

Khan, M. I., Shin, J. H., & Kim, J. D. (2018a). The promising future of microalgae: Current status, challenges, and optimization of a sustainable and renewable industry for biofuels, feed, and other products. In *Microbial Cell Factories* (vol. 17, Issue 1). BioMed Central Ltd. https://doi.org/10.1186/s12934-018-0879-x

Khan, M. I., Shin, J. H., & Kim, J. D. (2018b). The promising future of microalgae: Current status, challenges, and optimization of a sustainable and renewable industry for biofuels, feed, and other products. *Microbial Cell Factories*, *17*(1), 36.

Khanum, F., Giwa, A., Nour, M., Al-Zuhair, S., & Taher, H. (2020). Improving the economic feasibility of biodiesel production from microalgal biomass via high-value products coproduction. *International Journal of Energy Research*, *44*(14), 11453–11472. https://doi.org/10.1002/er.5768

Khoo, K. S., Chew, K. W., Yew, G. Y., Leong, W. H., Chai, Y. H., Show, P. L., & Chen, W. H. (2020). Recent advances in downstream processing of microalgae lipid recovery for biofuel production. In *Bioresource Technology* (vol. 304). Elsevier Ltd. https://doi.org/10.1016/j.biortech.2020.122996

Kimble, M., Pasdeloup, M.-V., & Spencer, C. (2008). *Sustainable Bioenergy Development in UEMOA Member Countries*. United Nations Foundation Press Center.

Kneifel, J. D., & Lavappa, P. (2018). *BEES (Building for Environmental and Economic Sustainability) Software*. www.nist.gov/services-resources/software/bees1/3www.nist.gov/services-resources/software/beesSOFTWARE

Kok, B., Malcorps, W., Tlusty, M. F., Eltholth, M. M., Auchterlonie, N. A., Little, D. C., Harmsen, R., Newton, R. W., & Davies, S. J. (2020). Fish as feed: Using economic allocation to quantify the Fish in—Fish-out ratio of major fed aquaculture species. *Aquaculture, 528*. https://doi.org/10.1016/j.aquaculture.2020.735474

Kolosz, B. W., Luo, Y., Xu, B., Maroto-Valer, M. M., & Andresen, J. M. (2020). Life cycle environmental analysis of “drop in” alternative aviation fuels: A review. In *Sustainable Energy and Fuels* (vol. 4, Issue 7, pp. 3229–3263). Royal Society of Chemistry. https://doi.org/10.1039/c9se00788a

Kumar, B., Agrawal, K., Bhardwaj, N., Chaturvedi, V., & Verma, P. (2019). Techno-economic assessment of microbe-assisted wastewater treatment strategies for energy and value-added product recovery. In *Microbial Technology for the Welfare of Society* (pp. 147–181). Springer. https://doi.org/10.1007/978-981-13-8844-6_7

Kumar, B., Bhardwaj, N., Agrawal, K., Chaturvedi, V., & Verma, P. (2020a). Current perspective on pretreatment technologies using lignocellulosic biomass: An emerging biorefinery concept. In *Fuel Processing Technology* (vol. 199). Elsevier B.V. https://doi.org/10.1016/j.fuproc.2019.106244

Kumar, B., Bhardwaj, N., Agrawal, K., & Verma, P. (2020b). Bioethanol Production: Generation-Based Comparative Status Measurements. In M. and M. P. K. and G. V. K. Srivastava Neha and Srivastava (Ed.), *Biofuel Production Technologies: Critical Analysis for Sustainability* (pp. 155–201). Springer. https://doi.org/10.1007/978-981-13-8637-4_7

Kumar, B., & Verma, P. (2020). Techno-economic assessment of biomass-based integrated biorefinery for energy and value-added product. In P. Verma (Ed.), *Biorefineries: A Step Towards Renewable and Clean Energy* (pp. 581–616). Springer. https://doi.org/10.1007/978-981-15-9593-6_23

Kumar, B., & Verma, P. (2021). Life cycle assessment: Blazing a trail for bioresources management. *Energy Conversion and Management: X, 10*. https://doi.org/10.1016/j.ecmx.2020.100063

Kumar, S., Shrestha, P., & Salam, P. A. (2013). A review of biofuel policies in the major biofuel producing countries of ASEAN: Production, targets, policy drivers and impacts. *Renewable and Sustainable Energy Reviews, 26*, 822–836.

Kylili, A., Christoforou, E., & Fokaides, P. A. (2016). Environmental evaluation of biomass pelleting using life cycle assessment. *Biomass and Bioenergy, 84*, 107–117. https://doi.org/10.1016/j.biombioe.2015.11.018

LCA_GaBi. (2023). *LCA for Experts (GaBi) LCA for Experts Software built on the world's most robust LCA databases. Achieve Your Sustainability Goals with Sphera's Best-In-Class Product Sustainability Software*. https://sphera.com/life-cycle-assessment-lca-software/

Liang, S., Gu, H., & Bergman, R. D. (2017). Life cycle assessment of cellulosic ethanol and biomethane production from forest residues. *BioResources, 12*(4), 7873–7883.

Lima, M. G. B., & Gupta, J. (2014). The extraterritorial dimensions of biofuel policies and the politics of scale: live and let die? *Third World Quarterly, 35*(3), 392–410. www.jstor.org/stable/24522149

Linares, L. C. F., Falfán, K. Á. G., & Ram\'\irez-López, C. (2017). Microalgal biomass: A biorefinery approach. *Biomass Volume Estimation and Valorization for Energy*, 293.

Luz, L. M. da, Francisco, A. C. de, Piekarski, C. M., & Salvador, R. (2018). Integrating life cycle assessment in the product development process: A methodological approach. *Journal of Cleaner Production, 193*, 28–42. https://doi.org/10.1016/j.jclepro.2018.05.022

Maga, D. (2017). Life cycle assessment of biomethane produced from microalgae grown in municipal waste water. *Biomass Conversion and Biorefinery, 7*(1), 1–10.

Martinot, E., Sawin, J. L., & others. (2005). *Renewables 2005: Global Status Report*. Worldwatch Institute.

Mat Aron, N. S., Khoo, K. S., Chew, K. W., Show, P. L., Chen, W.-H., & Nguyen, T. H. P. (2020). Sustainability of the four generations of biofuels – A review. *International Journal of Energy Research, 44*(12), 9266–9282. https://doi.org/https://doi.org/10.1002/er.5557

McHugh, D. J., & FAO UN. (2003). *A Guide to the Seaweed Industry*. Food and Agriculture Organization of the United Nations. www.fao.org/3/y4765e/y4765e00.htm

Mehariya, S., Goswami, R. K., Karthikeysan, O. P., & Verma, P. (2021a). Microalgae for high-value products: A way towards green nutraceutical and pharmaceutical compounds. *Chemosphere, 280*. https://doi.org/10.1016/j.chemosphere.2021.130553

Mehariya, S., Goswami, R. K., Verma, P., Lavecchia, R., & Zuorro, A. (2021b). Integrated approach for wastewater treatment and biofuel production in microalgae biorefineries. In *Energies* (vol. 14, Issue 8). MDPI AG. https://doi.org/10.3390/en14082282

Michiels, F., & Geeraerd, A. (2020). How to decide and visualize whether uncertainty or variability is dominating in life cycle assessment results: A systematic review. In *Environmental Modelling and Software* (vol. 133). Elsevier Ltd. https://doi.org/10.1016/j.envsoft.2020.104841

Moazeni, F., & Khazaei, J. (2020). Optimal operation of water-energy microgrids; A mixed integer linear programming formulation. *Journal of Cleaner Production, 275*. https://doi.org/10.1016/j.jclepro.2020.122776

Mobarezkhoo, H., Saidi-Mehrabad, M., & Sahebi, H. (2022). Life cycle evaluation of microalgae based biorefinery supply network: A sustainable multi-criteria approach. *Algal Research, 66*. https://doi.org/10.1016/j.algal.2022.102816

Molino, A., Mehariya, S., Di Sanzo, G., Larocca, V., Martino, M., Leone, G. P., Marino, T., Chianese, S., Balducchi, R., & Musmarra, D. (2020). Recent developments in supercritical fluid extraction of bioactive compounds from microalgae: Role of key parameters, technological achievements and challenges. *Journal of CO_2 Utilization, 36*, 196–209.

Montazeri, M., Soh, L., Pérez-López, P., Zimmerman, J. B., & Eckelman, M. J. (2016). Time-dependent life cycle assessment of microalgal biorefinery co-products. *Biofuels, Bioproducts and Biorefining, 10*(4), 409–421. https://doi.org/https://doi.org/10.1002/bbb.1649

Mu, D., Min, M., Krohn, B., Mullins, K. A., Ruan, R., & Hill, J. (2014). Life cycle environmental impacts of wastewater-based algal biofuels. *Environmental Science & Technology, 48*(19), 11696–11704.

Nagarajan, D., Lee, D.-J., Chen, C.-Y., & Chang, J.-S. (2020). Resource recovery from wastewaters using microalgae-based approaches: A circular bioeconomy perspective. *Bioresource Technology, 302*, 122817.

Nanaki, E. A., & Koroneos, C. J. (2012). Comparative LCA of the use of biodiesel, diesel and gasoline for transportation. *Journal of Cleaner Production, 20*(1), 14–19. https://doi.org/10.1016/j.jclepro.2011.07.026

Ndayisenga, F., Yu, Z., Yu, Y., Lay, C.-H., & Zhou, D. (2018). Bioelectricity generation using microalgal biomass as electron donor in a bio-anode microbial fuel cell. *Bioresource Technology, 270*, 286–293.

Neri, E., Cespi, D., Setti, L., Gombi, E., Bernardi, E., Vassura, I., & Passarini, F. (2016). Biomass residues to renewable energy: A life cycle perspective applied at a local scale. *Energies, 9*(11), 922.

Nilsson, A. E., Bergman, K., Gomez Barrio, L. P., Cabral, E. M., & Tiwari, B. K. (2022). Life cycle assessment of a seaweed-based biorefinery concept for production of food, materials, and energy. *Algal Research, 65*. https://doi.org/10.1016/j.algal.2022.102725

One Click LCA. (2023). *One Click LCA*. www.oneclicklca.com/

openLCA. (2023). *The Idea _ openlca.org*. www.openlca.org/the-idea/

openLCA Features. (2023). *openLCA Features _ openLCA.org*. www.openlca.org/openlca/openlca-features/#

Osiichuk, M., & Shepotylo, O. (2020). Conflict and well-being of civilians: The case of the Russian-Ukrainian hybrid war. *Economic Systems*, *44*(1). https://doi.org/10.1016/j.ecosys.2019.100736

Othoniel, B., Rugani, B., Heijungs, R., Benetto, E., & Withagen, C. (2016). Assessment of life cycle impacts on ecosystem services: Promise, problems, and prospects. In *Environmental Science and Technology* (vol. 50, Issue 3, pp. 1077–1092). American Chemical Society. https://doi.org/10.1021/acs.est.5b03706

Panesar, D. K., Seto, K. E., & Churchill, C. J. (2017). Impact of the selection of functional unit on the life cycle assessment of green concrete. *International Journal of Life Cycle Assessment*, *22*(12), 1969–1986. https://doi.org/10.1007/s11367-017-1284-0

Passarini, F., Nicoletti, M., Ciacci, L., Vassura, I., & Morselli, L. (2014). Environmental impact assessment of a WtE plant after structural upgrade measures. *Waste Management*, *34*(4), 753–762.

Passell, H., Dhaliwal, H., Reno, M., Wu, B., Amotz, A. Ben, Ivry, E., Gay, M., Czartoski, T., Laurin, L., & Ayer, N. (2013). Algae biodiesel life cycle assessment using current commercial data. *Journal of Environmental Management*, *129*, 103–111.

Paul Abishek, M., Patel, J., & Prem Rajan, A. (2014). Algae oil: A sustainable renewable fuel of future. *Biotechnology Research International*, *2014*.

Pittman, J. K., Dean, A. P., & Osundeko, O. (2011). The potential of sustainable algal biofuel production using wastewater resources. *Bioresource Technology*, *102*(1), 17–25.

PRé_Consultants. (2019). SimaPro. In *Amersfoort, The Netherlands, 2015*. PRé_Consultants.

Pulz, O., & Gross, W. (2004). Valuable products from biotechnology of microalgae. *Applied Microbiology and Biotechnology*, *65*(6), 635–648.

Raman, J. K., & Gnansounou, E. (2015). LCA of bioethanol and furfural production from vetiver. *Bioresource Technology*, *185*, 202–210.

Righi, S. (2019). Life Cycle Assessments of waste-based biorefineries—A critical review. In *Life Cycle Assessment of Energy Systems and Sustainable Energy Technologies* (pp. 139–154). Springer.

Rodríguez, R., Espada, J. J., Moreno, J., Vicente, G., Bautista, L. F., Morales, V., Sánchez-Bayo, A., & Dufour, J. (2018). Environmental analysis of Spirulina cultivation and biogas production using experimental and simulation approach. *Renewable Energy*, *129*, 724–732. https://doi.org/10.1016/j.renene.2017.05.076

Saini, K. C., Yadav, D. S., Mehariya, S., Rathore, P., Kumar, B., Marino, T., Leone, G. P., Verma, P., Musmarra, D., & Molino, A. (2021). Overview of extraction of astaxanthin from Haematococcus pluvialis using CO_2 supercritical fluid extraction technology vis-a-vis quality demands. In *Global Perspectives on Astaxanthin: From Industrial Production to Food, Health, and Pharmaceutical Applications* (pp. 341–354). Elsevier. https://doi.org/10.1016/B978-0-12-823304-7.00032-5

San Juan, J. L. G., Aviso, K. B., Tan, R. R., & Sy, C. L. (2019). A multi-objective optimization model for the design of biomass co-firing networks integrating feedstock quality considerations. *Energies*, *12*(11). https://doi.org/10.3390/en12122252

Saravanan, A. P., Mathimani, T., Deviram, G., Rajendran, K., & Pugazhendhi, A. (2018a). Biofuel policy in India: A review of policy barriers in sustainable marketing of biofuel. *Journal of Cleaner Production*, *193*, 734–747.

Saravanan, A. P., Mathimani, T., Deviram, G., Rajendran, K., & Pugazhendhi, A. (2018b). Biofuel policy in India: A review of policy barriers in sustainable marketing of biofuel. In *Journal of Cleaner Production* (vol. 193, pp. 734–747). Elsevier Ltd. https://doi.org/10.1016/j.jclepro.2018.05.033

Schenk, P. M., Thomas-Hall, S. R., Stephens, E., Marx, U. C., Mussgnug, J. H., Posten, C., Kruse, O., & Hankamer, B. (2008). Second generation biofuels: High-efficiency microalgae for biodiesel production. *Bioenergy Research, 1*(1), 20–43.

Seidel, C. (2016). The application of life cycle assessment to public policy development. *The International Journal of Life Cycle Assessment, 21*(3), 337–348.

Sharifi, A. (2020). Trade-offs and conflicts between urban climate change mitigation and adaptation measures: A literature review. In *Journal of Cleaner Production* (vol. 276). Elsevier Ltd. https://doi.org/10.1016/j.jclepro.2020.122813

Sheehan, J., Camobreco, V., Duffield, J., Graboski, M., & Shapouri, H. (1998). An Overview of Biodiesel and Petroleum Diesel Life Cycles. *Jointly Prepared by U.S. Department of Agriculture and U.S. Department of Energy (NREL/TP-580–24772)*, pp. 1–60. www.doe.gov/bridge

Shuba, E. S., & Kifle, D. (2018). Microalgae to biofuels: 'Promising' alternative and renewable energy, review. In *Renewable and Sustainable Energy Reviews* (vol. 81, pp. 743–755). Elsevier Ltd. https://doi.org/10.1016/j.rser.2017.08.042

Sills, D. L., Paramita, V., Franke, M. J., Johnson, M. C., Akabas, T. M., Greene, C. H., & Tester, J. W. (2013). Quantitative uncertainty analysis of life cycle assessment for algal biofuel production. *Environmental Science & Technology, 47*(2), 687–694. https://doi.org/10.1021/es3029236

SimaPro. (2023). *SimaPro_About*. https://simapro.com/about/

Sing, S. F., Isdepsky, A., Borowitzka, M. A., & Moheimani, N. R. (2013). *Production of Biofuels from Microalgae* (pp. 47–72). https://doi.org/10.1007/s11027-011-9294-x

Siqueira, S. F., Deprá, M. C., Zepka, L. Q., & Jacob-Lopes, E. (2018). Life Cycle Assessment (LCA) of third-generation biodiesel produced heterotrophically by phormidium autumnale. *The Open Biotechnology Journal, 12*(1).

Smyth, B. M., Ó Gallachóir, B. P., Korres, N. E., & Murphy, J. D. (2010). Can we meet targets for biofuels and renewable energy in transport given the constraints imposed by policy in agriculture and energy? *Journal of Cleaner Production, 18*(16–17), 1671–1685. https://doi.org/10.1016/j.jclepro.2010.06.027

Soh, L., Montazeri, M., Haznedaroglu, B. Z., Kelly, C., Peccia, J., Eckelman, M. J., & Zimmerman, J. B. (2014). Evaluating microalgal integrated biorefinery schemes: empirical controlled growth studies and life cycle assessment. *Bioresource Technology, 151*, 19–27.

Somé, A., Dandres, T., Gaudreault, C., Majeau-Bettez, G., Wood, R., & Samson, R. (2018). Coupling input-output tables with macro-life cycle assessment to assess worldwide impacts of biofuels transport policies. *Journal of Industrial Ecology, 22*(4), 643–655.

Somers, M. D., & Quinn, J. C. (2019). Sustainability of carbon delivery to an algal biorefinery: A techno-economic and life-cycle assessment. *Journal of CO_2 Utilization, 30*, 193–204. https://doi.org/10.1016/j.jcou.2019.01.007

Sorda, G., Banse, M., & Kemfert, C. (2010). An overview of biofuel policies across the world. *Energy Policy, 38*(11), 6977–6988.

Sphera_PSSD. (2023). *Product Sustainability Software & Data_Sphera*. https://sphera.com/product-sustainability-software/

Su, Y., Zhang, P., & Su, Y. (2015). An overview of biofuels policies and industrialization in the major biofuel producing countries. In *Renewable and Sustainable Energy Reviews* (vol. 50, pp. 991–1003). Elsevier Ltd. https://doi.org/10.1016/j.rser.2015.04.032

Supakorn, P., Chonlada, Y., & Anchalee, S. (2021). Pigment production under cold stress in the green microalga Chlamydomonas reinhardtii. *Agriculture (Switzerland), 11*(6). https://doi.org/10.3390/agriculture11060564

Thomas, J. B. E., Sodré Ribeiro, M., Potting, J., Cervin, G., Nylund, G. M., Olsson, J., Albers, E., Undeland, I., Pavia, H., & Gröndahl, F. (2021). A comparative environmental life cycle assessment of hatchery, cultivation, and preservation of the kelp Saccharina latissima. *ICES Journal of Marine Science, 78*(1), 451–467. https://doi.org/10.1093/icesjms/fsaa112

Thomassen, G., Van Dael, M., & Van Passel, S. (2018). The potential of microalgae biorefineries in Belgium and India: An environmental techno-economic assessment. *Bioresource Technology*, *267*, 271–280. https://doi.org/10.1016/j.biortech.2018.07.037

Thornley, P., Gilbert, P., Shackley, S., & Hammond, J. (2015). Maximizing the greenhouse gas reductions from biomass: The role of life cycle assessment. *Biomass and Bioenergy*, *81*, 35–43.

Timmis, K. N., McGenity, T., Van Der Meer, J. R., & de Lorenzo, V. (2010). *Handbook of Hydrocarbon and Lipid Microbiology*. Springer.

Tonini, D., & Astrup, T. (2012). Life-cycle assessment of a waste refinery process for enzymatic treatment of municipal solid waste. *Waste Management*, *32*(1), 165–176.

Ubando, A. T., Anderson S. Ng, E., Chen, W. H., Culaba, A. B., & Kwon, E. E. (2022). Life cycle assessment of microalgal biorefinery: A state-of-the-art review. In *Bioresource Technology* (vol. 360). Elsevier Ltd. https://doi.org/10.1016/j.biortech.2022.127615

Ubando, A. T., Rivera, D. R. T., Chen, W.-H., & Culaba, A. B. (2019). A comprehensive review of life cycle assessment (LCA) of microalgal and lignocellulosic bioenergy products from thermochemical processes. *Bioresource Technology*, *291*, 121837. https://doi.org/10.1016/j.biortech.2019.121837

Vance, C., Sweeney, J., & Murphy, F. (2022). Space, time, and sustainability: The status and future of life cycle assessment frameworks for novel biorefinery systems. In *Renewable and Sustainable Energy Reviews* (vol. 159). Elsevier Ltd. https://doi.org/10.1016/j.rser.2022.112259

Xu, H., Ou, L., Li, Y., Hawkins, T. R., & Wang, M. (2022). Life Cycle Greenhouse Gas Emissions of Biodiesel and Renewable Diesel Production in the United States. *Environmental Science and Technology*, *56*(12), 7512–7521. https://doi.org/10.1021/acs.est.2c00289

Yacobucci, B. D., & Bracmort, K. S. (2009). *Calculation of Lifecycle Greenhouse Gas Emissions for the Renewable Fuel Standard*. Library of Congress. Congressional Research Service. https://rosap.ntl.bts.gov/view/dot/17287

Index

R

S

U

V

W